Catalytic Amination for *N*-Alkyl Amine Synthesis

Catalytic Amination for *N*-Alkyl Amine Synthesis

Feng Shi

Xinjiang Cui

ELSEVIER

ACADEMIC PRESS

An imprint of Elsevier

Academic Press is an imprint of Elsevier
125 London Wall, London EC2Y 5AS, United Kingdom
525 B Street, Suite 1800, San Diego, CA 92101-4495, United States
50 Hampshire Street, 5th Floor, Cambridge, MA 02139, United States
The Boulevard, Langford Lane, Kidlington, Oxford OX5 1GB, United Kingdom

Notices

Knowledge and best practice in this field are constantly changing. As new research and experience broaden our understanding, changes in research methods, professional practices, or medical treatment may become necessary.

Practitioners and researchers must always rely on their own experience and knowledge in evaluating and using any information, methods, compounds, or experiments described herein. In using such information or methods they should be mindful of their own safety and the safety of others, including parties for whom they have a professional responsibility.

To the fullest extent of the law, neither the Publisher nor the authors, contributors, or editors, assume any liability for any injury and/or damage to persons or property as a matter of products liability, negligence or otherwise, or from any use or operation of any methods, products, instructions, or ideas contained in the material herein.

Library of Congress Cataloging-in-Publication Data

A catalog record for this book is available from the Library of Congress

British Library Cataloguing-in-Publication Data

A catalogue record for this book is available from the British Library

ISBN 978-0-12-812284-6

For information on all Academic Press publications
visit our website at https://www.elsevier.com/books-and-journals

Publisher: John Fedor
Acquisition Editor: Emily M. McCloskey
Editorial Project Manager: Emily Thomson
Production Project Manager: Prem Kumar Kaliamoorthi
Cover Designer: Mark Rogers

Typeset by SPi Global, India

Contents

Preface

The chemistry of nitrogen-containing compounds is of significant importance in many crucial fields because they play key roles in the cases of preparation of agrochemicals, dyes, explosives, surfactants, additives, and so on. In addition, nitrogen derivatives are typical building blocks for life matter and bioactive compounds such as amino acids, nucleotides, vitamins, hormones, organic catalysts, and others. Among nitrogen-containing compounds, *N*-alkylamine is one of the most important compounds, and huge amount of attention has been given to this research area since the 1900s. Traditionally, *N*-alkylamines are synthesized using alkyl halides as alkylating reagents. However, this procedure is problematic due to amine over-alkylation, the toxic nature of alkyl halides, and the generation of inorganic salt waste. In order to achieve sustainable synthesis of *N*-alkylamines, there are several aspects of note. (1) The use of alkyl halides as alkylating reagents should be replaced by alcohol, amine, alkene, alkane, carbon dioxide, and other sustainable molecules; (2) the employment of amines as nitrogen sources can be replaced by nitrobenzene, nitrile, and ammonia to avoid multistep reactions; and (3) economic and efficient catalysts should be developed for the above transformations.

This handbook is intended to provide an overview of catalytic amination with amine, nitro-, and nitrile-containing molecules for *N*-alkylamine synthesis in the past 20–30 years or so, with emphasis placed not only on the amination reactions themselves, but also the development of the catalyst systems, i.e., from heterogeneous catalysis (1900–80) to homogeneous catalysis (1980–2010), and to the merging time of heterogeneous and homogeneous catalysis (2000–10). This handbook is organized in seven chapters to highlight the catalytic amination with different alkylating reagents, i.e., alcohol, amine, alkene/carbon monoxide, alkene/carbon monoxide/H_2, carbon dioxide, and alkane, and different nitrogen sources, i.e., amine, nitro-, and nitrile compounds. Although extensively valuable results were achieved in the last decades, the amination reactions with alkyl halides and aldehyde/ketone as alkylation reagents will not be included in this book as many positive reviews have been published in the last years.

It is a pleasure to be the editors of this handbook because it provides us with the opportunity to survey the field of catalytic amination for *N*-alkylamine synthesis and to honor the work of so many fine chemists. We gratefully acknowledge the Elsevier editorial staff, and extend special thanks to Katey Birtcher and Emily Thomson.

We are very grateful to all of our colleagues for their excellent contributions to this book. Contributors include: Xingchao Dai (Chapter 1), Xingchao Dai and Xinjing Cui (Chapter 2), Hongli Wang (Chapter 3), Hongli Wang and Xingchao Dai (Chapter 4), Yujing Zhang (Chapter 5), Shaofeng Pang and Hongli Wang (Chapter 6), and Yajuan Wu (Chapter 7).

We hope this book will be of interest to all those involved in this field, from graduate students to independent catalytic and organic researchers in both academic and industrial sectors. It is hoped that this extensive survey of catalytic amination for *N*-alkylamine synthesis will stimulate a wider general interest in the subject.

Feng Shi
Xinjiang Cui
State Key Laboratory for Oxo Synthesis and Selective Oxidation
Center for Green Chemistry and Catalysis
Lanzhou Institute of Chemical Physics
Chinese Academy of Sciences
Lanzhou, 730000, China

Chapter 1

N-Alkyl Amine Synthesis by Catalytic Alcohol Amination

1.1 INTRODUCTION

Amines have been attracting worldwide attention because of their key role in the bulk and fine chemical industries for production of polymers, dyes, pharmaceuticals, agrochemicals, surfactants, and biologically active compounds.[1] Owing to their significant importance, a variety of procedures for their synthesis, such as Hofmann alkylation,[2] Buchwald-Hartwig[3–4] and Ullmann[5] reactions, hydroamination,[6] hydroaminomethylation,[7] reduction of nitriles[8] and nitro[9] compounds, or reductive amination,[10] have been developed in the last century and excellent results were obtained. Besides the reactions mentioned above, the direct alcohol amination constitutes an attractive process because alcohol is readily available, and water is generated as the sole by-product.

As early as 1901, the first coupling of amines with alcohols was reported by Nef and the alkylation of aniline was successfully performed using different sodium alkoxides.[11] Following Nef's work, the research on alcohol amination was mainly focused on the reaction of amine with alkoxide compounds and base-catalyzed alcohol amination.[12] Unfortunately, these transformations usually require harsh reaction conditions such as high temperatures (>200°C), high pressure, long reaction time, and with the addition of excess base. In 1981, the homogeneous transition metal-catalyzed alcohol amination was reported by Grigg and Watanabe successively[13–14] and the alcohol amination began to draw worldwide attention. Since then, various transition metals, such as Ru, Ir, Rh, Pt, Pd, Au, Ag, Co, Mn, Ni, Cu, and Fe, have been employed as homogeneous or heterogeneous catalysts in the alcohol amination and have yielded good results. Additionally, carbon materials and organic molecular-catalyzed alcohol aminations have been achieved in the recent years, and the catalytic performances were comparable to the transition metal catalysts.

Although numerous approaches for direct alcohol amination have been explored, these successful catalytic systems essentially operate through either nucleophilic substitution (S_N2 and S_N1) or the borrowing hydrogen mechanism (BH). In general, the S_N type alcohol amination requires π-activated alcohols, such as benzylic, propargylic, or allylic alcohols.[15–26] However, exceptional cases do exist and a very recent example demonstrates that even nonactivated

Catalytic Amination for *N*-Alkyl Amine Synthesis. https://doi.org/10.1016/B978-0-12-812284-6.00001-9

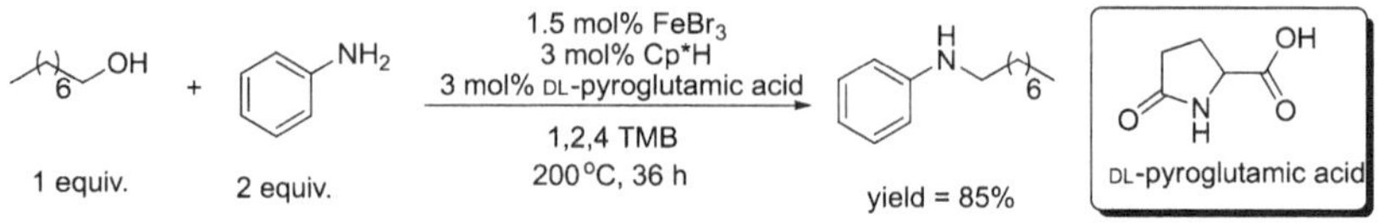

SCHEME 1.1 S_N type alcohol amination using the nonactivated alcohols as substrates.

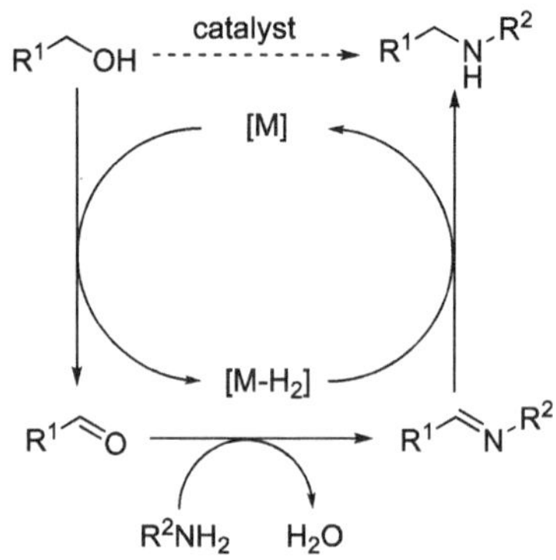

SCHEME 1.2 The alcohol amination with borrowing hydrogen mechanism.

alcohols, such as 1-octanol, can be converted with aniline to give the *N*-alkyl anilines in good yields (Scheme 1.1).[27]

Although the reaction temperature is high (200°C), some allylic alcohols have been successfully converted at 100°C. As another possible pathway, the so-called borrowing hydrogen (BH) mechanism (Scheme 1.2),[28–37] can be efficiently applied in the amination of the nonactivated alcohols and form a good complementation with the S_N mechanism. Here, the alcohol is first activated by dehydrogenation to give an aldehyde or ketone, and the generated in situ carbonyl compound then undergoes a condensation reaction with the amine nucleophile. Subsequently, the resulting amine reacts with the initially generated hydrogen to yield the desired amine product.

1.2 HOMOGENEOUS CATALYSTS SYSTEM

Although the first example of alcohol amination was performed in a homogeneous solution in the absence of transition-metal catalyst in 1901,[38] the homogeneous version using transition metal catalysts was not reported until 1981.[13–14] After that, homogeneous catalysts entered a growth period and various transition metals including the noble metals, such as Ru, Ir, Rh, Pd, and nonnoble metals, such as Cu, Ni, Fe, have been explored as homogenous catalysts for the *N*-alkylation of amines with alcohols, and encouraging results were obtained.

1.2.1 Ru Catalyst System

As early as 1981, Watanabe and coworkers reported the first homogeneous Ru-catalyzed *N*-alkylation of amines with alcohols. *N*-Alkylanilines and 2,3-alkylquinolines were obtained from aniline with saturated and 2,3-unsaturated primary alcohols in good yields, in the presence of [$RuCl_2$ $(PPh_3)_3$] catalyst at 180°C (Scheme 1.3).[14] Saturated alcohols combined with aniline in this procedure gave *N,N*-dialkyl and *N*-alkylaniline, and methanol had low reactivity for this reaction. Inversely, the reaction of 2,3-unsaturated alcohols, such as ally alcohol and crotyl alcohol, gave 2,3-alkylquinoline in good yields.

Since the initial achievements of 1981, the use of [$RuCl_2$ $(PPh_3)_3$] catalyst was further extended to a number of alcohol aminations including *N*-methylation reactions of aliphatic amines with methanol,[39] *N*-alkylation of symmetrical secondary amines with long-chain alcohols,[40–41] *N*-dialkylation of aniline with primary alcohols,[42] *N*-alkylation of amines with diols,[43,44] cyclization of aromatic amines with diols,[45] cyclization of amino alcohols,[46] cyclization of alkynildiols with primary amines,[47] and *N*-alkylation of carboxamides with primary alcohols.[48]

Additionally, some other ruthenium complexes have also been explored to catalyze the direct alcohol amination. For example, $RuH_2(PPh_3)_4$ has proven to be efficient for the reaction of aliphatic primary amines and primary alcohols to the corresponding secondary amines and moderate to high yields were obtained (Scheme 1.4).[49] The well-defined catalyst (η^4-1,5-cyclooctadiene)(η^6-1,3,5-cyclooctatriene) ruthenium [Ru(COD)(COT)] has also displayed a high activity in the selective alkylation of aminopyridine derivatives (Scheme 1.5).[50] However, it was ineffective in the alkylation of aniline.

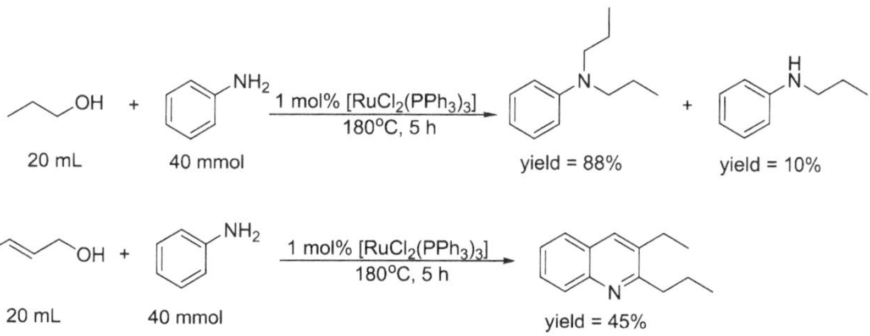

SCHEME 1.3 The Ru-catalyzed *N*-alkylation of amine with primary alcohol.

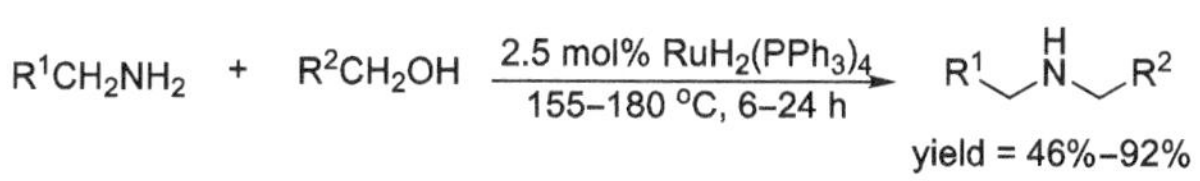

SCHEME 1.4 The *N*-alkylation of aliphatic primary amines with primary alcohols catalyzed by $RuH_2(PPh_3)_4$.

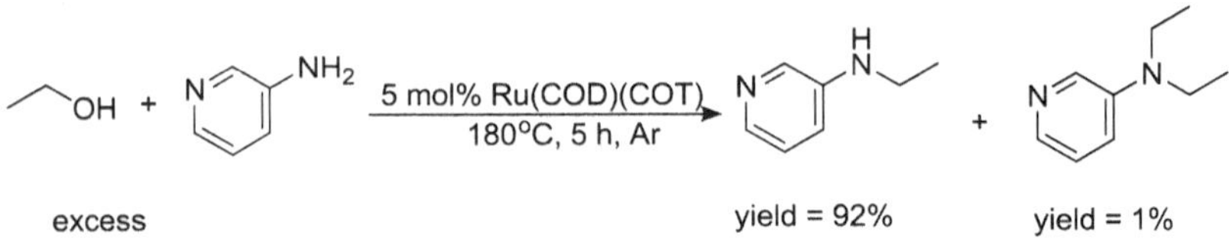

SCHEME 1.5 The *N*-alkylation of heteroaromatic amines with primary alcohols catalyzed by [Ru (COD)(COT)].

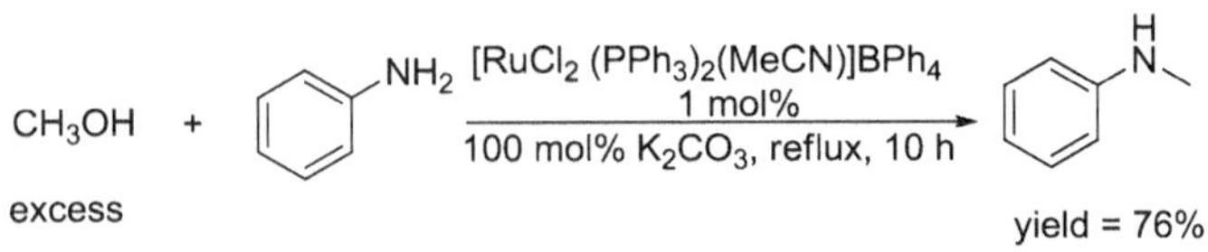

SCHEME 1.6 The monoalkylation of aniline derivatives with primary alcohols catalyzed by $[RuCl_2\ (PPh_3)_2\text{-}(MeCN)]BPh_4$.

The cationic ruthenium complex $[RuCl_2\ (PPh_3)_2(MeCN)]BPh_4$ combined with a base has been used to catalyze the *N*-methylation of aniline with methanol under solvent-free conditions (Scheme 1.6).[51] Different from the above results, the ruthenium(II) half-sandwich complex $[RuCl(\eta^5\text{-}C_5H_5)(PPh_3)_2]$ showed excellent activity for the *N*-methylation of aliphatic primary and secondary alkylamines with methanol, and the quantitative yields could be obtained.[52]

The above examples mainly focus on the amination of primary alcohols, however the Ru-catalyzed amination of secondary alcohols is still challenging. In 2000, the amination of secondary propargylic alcohols catalyzed by thiolate-bridged diruthenium complexes $[Cp^*RuCl(\mu_2\text{-}SMe)_2RuCp^*Cl]$ was reported by Nishibayashi and Hidai et al.[53] Researchers obtained 72% yield when the reaction of *p*-nitroaniline and 1-phenylprop-2-yn-1-ol was performed in the presence of $[Cp^*RuCl(\mu_2\text{-}SMe)_2RuCp^*Cl]$ (5 mol%) and NH_4BF_4 (10 mol%) at 60°C. A lower yield (64%) was achieved in the case of *N*-methylbenzamide. However, acetamide gave a similar yield (73%).[54] By further optimizing the reaction conditions, much better substrate generality was observed. A series of amines, such as aromatic amines, amides, sulfonamides, and lactams, and propargylic alcohols with different functional groups could be well tolerated.[55] In addition, the catalytic performance of (Se, Te)-bridged diruthenium complexes (neutral and cationic) in the amination of propargylic alcohols were also tested and compared with the corresponding S-bridged diruthenium complexes. The results revealed that both S- and Se-bridged complexes were quite effective for the amination of propargylic alcohols, while Te-bridged complexes did not show any catalytic activity.[56] However, the above process is only active for the activated secondary alcohols.

In 2006, Beller's group reported an efficient ruthenium catalyst for the amination of nonactivated secondary alcohols. A prototypical secondary alcohol substrate, 1-phenylethanol, gives high yields of amine products with simple

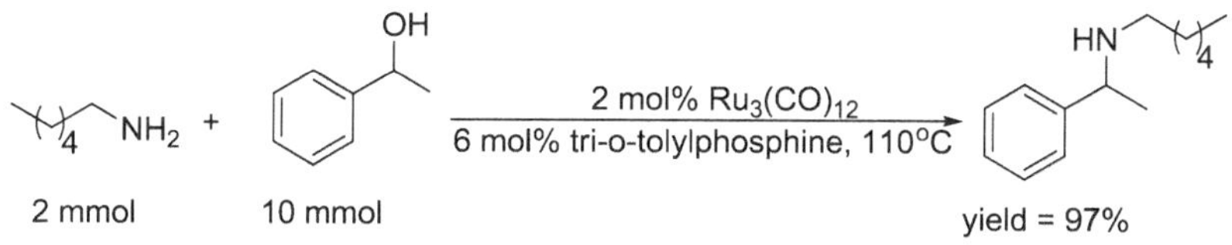

SCHEME 1.7 The Ru-catalyzed amination of secondary alcohols.

aliphatic amines using Ru_3CO_{12} in combination with tri-o-tolylphosphine or *n*-butyl-di-1-adamantylphosphine, but not with a simple arylamine such as aniline (Scheme 1.7).[57] The activity for the amination of secondary alcohols could be improved by using catalyst $Ru_3(CO)_{12}$/*N*-phenyl-2-(dicyclohexylphosphanyl) pyrrole.[58]

Clearly, a variety of tertiary amines are of pharmaceutical interest,[59] especially piperazine derivatives.[60] However, only a few examples are known for catalytic *N*-alkylations of secondary amines.[13,41,61] In 2007, Williams et al. reported a catalytic system of a commercially available $[Ru(cymene)Cl_2]_2$ with a bidentate phosphine ligand dppf, which is effective for the synthesis of secondary amines from aniline and primary alcohols including benzyl alcohols and ethanol (Scheme 1.8).[62] This method was also effective for the synthesis of tertiary amines by the reaction of secondary amines and primary alcohols.[63] The *N*-alkylation of sulfonamides with primary alcohols required higher temperature (150°C) and the addition of K_2CO_3.[64]

There is no general ruthenium-catalyzed synthesis of tertiary amines from secondary alcohols. In 2008, Beller's group revealed that a combination of $[Ru_3(CO)_{12}]$ and *N*-phenyl-2-(dicyclohexylphosphanyl)pyrrole could effectively catalyze the synthesis of tertiary amines from cyclic amines with different secondary alcohols in high yield (Scheme 1.9).[65] Unfortunately, it turned out that all of the substrates performed in their own way and each reaction needed its own set of optimized conditions.

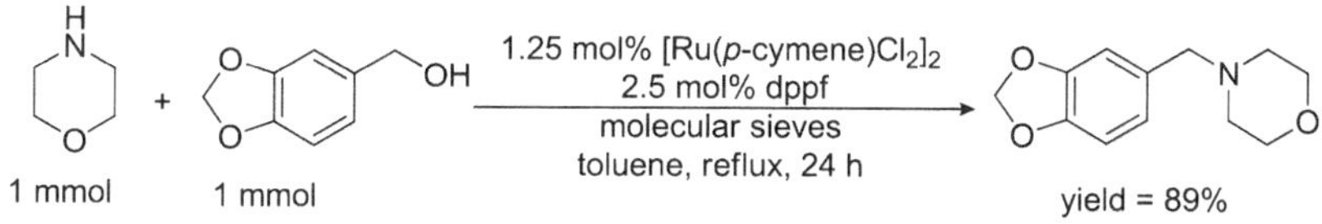

SCHEME 1.8 The alkylation of morpholine with piperonyl alcohol catalyzed by $[Ru(p\text{-cymene})Cl_2]_2$.

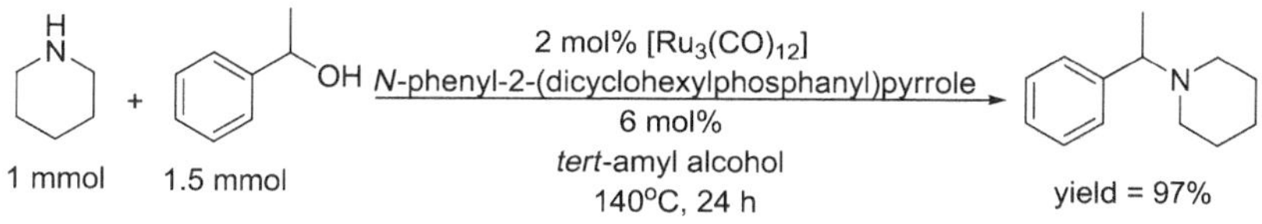

SCHEME 1.9 The alkylation of piperidine with 1-phenylethanol catalyzed by $[Ru_3(CO)_{12}]$.

Although the amination of alcohols with primary and secondary amines has been well studied, less is known of tertiary amines. Recently, Deng and coworkers reported the formation of unsymmetrically substituted tertiary amines from primary alcohols and aliphatic tertiary amines catalyzed by hydrated $RuCl_3{\cdot}3H_2O$.[66] The selectivity for the tertiary amine product strongly depended on the solvent. For instance, conversion of tributylamine and *p*-fluorobenzyl alcohol afforded monosubstituted product in chlorobenzene, whereas the formation of bis-substituted product was favored under positive conditions (Scheme 1.10).

Indole molecules have become an important component in several current pharmaceutical drugs due to their biological activity.[67] Among the numerous known bioactive indole molecules, various *N*-alkylated derivatives are known.[68] However, the homogeneously catalyzed *N*-alkylation of indoles with alcohols has not been investigated until 2010 because of its poor nucleophilicity. In 2010, Beller and Williams's groups reported the *N*-alkylation of indoles with alcohols (Scheme 1.11).[69] In the presence of the Shvo catalyst, an in situ transfer of hydrogenation indole and the alcohol occurred. The resulting indoline and aldehyde undergo a condensation reaction, and the *N*-alkylated indole is formed by an intramolecular isomerization. Applying primary alcohols, the selectively *N*-alkylated products were obtained in good to excellent yields.

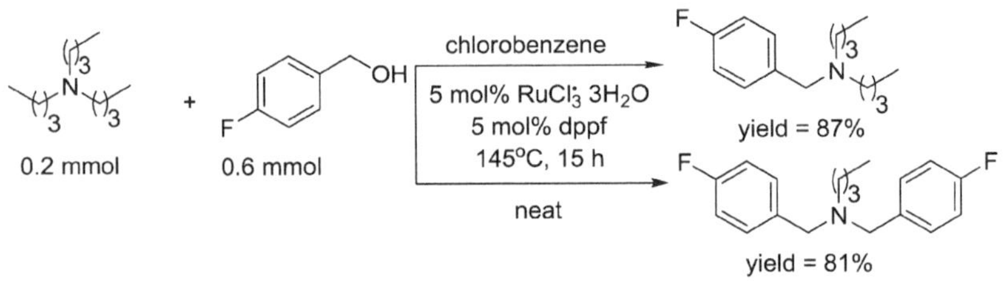

SCHEME 1.10 The amination of alcohols with tertiary amines.

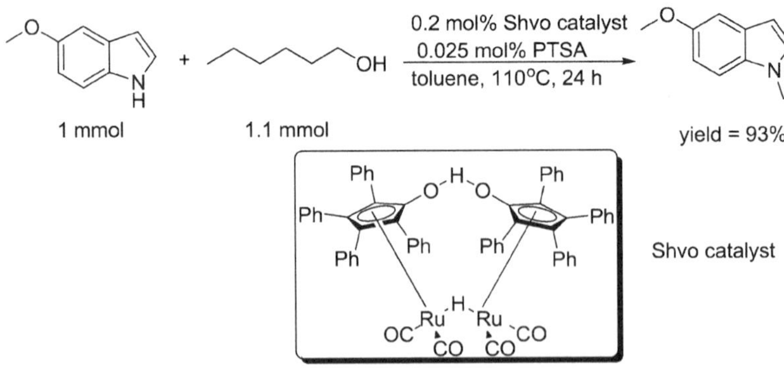

SCHEME 1.11 The Ru-catalyzed *N*-alkylation of indoles with alcohols.

A new Ru-(II)-CNN (C = carbon; N = nitrogen) pincer complex catalyzed the selective monoalkylation of (hetero)aromatic amines with a wide range of primary alcohols (including pyridine-, furan-, and thiophene-substituted alcohols) was reported by Martín-Matute et al. in 2012.[70] The catalyst system is of high efficiency and only equivalent amount of amines and alcohols is enough for high yields. However, the equivalent strong base (KO*t*Bu) is necessary. Of note the formation of tertiary amine was not observed, making this ruthenium system an excellent catalyst for the synthesis of *sec*-amines.

An efficient catalytic system for the synthesis of 1,2,5-substituted pyrroles from 1,4-alkynediols and amines was developed by Williams et al. using $Ru(PPh_3)_3(CO)H_2$ (2.5 mol%) with the bidentate phosphine Xantphos (2.5 mol%).[71] The reaction involves isomerization of the 1,4-alkynediol into a 1,4-diketone and subsequent Paal-Knorr cyclization into the corresponding pyrrole. Optimization of the reaction conditions showed that addition of amine after 30 min of heating the catalyst, ligand and starting material in toluene was ideal.

The ruthenium-catalyzed synthesis of 2- and 3-substituted quinolines from anilines and 1,3-diols was also reported by Madsen in 2011 (Scheme 1.12).[72] The reaction results in mesitylene solution with catalytic amounts of $RuCl_3{\cdot}xH_2O$, PBu_3, and $MgBr_2{\cdot}OEt_2$. In the 1,3-diol molecule, a substituent was allowed in the 1- or 2-position, and the best results were obtained with 2-alkyl substituted 1,3-diols to afford 3-alkylquinolines in 61% yield.

In 2014, Enyong et al. reported the *N*-alkylation of secondary aliphatic amines at room temperature using the catalytic system of $[Ru(p\text{-cymeneCl}_2]_2$ (6 mol%), *t*-BuOK (24 mol%), amino amide ligand (12 mol%), and alcohols as solvent (Scheme 1.13).[73] This is the first example of Ru-catalyzed *N*-alkylation of amines with primary aliphatic alcohols at room temperature.

In 2015, the *N*-alkylation of aniline using equimolar amounts of benzylic alcohol at room temperature was achieved by Taddei et al. using a catalytic system of $[Ru(cod)Cl_2]n$ (2.5 mol%), PTA (1,3,5-triaza-7-phosphaadamantane) (5 mol%), and *t*-BuOK (2 equiv.), and *N*-benzylaniline was obtained with 70% yield (Scheme 1.14).[74]

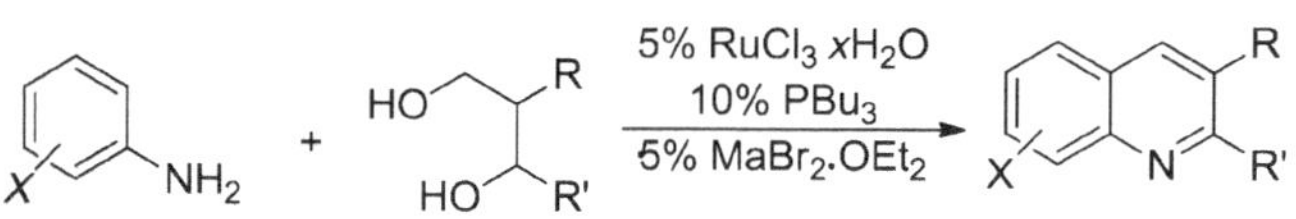

SCHEME 1.12 The synthesis of substituted quinolines from anilines and 1,3-diols.

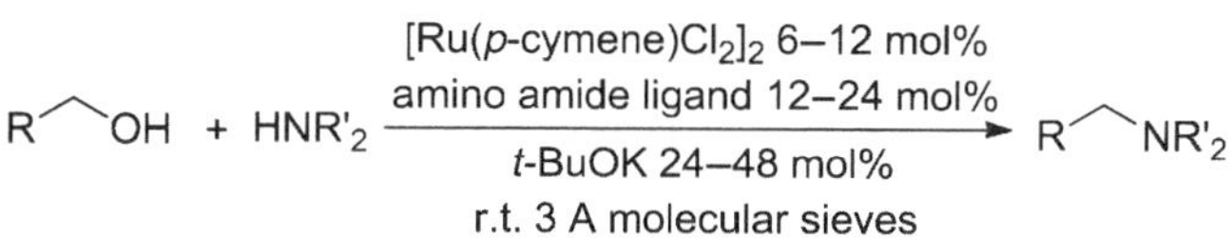

SCHEME 1.13 The Ru-catalyzed *N*-alkylation of amines with primary aliphatic alcohols at room temperature.

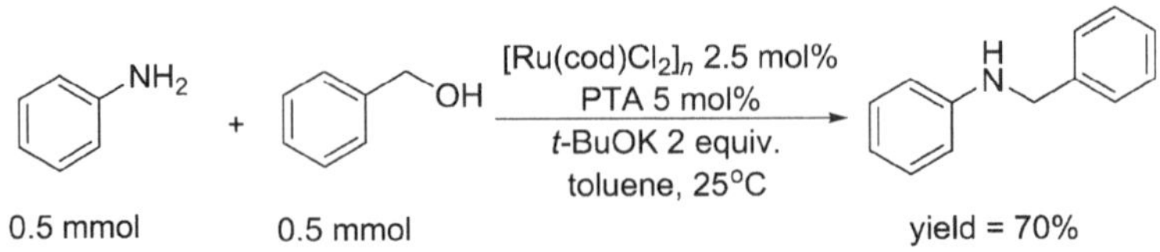

SCHEME 1.14 The *N*-alkylation of aniline with benzyl alcohol at room temperature.

In 2015, Seayad and coworkers presented an in situ-generated Ru-based complex from $[RuCp^*Cl_2]_2$ and dpePhos ligand, which is effective in the presence of 5 mol% of LiO*t*Bu for the *N*-methylation of amines using methanol as the methylating agent at moderate conditions, following a hydrogen borrowing strategy (Scheme 1.15).[75] This simple catalyst system provides selective *N*-monomethylation of substituted primary anilines and sulfonamides as well as *N,N*-dimethylation of primary aliphatic amines in excellent yields at 40–100°C with good tolerance to reducible functional groups. The catalytic intermediate Cp*Ru(dpePhos)H was isolated and was shown to be active for methylation in the absence of base.

In 2016, Takacs et al. reported a new ruthenium complex for the amination of primary and secondary alcohols, and the regioselective mono- and sequential diamination of diols were realized via the borrowing hydrogen pathway (Scheme 1.16).[76] An eight-membered ring heterocycle could be synthesized by the cross-coupling of diamines with diols.

The basis of synthetic amine chemistry and amine industry relies on ammonia, and all the organic amines are produced directly or indirectly from this commodity. Among the methods for commercial production of amines,[1,77] by far the largest and most utilized are based on the reaction of alcohols with ammonia. However, the heterogeneous processes suffer from the requirement of very high temperatures and pressures and lead to mixtures of amines, as well as the

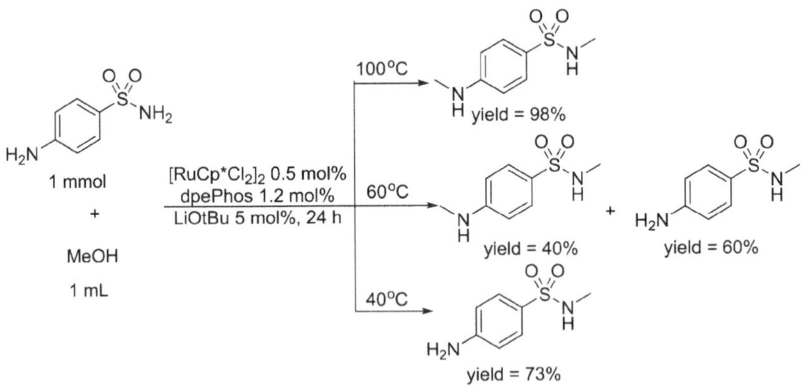

SCHEME 1.15 The selective *N*-monomethylation and *N,N*-dimethylation of amines using methanol.

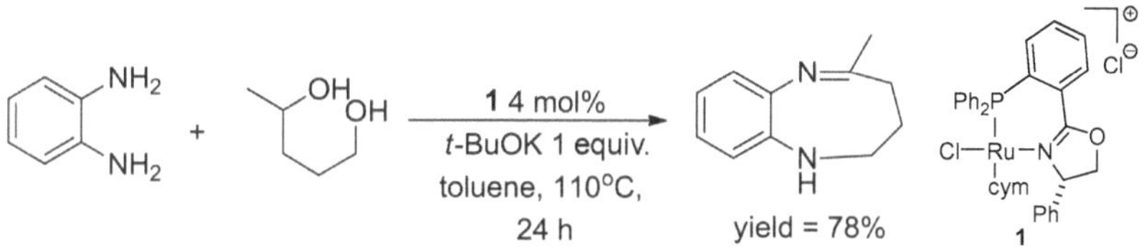

SCHEME 1.16 The regioselective mono- and sequential diamination of diols.

formation of alkene and alkane as side products.[77] Thus, the selective catalytic synthesis of primary amines directly from alcohols and ammonia with elimination of water under relatively mild conditions, without producing waste is highly desirable.

In 2008, Gunanathan and Milstein reported the selective synthesis of primary amines from alcohols and inexpensive ammonia gas in the presence of a ruthenium PNP pincer complex (Scheme 1.17).[78] Primary benzylic amines were synthesized from the corresponding primary alcohols, resulting in good to excellent yields. However, the conversion of nonbenzylic alcohols, such as 2-phenylethanol, resulted in a mixture of primary and secondary amines. The selectivity toward primary amines was increased by running the reaction in water or a mixture of water and organic solvents. Notably, the reaction requires only 0.1 mol% catalyst loading, which indicates the high activity of the Milstein catalyst. However, it is obvious that only primary alcohols and water-insoluble alcohols were efficiently aminated with this catalytic system. The homogeneously catalyzed amination of secondary alcohols with ammonia to give primary amines was independently reported in 2010 by both Beller's group[79] and Vogt's group.[80] Both groups successfully converted a number of secondary alcohols into the corresponding amines by using the combination of $[Ru_3(CO)_{12}]$ and commercially available CataCXiumPCy.

In 2016, the ruthenium-triphos and diphosphine-catalyzed amination of aliphatic alcohols with ammonia was reported (Scheme 1.18).[81] Primary and secondary amines can be selectively synthesized by changing the type of ligands. But tertiary amines cannot be generated as the main product.

In the last few years, there has been increasing interest in producing and using chemical products based on functionalized alcohols from renewable

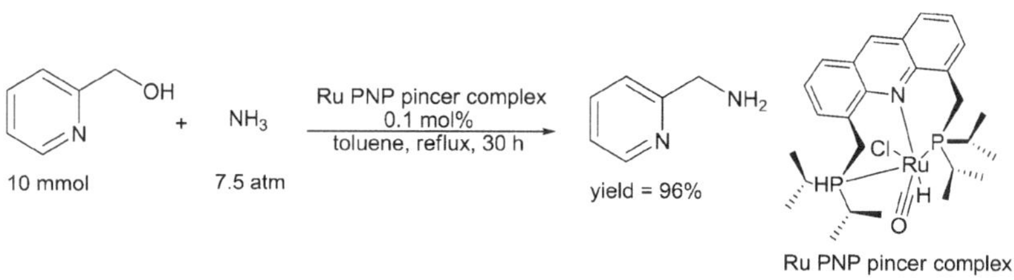

SCHEME 1.17 The synthesis of primary amines from alcohols and ammonia gas.

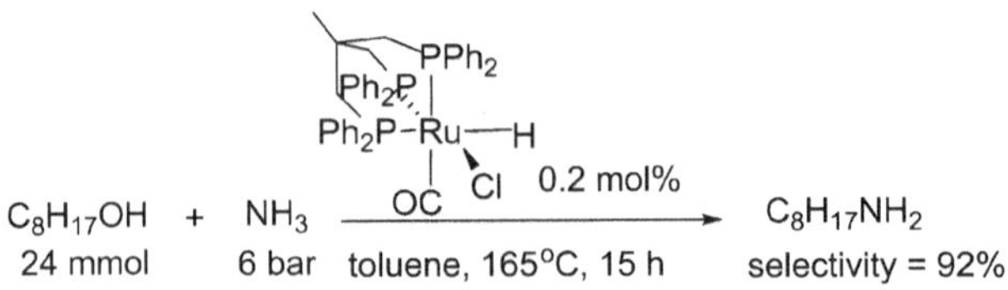

SCHEME 1.18 The synthesis of primary amines from aliphatic alcohols and ammonia.

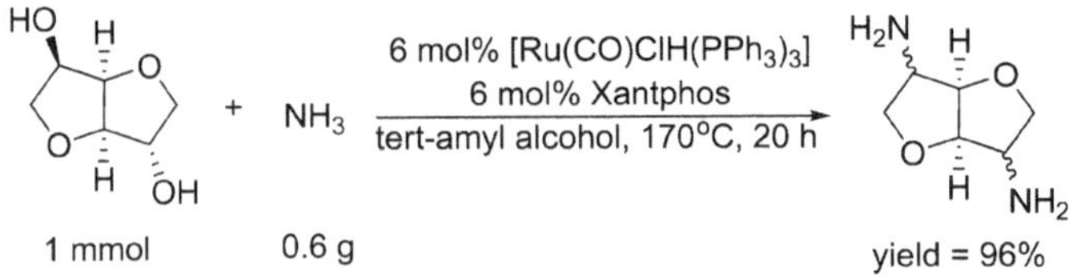

SCHEME 1.19 The catalytic diaminations of biomass-derived primary and secondary diols with ammonia.

resources, especially biomass-derived carbohydrate. However, aminations of such challenging substrates with ammonia have not been explained until 2011.[82] Beller's group reported the first catalytic diaminations of primary and secondary diols with ammonia using commercially available [Ru(CO)ClH(PPh_3)$_3$]/Xantphos catalysts, and the corresponding diamines were obtained in good yields and high selectivity (Scheme 1.19).[83] This transformation was successfully applied for the selective diamination of isosorbide, which is easily obtained from glucose. The resulting diamines are of industrial interest as new monomers for polyamides. In addition, other primary as well as secondary alcohols and diols, including hydroxy-substituted esters, can be efficiently converted to the corresponding primary amines or diamines with good to excellent yields. Recently, a similar protocol has been published by the Vogt group.[84] Various bio-based monoalcohols and diols could be selectively converted with ammonia to synthesize primary amines in the presence of $Ru_3(CO)_{12}$ and the acridine-based diphosphine. Noteworthy, the catalyst has a high thermostability and can be reused in at least six consecutive runs.

In 2011, Beller's group presented the catalytic amination of α-hydroxy amides with a variety of amines including anilines, primary and secondary aliphatic amines, and ammonia in the presence of a commercially available [$Ru_3(CO)_{12}$]/DCPE catalyst system without special equipment (Scheme 1.20).[85] A wide range of α-amino acid amides is synthesized from amination of α-hydroxy amides or esters.

1.2.2 Ir Catalyst System

In 2002, Fujita and coworkers reported the oxidative N-heterocyclization of amino alcohols catalyzed by iridium complexes.[86] Indole derivatives and

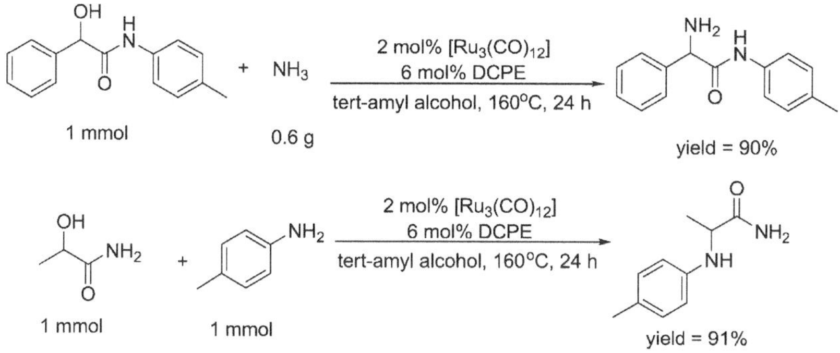

SCHEME 1.20 The catalytic aminations of α-hydroxy amides with amines/ammonia.

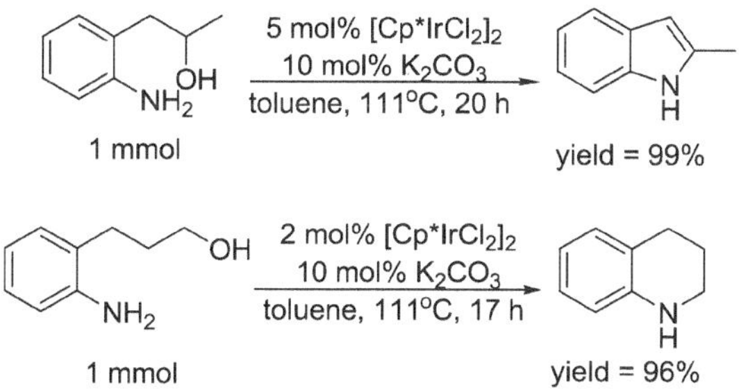

SCHEME 1.21 The Ir-catalyzed oxidative *N*-heterocyclization of amino alcohols.

1,2,3,4-tetrahydroquinolines are synthesized in good to excellent yields by means of a $[Cp^*IrCl_2]_2/K_2CO_3$ catalytic system (Scheme 1.21).

Since 2002, the catalytic system of $[IrCp^*Cl_2]_2$ has been applied in a number of alcohol amination reactions[87] including cyclization reactions of diols with amines[88] and amination of secondary alcohols.[89] In 2008, Fujita et al. further improved the protocol by changing the base to $NaHCO_3$ (Scheme 1.22).[90] The employment of equimolar amounts of amine and alcohol is sufficient to obtain high yields of the desired products, and the reaction of equimolar

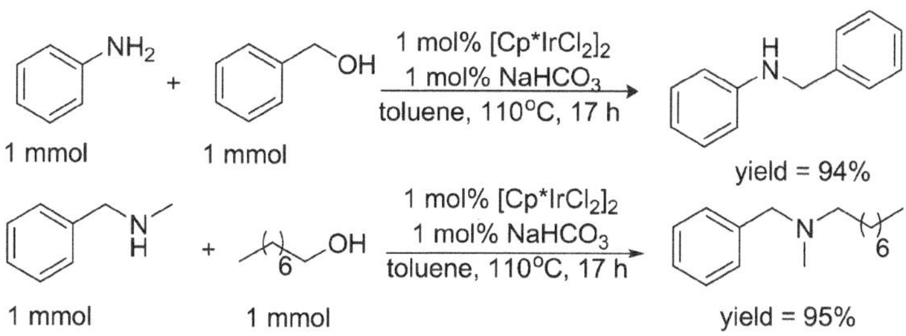

SCHEME 1.22 The amination of alcohols catalyzed by $[Cp^*IrCl_2]_2/NaHCO_3$.

amounts of aniline and benzyl alcohol in the presence of $[Cp^*IrCl_2]_2$ (1.0 mol% Ir) and $NaHCO_3$ (1.0 mol%) in toluene at 110°C gives *N*-benzylaniline in 94% yield. A range of primary and secondary amines was *N*-alkylated with primary and secondary alcohols at a comparatively mild temperature of 110°C, and a wide variety of secondary and tertiary amines can be synthesized with good to excellent yields.

In general, the protocols using $[IrCp^*Cl_2]_2$ require a base and DFT calculations have indicated that the coordinated base is part of the catalytically active species.[91]

In 2010, the group of Williams demonstrated that primary amines could be directly alkylated with primary alcohols using 1 mol% $[IrCp^*I_2]_2$ catalyst in the absence of base or other additives (Scheme 1.23).[92] A range of secondary amines were synthesized with good to excellent yields. The reaction conditions were comparable to Fujita's protocol. Interestingly, the reaction was run in water. After that, the authors further extended the application of this catalytic system.[93] The conversion of primary amines into *N*-heterocycles by the reaction with diols was achieved and the *N*-alkylation of sulfonamides was also contained, although base was required. Noteworthy, the use of ionic liquid as solvent was especially beneficial for the synthesis of tertiary amines, and this was the first time that ionic liquid was used as the solvent for borrowing hydrogen methodology.

A similar protocol without the addition of base was presented by Fujita and coworkers.[94,95] A wide variety of secondary and tertiary amines were synthesized by the *N*-alkylation reactions of theoretical equivalents of amines with alcohols in the presence of water-soluble and air-stable iridium-ammine iodide complex, $[Cp^*Ir(NH_3)_3][I]_2$, in water under air without a base.[95] It is worth noting that this method is also effective for *N*-alkylation of aqueous ammonia with alcohol, and the tertiary and secondary amines can be synthesized by the multi-alkylation of aqueous ammonia with theoretical equivalents of primary and secondary alcohols (Scheme 1.24).[94]

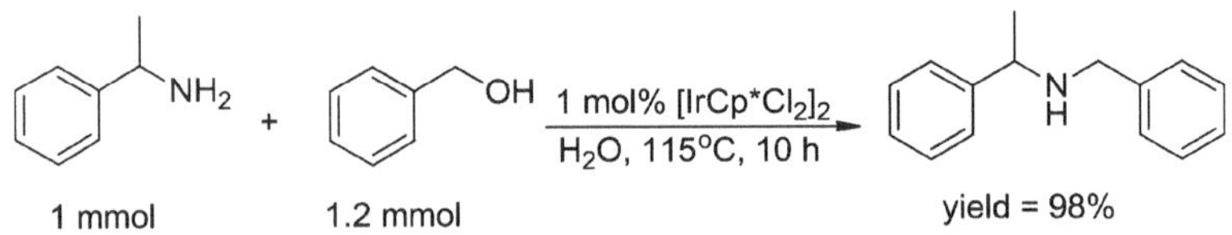

SCHEME 1.23 The amination of alcohols catalyzed by $[IrCp^*Cl_2]_2$ without base.

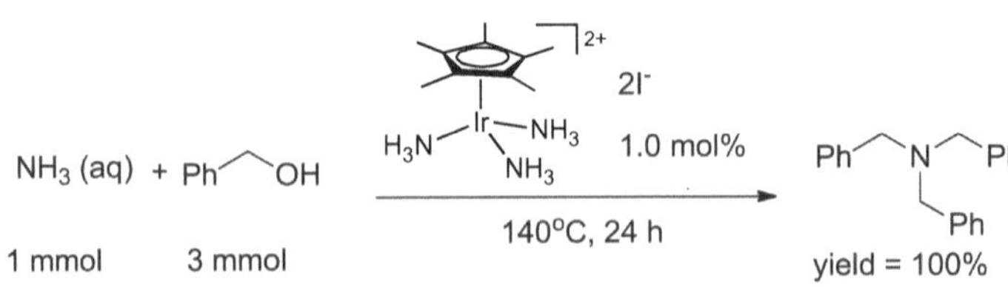

SCHEME 1.24 The *N*-alkylation of aqueous ammonia with alcohol.

In 2008, Peris reported another protocol without base addition.[96] Several secondary amines could be synthesized by the *N*-alkylation of primary amines with excess of primary and secondary alcohols in the presence of [$IrCl_2Cp^*$ (I^{nBu})] (I^{nBu} = 1,3-di-nbutylimidazolylidene). But good yields were obtained only in the case of aniline and benzylamine.

In 2012, Martín-Matute et al. prepared iridium complexes with chelating [NHC-alcohol] ligands, which are effective for the alkylation of anilines with alcohols under base-free conditions (Scheme 1.25).[97] A large variety of aromatic amines could be alkylated with primary and secondary alcohols to afford higher-order amine products with up to >99% yield over bis-cationic Ir^{III} complexes [Cp*(NHC-OH)Ir (MeCN)]$^{2+}$2[BF_4^-] and with a 1:1 ratio of reactants in short reaction times (2–16h). Quantitative yields were also obtained at 50°C, but longer reaction times (48–60h) were needed.

In 2009, Ishii et al. reported the *N*-heterocyclization of naphthylamine with 1,2-and 1,3-propanediol.[98] In the presence of $IrCl_3{\cdot}3H_2O$ and the BINAP ligand, 1,2- and 1,3-diols were converted into benzoquinolines and benzoindoles derivatives in moderate to excellent yields (Scheme 1.26). The reaction is believed to start with the *N*-alkylation of naphthylamine with the diol through borrowing hydrogen amination followed by C—H activation of the aromatic ring.

In the same year, the group of Crabtree developed an air-stable iridium complex bearing a chelating pyrimidine-functionalized *N*-heterocyclic carbene.[99] It was found to be active for *N*-alkylation of amines with primary alcohols. However, good yields were only achieved for benzyl alcohol and the reaction required 50mol% $NaHCO_3$.

Recently, Kempe's group developed a novel P,N-ligand-stabilized iridium catalyst for the selective monoalkylation of (hetero)aromatic amines with

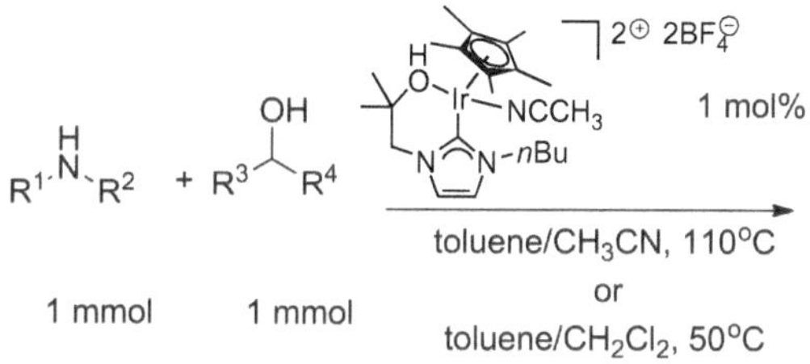

SCHEME 1.25 The *N*-alkylation of anilines with alcohol catalyzed [Cp*(NHC-OH)Ir (MeCN)]$^{2+}$2[BF_4^-].

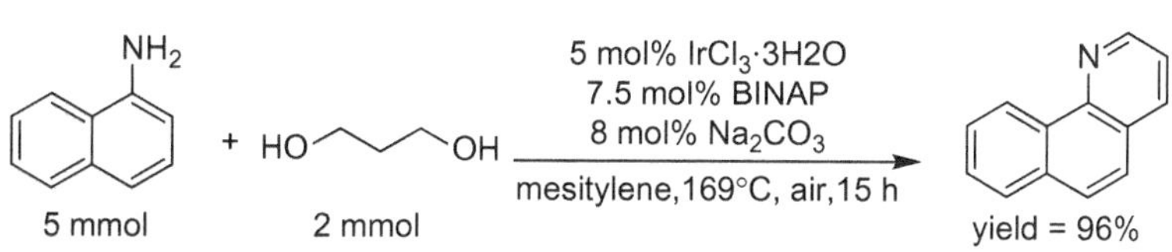

SCHEME 1.26 The *N*-heterocyclization of naphthylamine with diols.

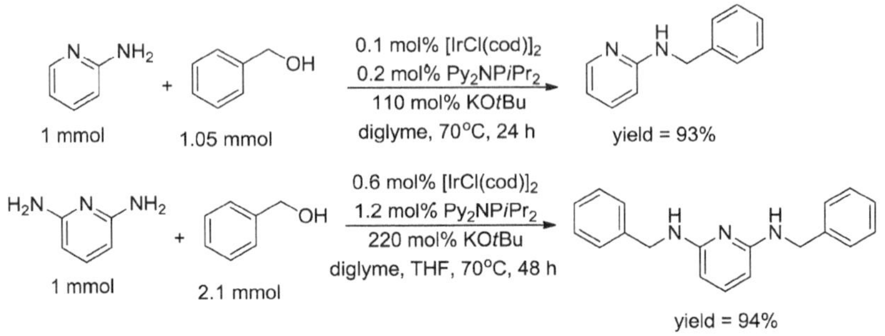

SCHEME 1.27 The *N*-alkylation of heteroaromatic amines and diamines with alcohols.

primary alcohols.[100] A series of symmetrically and unsymmetrically alkyl diamines were synthesized.[101,102] Further, they demonstrated this catalyst to be active at 70°C with catalyst loadings as low as 0.1 mol% Ir, although equivalent strong base KO*t*Bu was required (Scheme 1.27).[103] These mild reaction conditions could not be applicable to the alkylation of anilines. For these compounds, it was necessary to increase the catalyst loading to 0.6 mol% to obtain comparable results. Therefore, an improved molecular defined catalyst was developed by Kempe in 2010.[104] It showed significantly better performance for the alkylation of anilines with alcohols than the previously reported system. However, the drawback of Kempe's iridium systems is obvious, i.e., they require stoichiometric amounts of base, which lowers the atom efficiency and requires aqueous workup.

In 2010, Börner, Andrushko, and coworkers reported the amination of aliphatic primary alcohols and diols in the presence of the chlorodihydride complex $[IrH_2Cl(iPr_2PC_2H_4)_2NH]$ (Scheme 1.28).[105] The selective monoamination of diols with secondary amines should be highlighted. Ethylene glycol reacted smoothly with diethylamine, dimethylamine, and piperidine, and the monoamination products were obtained with good to excellent yields. In the case of pyrrolidine, it is necessary to add the base additive NaOtBu and increase the reaction temperature to 140°C to gain comparable results.

A breakthrough in the *N*-monomethylation of amines with methanol as a methylating agent has been achieved by Li et al. in 2012 (Scheme 1.29).[106] The direct *N*-monomethylation of arylsulfonamides and regioselective *N*-monomethylation of amino-azoles with methanol was accomplished with commercially available $[Cp^*IrCl_2]_2$ (0.1–0.4 mol%) and NaOH (1 equiv.) at 150°C.

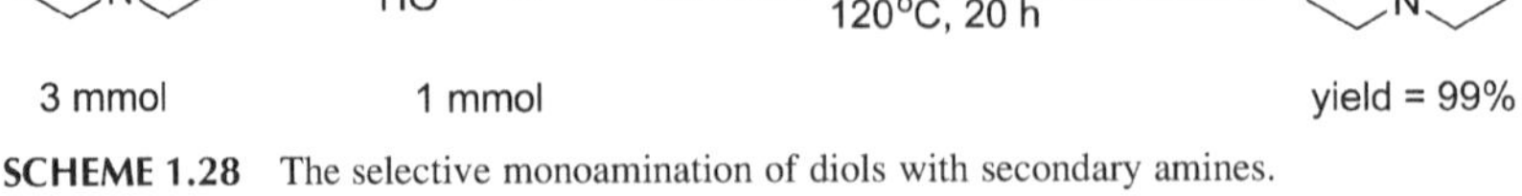

SCHEME 1.28 The selective monoamination of diols with secondary amines.

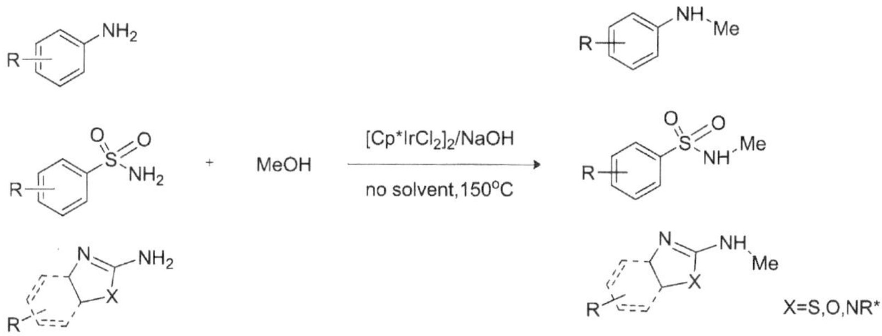

SCHEME 1.29 The *N*-monomethylation of arylsulfonamides and regioselective *N*-monomethylation of amino-azoles with methanol.

In 2012, Bruneau reported the three-component preparation of diverse *N*-arylpiperidine derivatives using easily accessible diols, primary amines, and aldehydes as starting materials, a new iridium complex featuring a phosphanesulfonate chelate as catalyst and formic acid (2equiv,) as a reducing agent (Scheme 1.30).[107] This method consists of three steps: step 1, anilines were reacted with diols without any bases, in the presence of Ir complex catalyst (3mol%) at 150°C for 16h; step 2, after cooling, aldehydes (1.1equiv.) were added and the reaction was heated for an additional 19h at 150°C; step 3, formic acid was added as a reducing agent to ensure complete reduction of the unsaturated intermediates and the reaction continued to perform at 150°C for 2h under inert atmosphere to achieve the desired *N*-arylpiperidine with moderate to good yields.

In 2013, a bidentate iridium NHC-phosphine complex was developed by Andersson,[108] which is highly effective for the *N*-monoalkylation of aromatic amines with a wide range of primary alcohols. High isolated product yields could be obtained under very mild conditions (0.5–1.0mol% catalyst loading, 50°C). Even at room temperature and under solvent-free conditions, this reaction could also be performed smoothly with a longer reaction time (48h).

On the basis of pyrrole synthesis which was previously established, Kempe developed a new [2+4] pyridine synthesis catalytic system in 2013.[109] A primary or secondary alcohol can act as a C_2 building block that reacts with

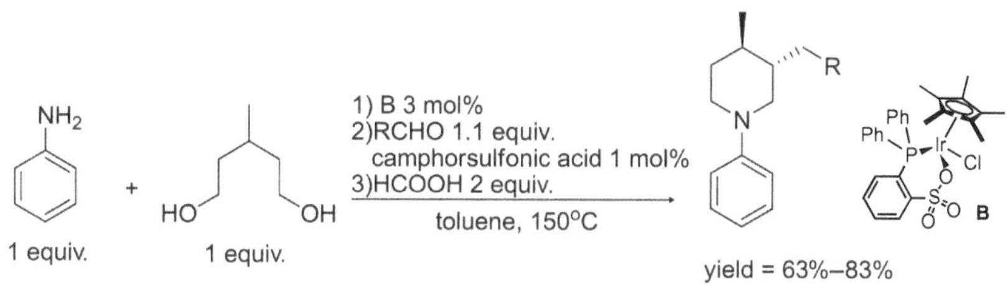

SCHEME 1.30 The three-component preparation of *N*-arylpiperidine derivatives.

a 1,3-amino alcohol in the presence of an iridium-based catalyst to provide a variety of regioselectively substituted pyridines, including challenging unsymmetrically substituted pyridines. In addition, the synthesis of quinolines from secondary or primary alcohols and amino alcohols is also reported by Kempe in 2014 using a novel family of iridium catalysts stabilized by P,N-ligands.[110]

Cp*iridium(III) half-sandwich complexes with amino acidato ligands are highly active and selective catalysts for the alkylation of amines with alcohols.[111] This method does not require basic additives or harsh reaction conditions for high activity and works both in an organic solvent and in water. For example, the reaction of 1-octylamine and 1-hexanol afforded the corresponding secondary amine with 93% yield in toluene and 96% yield in water.

In 2015, Crabtree reported a series of iridium bis-NHC compounds, which are active and highly selective for *N*-monomethylation of anilines with methanol through a "borrowing hydrogen" reaction.[112] But 1–5 equiv. of KOH and *MW* heating at 120°C are necessary.

The groups of Stephens and Marr reported the one-pot microbial and chemocatalytic reaction to transform waste glycerol into valuable secondary amines in 2009.[113] At first, the fermentation of glycerol by Clostridium butyricum formed 1,3-propanediol. After fermentation, the aqueous solution containing 1,3-propanediol was transferred to a sealed tube, and aniline, iridium catalyst, toluene, and K_2CO_3 were added. The amination of 1,3-propanediol produced by Clostridium butyricum grown on crude glycerol from biodiesel manufacturing was also demonstrated, and the reaction afforded only 15% conversion but the 82% selectivity for the monoaminated product was remarkable (Scheme 1.31).

The amination of carbohydrates was presented by Cumpstey, Martin-Matute, and coworkers in 2011 (Scheme 1.32).[114] Primary carbohydrate amines at primary and secondary carbons are alkylated by alcohols in the presence of $[Ir^*CpCl_2]_2$. When primary carbohydrate alcohols are used as the coupling partners and in the presence of Cs_2CO_3, amine-linked pseudodisaccharides are obtained. Secondary carbohydrate alcohols are unaffected under these conditions, which allow regioselective reactions.

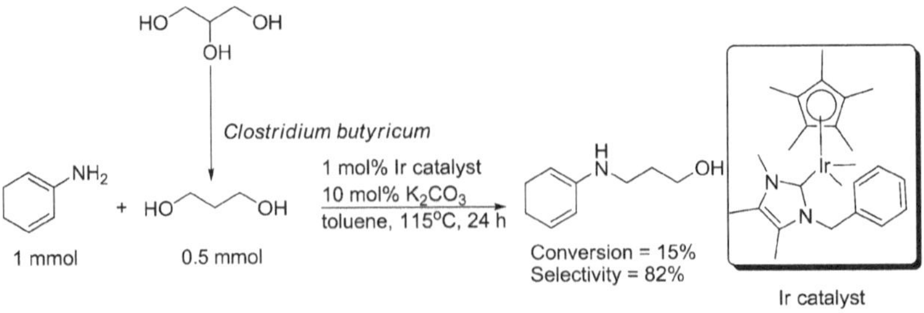

SCHEME 1.31 The combination of bio- and chemocatalysis for alcohol amination.

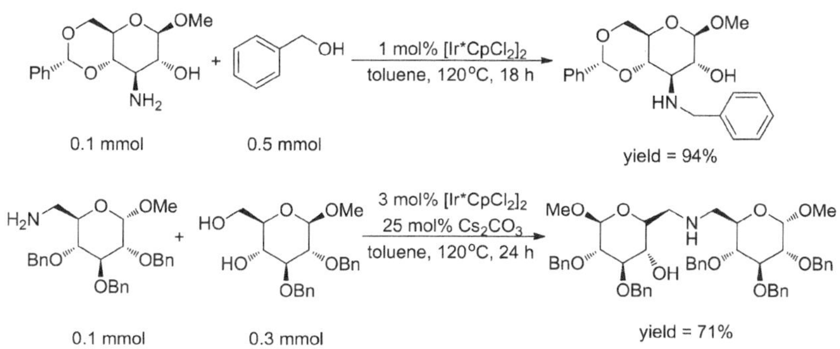

SCHEME 1.32 The catalytic amination of carbohydrates.

In 2007, Hartwig reported the important enantioselective aminations of terminal allylic alcohols by using the cyclometalated iridium catalysts prepared in situ from $[Ir(COD)Cl]_2$ and the chiral phosphoramidite ligands.[115] A series of branched allylic amine products were obtained with high regioselectivity and enantioselectivity, but the equivalent amount of a niobium alkoxide or catalytic amounts of triphenylborane was still required to activate the hydroxyl group in the alcohols.

A breakthrough in the field was provided by Roggen and Carreira, who documented the efficient Ir-catalyzed stereospecific allylic amination of unactivated secondary (2 degrees)-allylic alcohols by means of sulfamic acid as an ammonia equivalent, providing the primary amines with good to excellent enantioselectivity (Scheme 1.33).[116,117] The 1 degree allylic amines can be isolated directly in moderate to good yields. Further study revealed that the enantioselective allylic amination of racemic secondary allylic alcohols could also be achieved by changing the catalyst to $[Ir(coe)_2Cl]_2$ and optimizing the reaction conditions.[118,119] It is particularly noteworthy that the use of *N,N*-dimethylformamide was proposed to be better for the activation of the secondary allylic alcohols by reacting with the sulfamic acid.[120]

In 2013, Zhao et al. developed the enantioselective amination of alcohols on the basis of the borrowing hydrogen methodology by using a chiral Ir complex in cooperation with a chiral phosphoric acid, and a wide range of chiral amines were synthesized with good to excellent enantioselectivity (Scheme 1.34).[121]

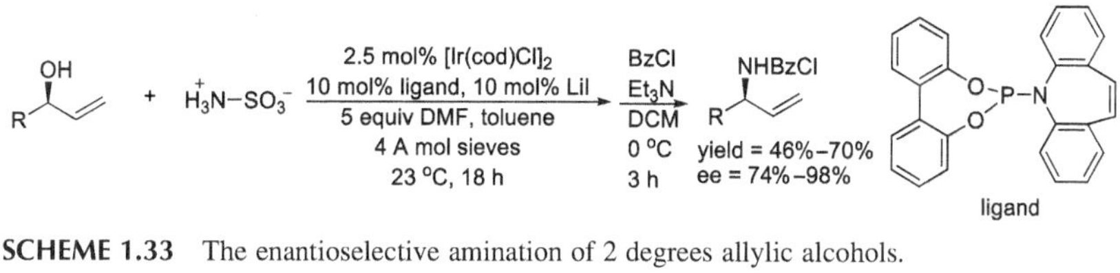

SCHEME 1.33 The enantioselective amination of 2 degrees allylic alcohols.

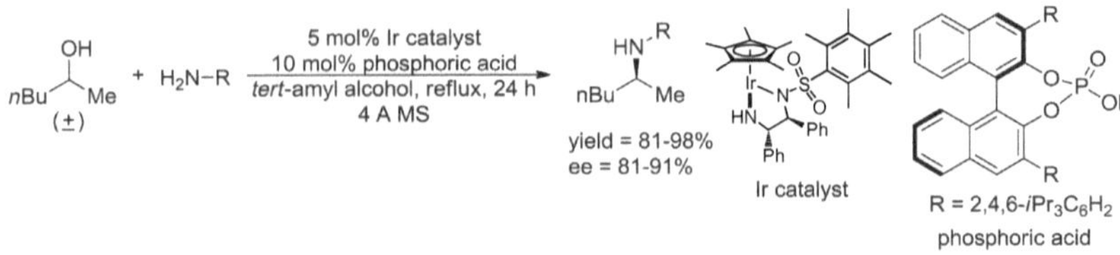

SCHEME 1.34 The enantioselective amination of alcohols.

Various anilines including those bearing substituents of an electron-donating or electron-withdrawing nature were well tolerated. It is particularly notable that heterocyclic amines, such as 5-aminoindole, can be successfully coupled with an alcohol to yield the chiral amine in good yield and enantioselectivity. However, in the case of other amines, such as benzylamine, tosylamide, or secondary amines, this method failed to yield the product with good efficiency or selectivity. The current method also exhibited potential activity for the intramolecular amination of alcohols and the chiral amines could be synthesized with good yield (73%) and promising enantioselectivity (68% ee). Recently Zhao et al. further extended the application of this catalytic system for the dynamic kinetic asymmetric amination of α-branched alcohols. Under the optimized reaction conditions, the mixture of four isomers in the alcohol substrate was converted to diastereo- and enantioenriched acyclic chiral amines with excellent selectivity and moderate to good yields, although the reaction time was usually long, up to 96h.[122]

1.2.3 Pd Catalyst System

As early as 1970, the amination of allylic alcohols with primary or secondary amines was reported by Atkins et al.[123] Up to 95% yield of the desired tertiary amine was obtained when allylic alcohols reacted with diethylamine in the presence of palladium acetylacetonate (0.5mol%) and triphenylphosphine (0.5mol%) at 50°C. However, a mixture of mono- and disubstituted amines was provided when a primary amine and *n*-butylamine were used as the starting materials. In 2006, Yang et al. further revealed that the combination of palladium acetylacetonate and triphenylphosphine was also effective for the allylation of aromatic amines in water and the addition of a 1-adamantanecarboxylic acid (1-$AdCO_2H$) could remarkably enhance both the reaction rate and yield.[124]

The intramolecular amination of allylic alcohols was achieved by Hirai with 30mol% $PaCl_2(MeCN)_2$.[125,126] The chiral 2-functionalized piperidine could be synthesized with the optically active urethane as substrates. It is noteworthy that this catalytic system allowed the reaction performing at room temperature without any organic ligands or additives.

In 1995, Masuyama et al. reported that the Pd(0) catalyst could catalyze the amination of allylic alcohols. Both primary and secondary amines were tolerated and 15%–75% yields were obtained in the presence of $Pd(PPh_3)_4$ (2mol%)

and a stoichiometric amount of $SnCl_2$ and Et_3N.[127] A better result was obtained when the stoichiometric amount of Et_3B was used.[128] Interestingly, the undesirable additives could be avoided when the reaction was performed under normal pressure of CO_2[129] or in the presence of $Pd(POC_6H_5)_3)_4$.[130]

In 1999 Yang et al. demonstrated the *N*-allylation of anilines with the direct use of allylic alcohol. About 48%–78% yields of desirable mono-allylanilines were obtained in the presence of $Pd(OAc)_2$ (1 mol%), PPh_3(4 mol%), and $Ti(OiPr)_4$ (25 mol%) at 50°C in benzene.[131] Better results could be achieved under reflux conditions or by increasing the amount of palladium catalyst to 2.5 mol%.[132] 10-mmol scale reaction could also be performed smoothly to get the desirable mono-allylanilines in 48%–88% yields.[133] The detailed investigation about the effect of the reaction temperature and the phosphine ligands revealed that the reaction process was temperature dependent, and that decreasing the size or the electron-withdrawing ability of the phosphine ligands was in favor of the formation of less highly substituted alkenes.[134] Further study found that this catalyst system was also effective for the amination of allylic alcohols with aminonaphthalenes and the monoallylation products were formed with good selectivity.[135]

When a higher amount of $Pd(OAc)_2$ (5 mol%) and the sodium diphenylphosphinobenzene-3-sulfonate (TPPMS) (10 mol%) were used, the catalyst system showed good functional group tolerance. The allylation of anthranilic acids with allylic alcohols performed smoothly at room temperature and good to excellent yield of mono-*N*-allylated products were formed (Scheme 1.35).[136] The unprotected amino acids also reacted well with 1,1-dimethylallyl alcohol to supply the mono-*N*-allylation products in good yields, although the high temperature (120°C) and equivalent amount of AcONa (2 equiv.) additive were necessary.[137] However, the above catalysts were not recyclable. In 2012, Bhanage et al. reported a recyclable homogeneous catalyst system $Pd(OAc)_2$/TPPTS (triphenylphosphine trisulfonate sodium salt), which was applicable to a wide range of hindered and functionalized aromatic/aliphatic amines, and the catalyst could be reused only by simple extraction separation.[138]

In 2012, Beller's group reported the amination of allylic alcohols with electron-deficient heterocycles bearing aryl or alkyl substituents using the combination of $Pd(OAc)_2$ (2.5 mol%) and 1,2-bis(diphenylphosphinomethyl) benzene (5 mol%). Under the optimized reaction conditions (100°C, toluene,

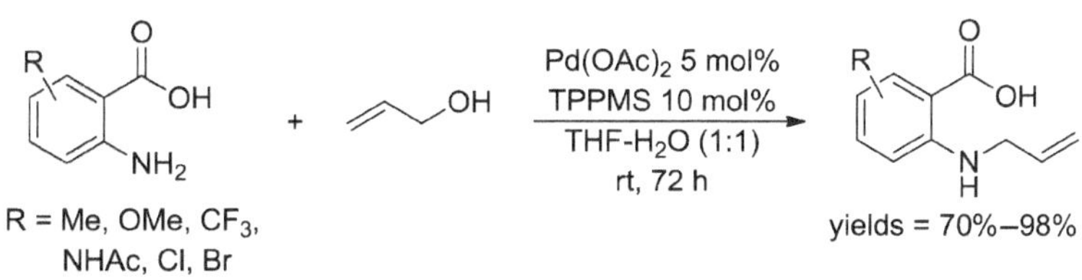

SCHEME 1.35 The Pd-catalyzed allylation of anthranilic acids.

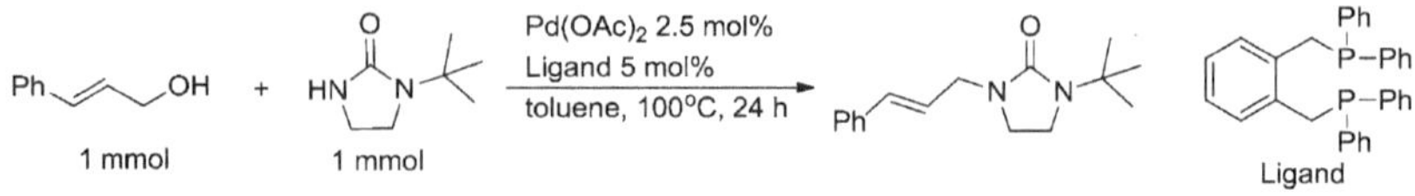

SCHEME 1.36 The Pd-catalyzed allylation of electron-deficient heterocycles.

24 h), good to excellent yields were obtained (Scheme 1.36).[139] However, the use of fancy phosphorus ligands was necessary in the above catalytic system. In the same year, Beller's group developed a catalytic system without phosphorus for the amination of allylic alcohols with aliphatic amines. The combination of $Pd(OAc)_2$ (2.5 mol%) with 1,10-phenanthroline (5 mol%) gave the best results.[140]

An example for the direct amination of allylic alcohols with anilines in the absence of activating agents was presented by Ozawa.[141–143] In the presence of (π-allyl)palladium complexes bearing sp^2-hybridized phosphorus ligands (diphosphinidenecyclobutene: DPCB-Y), different allylic alcohols could be converted with anilines at room temperature to afford the monoallylated products with excellent regio- and stereoselectivity (Scheme 1.37). Noteworthy, the optically active alcohol could also be converted into the corresponding monoallylated aniline without a notable loss of optical purity. Additionally, thorough investigation of the mechanism revealed that the DPCB-Y ligands bearing sp^2-hybridized phosphorus atoms might interact very flexibly with palladium by σ-donation and π-back-donation interaction, which was the origin of the high catalytic activity of DPCB-Y complexes.[144]

Following the work of Ozawa, some other (allyl)palladium complexes, such as $[PdCl(\eta^3\text{-allyl})]_2$,[145–149] $[Pd(P(OPh)_3)_2(\eta^3\text{-allyl})][OTf]$,[150] DPP-Xantphos $Pd(\eta^3\text{-allyl})][OTf]$ complex,[151] and $[Pd(cod)(\eta^3\text{-allyl})]BF_4$,[152] have been used as catalysts or catalyst precursors for the amination of allylic alcohols with or without the equivalent amount of additives.

In 2010 Qu et al. reported a cubane-type sulfido cluster $[(Cp^*Mo)_3S_4Pd(dba)][PF_6]$, which was able to effectively catalyze the amination of allylic alcohols with H_3BO_3 as an additive.[153] Various amines and allylic alcohols were explored and in the case of allylic alcohols bearing substituents at either α- or γ-position, only linear allylic amination products were obtained.

$Pd(Xantphos)Cl_2$ has been identified as an efficient catalyst for the catalytic amination of allylic alcohols, which allows the direct amination of allylic

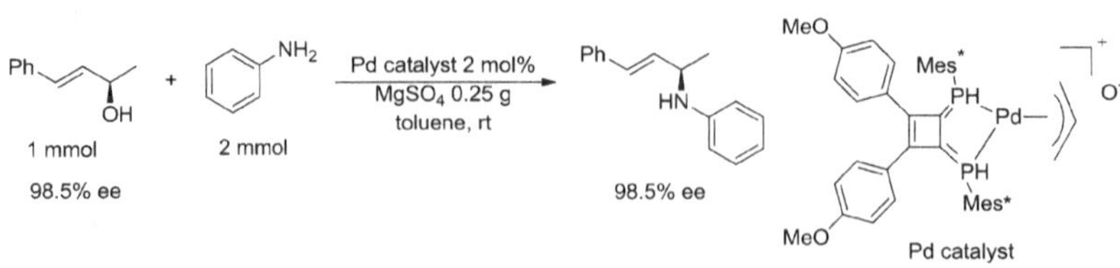

SCHEME 1.37 The Pd-catalyzed amination of allylic alcohols without activating agents.

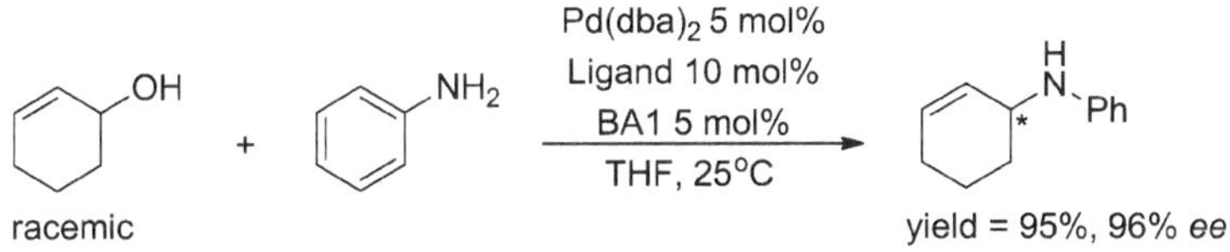

SCHEME 1.38 The Pd-catalyzed amination of allylic alcohols.

alcohols with amines to proceed at room temperature in the absence of additives.[154] A wide range of linear allylic amines can be prepared in good to excellent yields with high stereoselectivities.

In 2014, Beller's group reported efficient enantioselective amination of racemic allylic alcohols with a variety of functionalized amines by the combination of [$Pd(dba)_2$], chiral phosphoric acid, and phosphoramidite ligand.[155] The reaction proceeded even with less reactive alkyl allylic alcohols and furnished a number of products in almost quantitative yields with excellent enantioselectivity (Scheme 1.38).

1.2.4 Rh/Os/Re/Pt Catalyst Systems

As the first homogeneous example of transition metal-catalyzed *N*-alkylation of amines with alcohols, $RhH(PPh_3)_4$ complex was effective in the alkylation of primary amines with a large excess of primary alcohols (such as methanol, ethanol, or benzyl alcohol) to give the expected *N*-alkylated amines (Scheme 1.39).[13] The results were usually excellent, and only when the secondary amine pyrrolidine and aniline were used as starting materials, did the yield decrease to 74% and 39%, respectively.

By using the $RhH(PPh_3)_4$ catalyst, *N*-alkylation of sulfonamides with benzyl alcohols can be easily accomplished under aerobic conditions, and monoalkylated sulfonamides could be obtained in high yields (64%–99%), too.[156]

In 2011, Gusev et al. developed osmium pincer type complexes and employed them in the amination of primary alcohols with primary amines (Scheme 1.40).[157] Employing 0.1 mol% osmium pincer type complexes, good yields of the secondary amine products were achieved. The reactions were performed at a very high temperature (200°C), but interestingly only traces of tertiary amines were observed.

As the first transition metal complex in trace quantity to catalyze direct formation of ethers from alcohols, methylrhenium trioxide (MTO) also showed good catalytic performance for the amination of primary aromatic alcohols

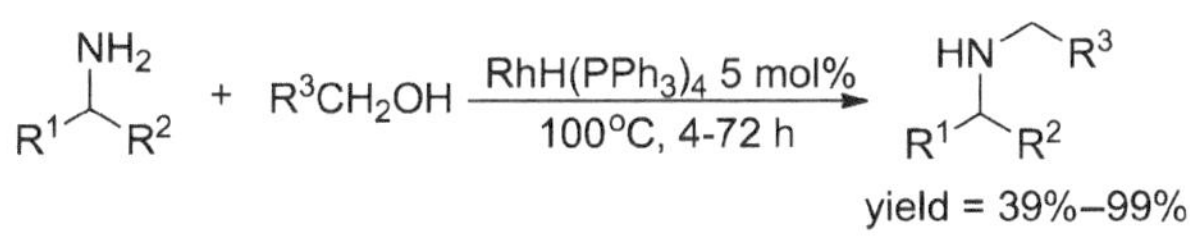

SCHEME 1.39 The Rh-catalyzed *N*-alkylation of amines with alcohols.

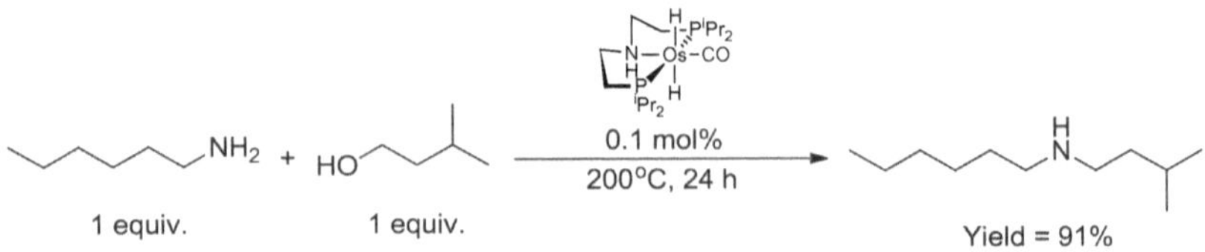

SCHEME 1.40 The Os-catalyzed *N*-alkylation of amines with alcohols.

to supply the desired secondary/tertiary amines at room temperature.[158] Good yields were obtained when aniline and hexylamine reacted with alcohols in excess (3 equiv.). However, when the alcohols were used in equal quantities, the yields of amine products were considerably reduced.

An air- and moisture-tolerant rhenium-oxo complex Re(dppm)OCl_3 has been used in the synthesis of propargylamines by the coupling of propargyl alcohols with amines in the presence of NH_4PF_6.[159] *p*-Nitroaniline, tosylamines, and carbamates reacted smoothly with propargylic alcohol to supply the corresponding propargylamines in good yields. However, the allylamine was not tolerated.

The rhenium heptahydride complex [$ReH_7(PCy_3)_2$] was found to be an effective homogeneous catalyst for amination of various alcohols through hydrogen transfer mechanism (Scheme 1.41).[160] In a carbon monoxide atmosphere, a variety of anilines are able to be converted to corresponding secondary amines with a range of benzyl alcohols with good to excellent yields. The exploration of the reaction mechanism revealed that CO could stabilize active rhenium species in the reaction process.

In 2007, the direct amination of allylic alcohols with anilines and aliphatic amines without the use of an activating reagent catalyzed by the combination of Pt(cod)Cl_2 and a large bite-angle ligand DPEphos was reported by Ohshima and Mashima et al. The use of the large bite-angle ligand DPEphos is essential for obtaining high catalyst activity. Moreover, a sterically congested active site created by the large bite-angle ligand led to selective monoallylation products in good to excellent yields.[161] Additionally, they demonstrated that the catalyst system was also effective for the synthesis of a variety of primary allylamines by the direct reaction of allylic alcohols with aqueous ammonia (Scheme 1.42).[162] DFT study reveals that the large bite angle in diphosphine chelate ligands could prevent the formation of cationic platinum hydride complexes on protonation, which eventually led to deactivation of the catalyst.[163] The mechanistic investigations revealed that the formation of the π-allylplatinum

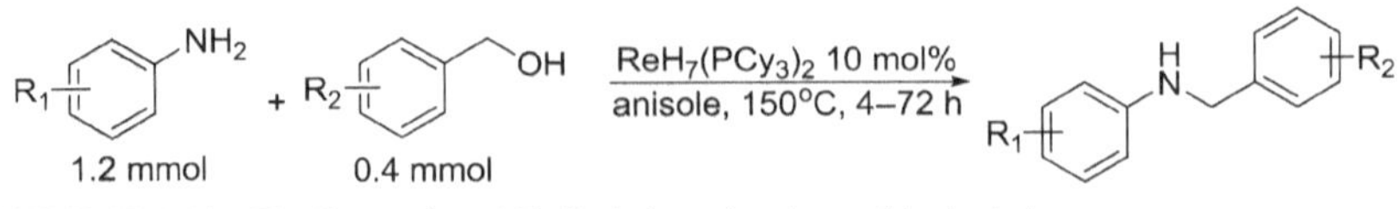

SCHEME 1.41 The Re-catalyzed *N*-alkylation of amines with alcohols.

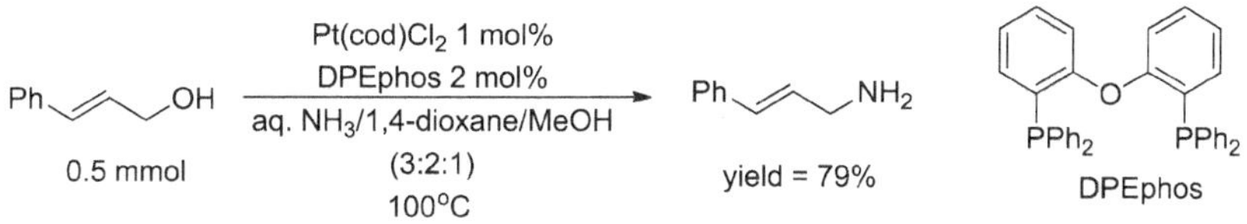

SCHEME 1.42 The Pt-catalyzed amination of allylic alcohols

complex through the elimination of water is an irreversible rate-determining step and that other processes in the catalytic cycle also are reversible, even at room temperature.[164]

1.2.5 Mo Catalyst System

The homogeneous molybdenum (V) complex-catalyzed amination of alcohols with sulfonamides and carbamates was described by Reddy in 2007.[165] This reaction proceeded as expected in CH_2Cl_2 at room temperature in the presence of 5 mol% $MoCl_5$ and secondary benzyl alcohols and propargyl alcohols gave desired products in good yields. However, primary benzyl alcohols and propargyl alcohols generated from aliphatic aldehydes were inert.

Following the work of Reddy, some molybdenum (VI) complexes have also been explored in the amination of alcohols. In 2009, Zhe et al. showed the combination of 10 mol% $MoO_2(acac)_2$ and 10 mol% NH_4PF_6 to be active for amination of allylic alcohols with formamides, sulfonamides, and nitro-substituted anilines.[166] The corresponding products were obtained in moderate to excellent yields at 65°C in acetonitrile. Additionally, they demonstrated that this catalyst system was also effective for amination of propargylic alcohol with sulfonamides and electron-withdrawing group-substituted aromatic amines.[167]

1.2.6 Co Catalyst System

In 2015, Kempe et al. reported the *N*-alkylation of aromatic amines by alcohols catalyzed by a cobalt complex stabilized by a PN_5P ligand (Scheme 1.43).[168] The catalyst allows the reaction to be carried out under mild conditions (80°C) with a relatively low catalyst loading (2 mol%), but strong base KO*t*Bu (1.2 equiv.) is required. In the next year, Kirchner reported a similar protocol.[169] A range of primary alcohols and aromatic amines were efficiently converted into mono-*N*-alkylated amines in good to excellent isolated yields over well-

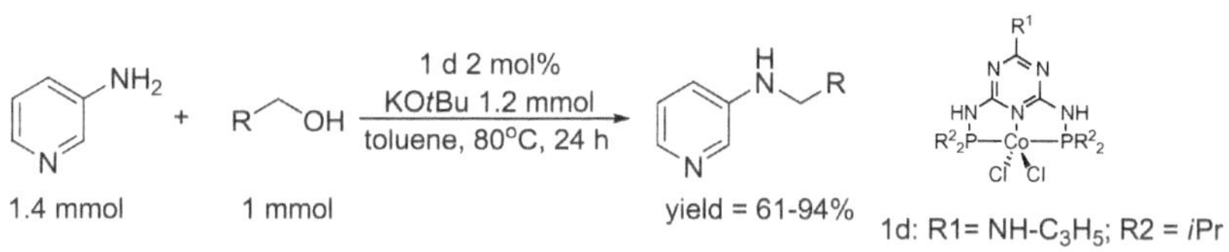

SCHEME 1.43 The cobalt complex-catalyzed *N*-alkylation of amines with alcohols.

defined Co(II) complexes stabilized by a PCP ligand (*N,N'*-bis-(diisopropyl-phosphino)-*N,N'*-dimethyl-1,3-diaminobenzene).

A similar cobalt(II)-catalyzed *N*-alkylation of amines with alcohols was achieved by Zhang and Zheng et al. in the same year.[170] A range of alcohols and amines including both aromatic and aliphatic substrates were efficiently converted to secondary amines in good-to-excellent yields over a well-defined cobalt(II) complex based on a pincer PNP ligand under base free conditions. Even cyclohexanol, a secondary alcohol, could react with aniline to afford secondary amine under the optimized reaction conditions, although only a 48% yield was obtained.

In 2016, Kirchner reported the cobalt (II) PCP pincer complexes with a 1,3-diaminobenzene backbone and employed it as a catalyst in the alkylation of amines by alcohols.[169] A range of primary alcohols and aromatic amines were efficiently converted into mono-*N*-alkylated amines in good to excellent isolated yields. The reaction could perform at 80°C but the addition of 1.3 equiv. amount of KO*t*Bu was necessary.

1.2.7 Cu Catalyst System

In 2009, Shi and Beller et al. reported the homogeneous copper-catalyzed *N*-alkylation of sulfonamides with alcohols.[171] High conversion (93%) and excellent selectivity (>99%) were obtained when the coupling reaction of *p*-toluenesulfonamide and benzyl alcohol was performed in the $Cu(OAc)_2$/K_2CO_3/air system (Scheme 1.44). HR-MS analysis revealed the formation of the bissulfonylated amidine as a new ligand, which stabilized the catalyst. The authors further demonstrated benzyl alcohols with different functional groups and thiophen-2-ol were well tolerated.[172] Further UV-visible measurements suggested interaction between the copper center and the bissulfonylated amine.

Ramón and Yus extended the application of $Cu(OAc)_2$/base system by replacing K_2CO_3 with KO*t*Bu in 2010.[173] The selective *N*-alkylation of amino derivatives with poor nucleophilic character, such as aromatic and heteroaromatic amines as well as sulfonamides, using primary alcohols as initial source of the electrophiles were achieved. In the case of sulfonamides, the alkylation process followed by a naphthalene-catalyzed reductive deprotection gives primary amines, which is an indirect alternative to the direct monoalkylation of ammonia. A study of the reaction using deuterium labeled reagents was

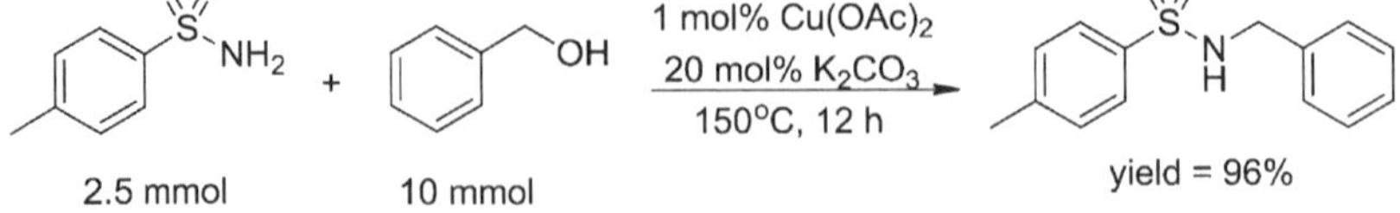

SCHEME 1.44 The homogeneous copper-catalyzed amination of alcohols.

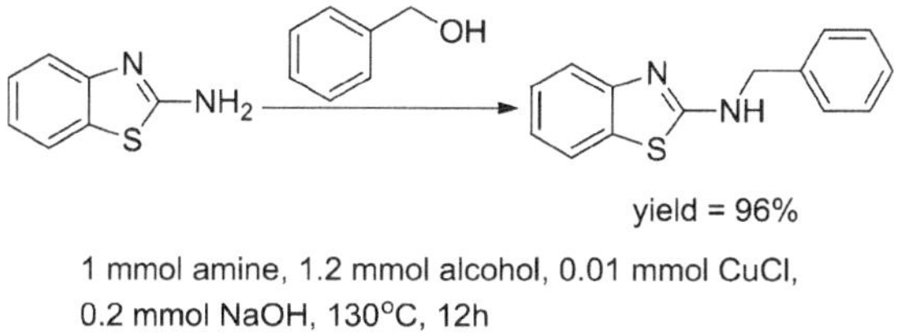

SCHEME 1.45 The regioselective *N*-alkylation of 2-aminobenzothiazoles with alcohols.

performed and the result indicated that dehydrogenation and hydrogenation steps did not take place on the same copper-atom coordination sphere, with the condensation step occurring out of the dehydrogenating catalytic species.[174]

In 2011, Li et al. demonstrated that regioselective *N*-alkylation of 2-aminobenzothiazoles to *N*-exosubstituted 2-(*N*-alkylamino)benzothiazoles occurred when using alcohols as alkylating agents in the presence of CuCl/NaOH (Scheme 1.45).[175] A wide range of 2-(*N*-alkylamino)benzothiazoles were synthesized with good to excellent yields.

1.2.8 Fe Catalyst System

The homogeneous iron-catalyzed *N*-alkylation of sulfonamides with benzylic alcohols through hydrogen-borrowing method was presented by Shi et al. in 2010 (Scheme 1.46).[176] Under optimized conditions, the $FeCl_2/K_2CO_3$ catalyst system is able to catalyze 21 different alkylation reactions and high yields, in general greater than 90%, are achieved in most cases. XPS analysis suggested a possible catalyst cycle between Fe(II) and Fe(0).

In 2011, a novel iron/amino acid catalyst system for direct amination of alcohols was reported by Saito et al. (Scheme 1.47).[27] The combination of $FeBr_3$ and DL-pyroglutamic acid showed broad substrate scope and functional

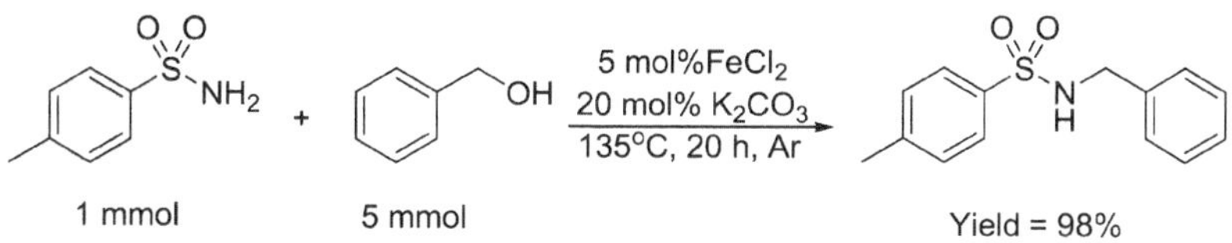

SCHEME 1.46 The iron-catalyzed *N*-alkylation of sulfonamides with alcohols through hydrogen-borrowing method.

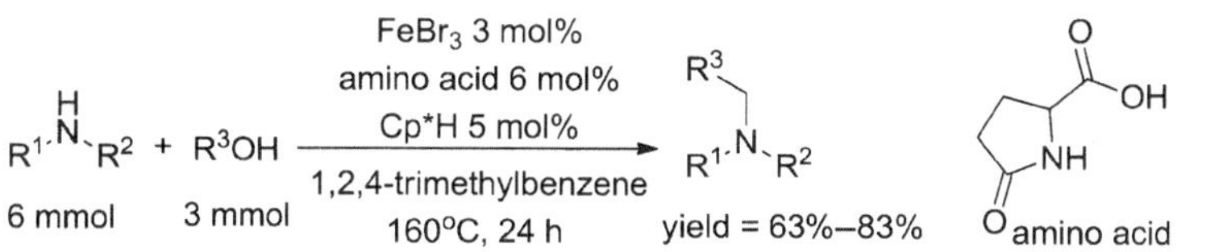

SCHEME 1.47 The selective *N*-monomethylation and dimethylation of aniline with methanol.

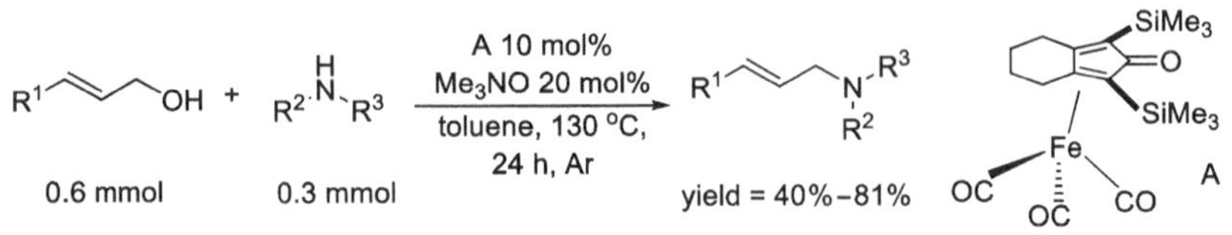

SCHEME 1.48 The Fe-catalyzed amination of allylic alcohols.

group tolerance with respect to each amine and alcohol component. Additionally, selective *N*-mono- or dimethyl products of aniline could be obtained as well, by varying the ratio of aniline and CH_3OH.

Iron phthalocyanine as an efficient catalyst for *N*-alkylation of amines with alcohols was reported by Singh in 2013.[177] This catalyst is effective for direct *N*-alkylation of aminobenzothiazoles, aminopyridines, and aminopyrimidines with alcohols and synthesis of 2-substituted benzimidazoles, benzothiazoles, and benzoxazoles from *N*-alkylation of *ortho*-substituted anilines (—NH_2, —SH, and —OH), but two equivalent NaOtBu are necessary.

In 2014, Barta's group reported a well-defined homogeneous Fe-based catalyst, which is effective for monoalkylation of anilines and benzyl amines with a wide range of alcohols, and formation of five-, six-, and seven- membered nitrogen heterocycles with diols.[178] Since then, a series of similar homogeneous Fe-based catalyst were synthesized and used in various alcohol aminations including amination of secondary alcohols,[179,180] amination of benzyl alcohol with aliphatic amines,[181] and amination of allylic alcohols (Scheme 1.48).[182]

In 2015, Kirchner reported a hydride iron (II) PNP pincer complex based on the 2,6-diamonopyridine scaffold, which are demonstrated to be active for the alkylation of aromatic, benzylic, and aliphatic primary amines with primary alcohols to give mono-*N*-alkylated amines. These reactions occurred at 140° C in toluene under base-free conditions but required the addition of molecular sieves.[183] The reaction temperature could be lowered to 80°C by the addition of strong base KO*t*Bu (1.3 equiv,) and replacing the catalyst with low-spin complexes [Fe(PNP)(CO)Br_2] bearing a carbonyl coligand.[184]

1.2.9 Ni Catalyst System

The amination of allylic alcohols with amines was investigated by Furukawa et al. in 1973.[185] In the presence of (n-$Bu_3P)_2NiBr_2$ catalyst, the allyl-substituted amines were obtained in high yields, but only morpholine was well tolerated.

The Ni(dppb)$_2$ complex in situ prepared form Ni(COD)$_2$ (COD = cycloocta-1,5-diene) and the diphosphine, have been proven to be intrinsically active catalysts for amination of allylic alcohols with diethylamine.[186] Initial activities observed with these types of catalysts are often higher than those of comparable palladium systems, but the Ni(dppb)$_2$ species suffer from poor stability during the reaction course.

1.2.10 Mn Catalyst System

Recently, Beller's group reported the homogeneous manganese catalyst for *N*-alkylation of amines with a variety of alcohols including methanol through hydrogen autotransfer methodology (Scheme 1.49).[187] A variety of substituted anilines are monoalkylated with different (hetero)aromatic and aliphatic alcohols even in the presence of other reducible functional groups over defined PNP manganese pincer complexes with the addition of 0.75 equiv. KO*t*Bu. The chemoselective monomethylation of primary anilines using an excess amount of methanol (1 mL) is also achieved by increasing the temperature to 100°C and the amount of KO*t*Bu to 1 equiv. Beller's group recently developed an improved protocol for selective *N*-methylation of aromatic amines with methanol by changing the manganese catalyst to lutidine-based manganese PNP-pincer complexes.[188] Compared to previously described manganese catalysts, this new complex allowed methylation with improved yields, but only with the addition of 0.5 equiv. KO*t*Bu.

1.2.11 Transition Metal-Free Catalyst System

Except for the amination of alcohols catalyzed by transition metals, some catalyst systems without transition metals have also been reported.[189–199] As early as 1993, Castedo et al. found that the reaction of benzyl alcohols and *N*-tosyl aminoacetaldehyde dimethyl acetal could be performed smoothly under Mitsunobu conditions[200] to supply the corresponding *N*-alkyl products.[198] Primary and secondary benzyl alcohols both were well converted in the presence of diethyl azodicarboxylate (DEAD) (2.5 equiv.) and PPh_3 (3 equiv.) at room temperature and moderate to good yields were obtained.

A highly efficient iodine-catalyzed allylic alkylation of amines with allylic alcohols was reported by Chan et al. in 2008.[197] A wide variety of sulfonamides and carbamates was tolerated and good to excellent yields were obtained in the presence of I_2 (5 mol%) and $CaSO_4$ (50 mg).

In 2011, Wang et al. demonstrated that the zinc-based ionic liquid ([CHCl][$ZnCl_2$]$_2$) as the reaction media and was able to catalyze the amination of secondary benzylic and allylic alcohols with some selected amines *via* a carbocation mechanism.[199] Of significance was the existence of carbocation detected by UV-vis spectroscopy.

A similar protocol was reported by Baeza and Nájera et al.[195] Fluorinated alcohols, such as 1,1,1,3,3,3-hexafluoroisopropanol (HFIP), 2,2,2-

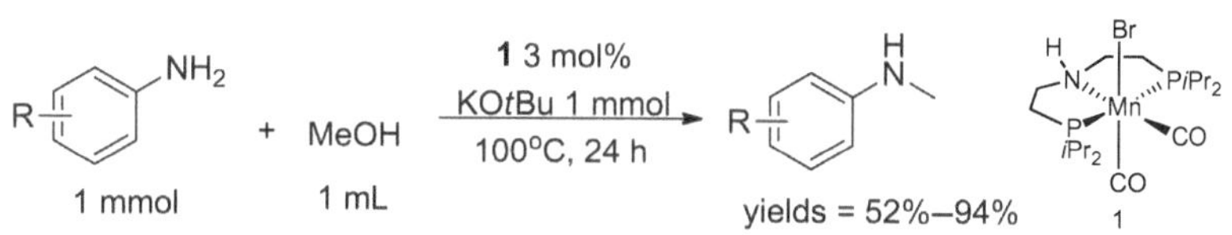

SCHEME 1.49 The Mn-catalyzed *N*-alkylation of amines with alcohols.

trifluoroethanol (TFE), as reaction media showed good catalytic performance for amination of allylic alcohols with different compounds including sulfonamides, carbamates, carboxamides, and amines, but a lower reaction temperature (50°C or 70°C) is required.

In 2011, Cossy et al. presented a one-pot, two-step method for amination of alcohols with primary and secondary amines via the combination of 2,2,6,6-tetramethyl-1-piperidinyloxyl and [bis(acetoxy)-iodo]benzene (TEMPO-BAIB) oxidation systems and $NaBH(OAc)_3$ reducing agents.[191] This method allows the chemoselective *N*-alkylation of amines with various nonactivated primary and secondary alcohols at room temperature and prevents epimerization of optically active substrates (amines or alcohols). For example, the amination of optically active alcohols (*S*)-solketal possessing a stereogenic center in the β-position was achieved with *tert*-butylamine, and 80% yield of secondary amine product was obtained.

In borrowing hydrogen type *N*-alkylation reactions, aldehydes were the products commonly obtained from activation of alcohols. Incidentally, Xu et al. found that the addition of external aldehydes could be used as catalysts for amination of alcohols with sulfonamides and amines (Scheme 1.50).[190] Although NMR analysis revealed that the aldehyde as catalyst could be regenerated and then recycled, the catalyst must be changed when changing of alcohols as substrates.

Recently, Shi's group demonstrated that a molecular-defined conjugated ketone was able to catalyze alcohol amination reaction through a hydrogen transfer mechanism.[189] Among the ketone derivatives screened, the conjugated Xanthone exhibited the best catalytic performance and under optimized reaction conditions, the yields of desired products reached 98%. Additionally, the reaction mechanism and kinetic study assumed carbonyl-hydroxyl groups are the catalytically active sites, and transfer hydrogenation reactions progress via the recycling of carbonyl and hydroxyl groups.

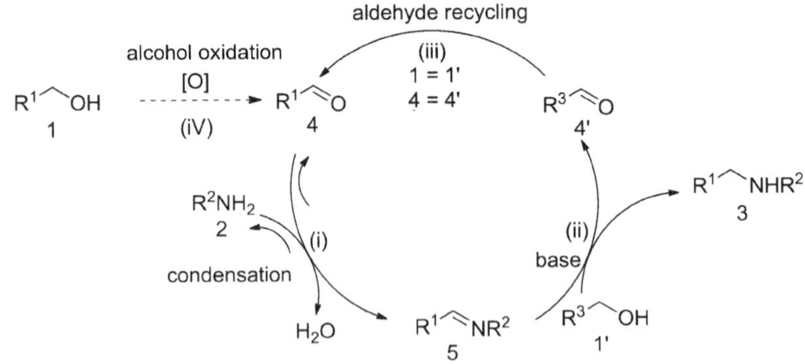

SCHEME 1.50 The possible mechanism for the aldehyde-catalyzed *N*-alkylation of amines with alcohols.

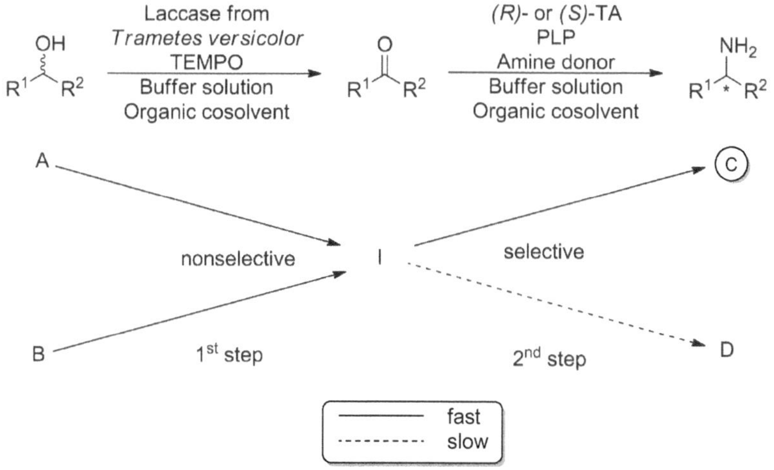

SCHEME 1.51 Synthesis of optically active amines from racemic alcohols through a one-pot sequential bienzymatic system using the Trametes versicolor laccase and a transaminase; A and B are the substrate enantiomers; I is the ketone intermediate; C and D are the product enantiomers.

A redox-neutral catalyst network for the asymmetric amination of *sec*-alcohols to corresponding α-chiral primary amines was successfully designed by Kroutil et al.[193] The artificial systems involved three to five redox enzymes and L-alanine was used as the amine donor. First, alcohols were oxidized to the corresponding ketone by an alcohol dehydrogenase. Then ketones were aminated by employing a]-transaminase. Under optimized conditions, (*S*)-octan-2-ol was converted to corresponding (*S*)-amine with 64% conversion and 96% ee in the presence of lactate dehydrogenase.

Gotor-Fernández and Lavandera et al. recently disclosed a one-pot/two-step bienzymatic asymmetric amination of secondary alcohols.[192] The approach is based on a sequential strategy involving the use of a laccase/TEMPO catalytic system for oxidation of alcohols into ketone intermediates, and then following transformation into optically enriched amines by using transaminases. A range of racemic (hetero)aromatic *sec*-alcohols with different substitutions have been converted into enantioenriched amines with good to excellent selectivities (90%–99% ee) and conversion values (67%–99%) (Scheme 1.51).

1.3 HETEROGENEOUS CATALYST SYSTEM

Although highly efficient homogeneous catalyst systems have been developed for *N*-alkylation of amines with alcohols, the expense and difficulty of recycling severely limit their industrial application. Therefore, it is highly desirable to develop an easily recoverable heterogeneous catalyst. In the past decades, various heterogeneous systems based on Cu, Ni, Fe, Mn, as well as noble metal catalysts, have been reported and excellent results were obtained Some are comparable to homogeneous noble metal catalysts.

1.3.1 Cu Catalyst System

Simple copper and copper oxide have been used as catalysts in the preparation of pyrazine from 2-aminoethanol by a hydrogen autotransfer process in a continuous flow reactor.[201] However, the yield for the aromatic compound was very low (6%), even at temperatures as high as 300°C. After a few hours, however, the copper catalyst lost its activity.

In order to stabilize the copper catalysts, other metals have been added as cocatalysts. One of them is the so-called copper-chromite catalyst ($CuCr_2O_4$-$BaCr_2O_4$). By using $CuCr_2O_4$-$BaCr_2O_4$ as a catalyst, several aliphatic amines have been converted to corresponding secondary or tertiary amines with modest yields using an equimolecular amount of a aliphatic primary or secondary alcohol in a hydrogen atmosphere (Scheme 1.52).[202] Additionally, the applications of a copper-chromite catalyst have been extended to various alcohol aminations including the transalkylation of triethylamine with primary alcohol,[203] amination of cyclohexanol with ammonia,[204] and double methylation of dodecylamine with methol.[205] It has been found that supported catalysts bearing copper and chromium can catalyze the alkylation of dimethylamine with dodecanol with practically quantitative yields.[206]

CuO supported on γ-Al_2O_3 (60 wt%) could be used as catalyst to prepare different *N,N*-dimethylphenylalkylamine with high yields.[207] A similar catalyst derived from CuO on γ-Al_2O_3 (63 wt%) has been used in the preparation of 2-methylaminoalkanols by reaction of dimethylamine with the corresponding aliphatic diol (Scheme 1.53).[208]

Several cyclic amines have been synthesized starting from the corresponding aminoalcohol by using copper supported either on γ-Al_2O_3 or on MgO in a continuous-flow, fixed-bed reactor.[209,210] The same results were obtained when the reaction was performed with hydrogen or nitrogen, but a deactivation of the catalyst was observed in the absence of hydrogen.

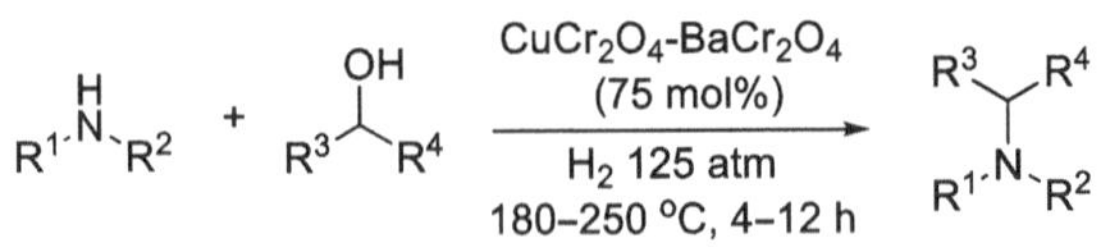

SCHEME 1.52 The *N*-alkylation of aliphatic amines with aliphatic alcohols catalyzed by $CuCr_2O_4$-$BaCr_2O_4$.

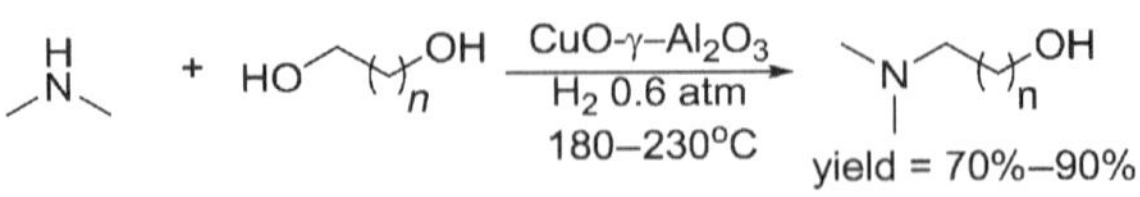

SCHEME 1.53 The *N*-alkylation of dimethylamine with aliphatic diol catalyzed by CuO-γ-Al_2O_3

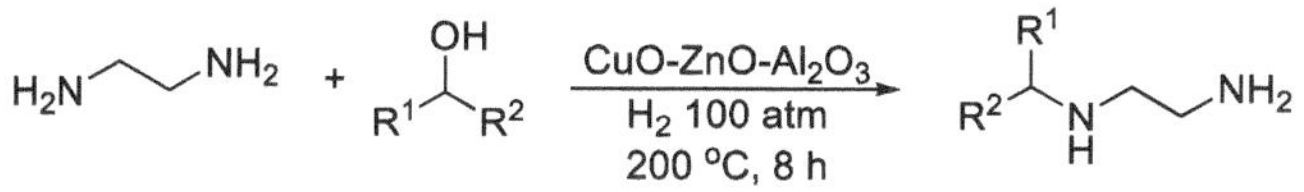

SCHEME 1.54 The *N*-alkylation of ethylenediamine with alcohols catalyzed by CuO-ZnO-Al_2O_3.

Commercially available CuO-ZrO-γ-Al_2O_3 has been used as catalyst for the methylation of *n*-butylamine in a hydrogen atmosphere at 185°C, and the corresponding *N*-methyl-1-butylamine was obtained with 54% yield.[211]

CuO-ZnO-Al_2O_3 catalyst prepared by coprecipitation and subsequent calcination has been used in the alkylation of ethylenediamine with different alcohols (Scheme 1.54). The use of ZnO-Al_2O_3 as catalyst did not produce the alkylated amine under the same reaction conditions, showing that the activity was due to copper species. Nevertheless, the synergy effect between different copper and zinc species was observed since the activity of CuO-ZnO-Al_2O_3 catalyst was much higher than the related CuO-Al_2O_3 one.[212]

A colloidal catalyst bearing copper, nickel, and barium has been used in the *N*-alkylation of dimethylamine with dodecylalcohol and the corresponding amine was obtained with a 96% yield when the stearic salt of Cu/Ni/Ba was used in a 5:1:1 ratio.[213] X-ray diffraction analysis of the reduced catalyst indicated that Cu metal, CuO, and Ni metal were colloidally dispersed in the system. A further addition of calcium stearate to the initial catalytic mixture increased yield up to 99% for the desired product.[214] Although the catalyst is a colloidal system, it could not be separated by simple filtration after the amination reaction due to its small particle size (1.5 nm); however its recovery could be performed by distillation.[215]

The aforementioned three-component colloidal catalyst has also been used in alkylation of dimethylamine with different diols. One such example is the reaction of dimethylamine with 1,6-hexanodiol, giving the corresponding *N,N,N′,N′*-tetramethylhexanediamine with 85% yield. Different alcohols, such as mixtures of commercial long-chain oxoalcohols, were also used as the alkylation reagents, giving the corresponding *N,N*-dimethylamines with good results (78%–89%).[216]

Copper-aluminum hydrotalcite (CuAl-HT)/K_2CO_3 has been employed in the *N*-alkylation of benzylamines and anilines with benzyl alcohols to supply the corresponding amines in good to excellent yields.[217] For example, the reaction 4-nitroaniline with benzyl alcohol produced corresponding secondary amine with 98% yield (Scheme 1.55). The catalyst was separated from the reaction mixture and was reused five times without losing its activity.

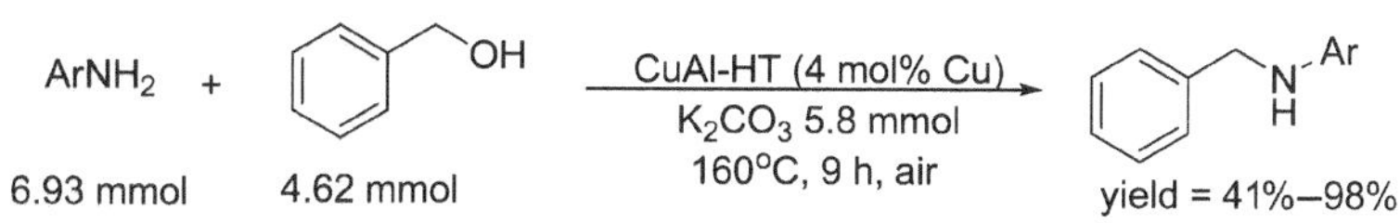

SCHEME 1.55 The *N*-alkylation of anilines with benzyl alcohols catalyzed by CuAl-HT.

CuO-NiO/γ-Al_2O_3 catalyst prepared via impregnation method has been used in the *N*-alkylation of morpholine with alcohols in a fixed-bed reactor.[218] The *N*-methylation of morpholine with methanol produced 95.3% conversion with a 93.8% selectivity of *N*-methylmorpholine under optimum reaction conditions. The present catalytic system is also applicable to other low-carbon primary alcohols. In contrast, secondary alcohols gave relatively low product selectivity because the resulting ketone is a poorer electrophile than an aldehyde.

Simple CuO NPs catalyst has been used for the direct *N*-alkylation of aniline with 1,4-phenylenedimethanol under optimized reaction conditions to produce *N,N'*-(1,4-phenylenebis(methylene))dianiline with 87% yield (Scheme 1.56).[219] The catalyst can be easily recovered and reused for four times without any significant loss of activity.

The supported Cu/Al_2O_3 catalyst has been used in *N*-alkylation of aniline with secondary and benzylic alcohols to afford the corresponding secondary amines without any additive.[220] For example, the reaction of aniline and secondary alcohols with the cyclohexanol structure gave the corresponding amine 100% conversion and 95% selectivity, while cyclooctanol and cyclododecanol produced the amine in a lower selectivity.

In 2015, Shi's group reported the selective synthesis of unsymmetric tertiary amines using a simple $CuAlO_x$ catalyst prepared using the hydrothermal method (Scheme 1.57).[221] This simple catalyst is effective for a wide variety of primary amines and alcohols substrates. For example, the reaction of benzyl alcohol, hexyl alcohol, and 4-butylaniline gave the corresponding unsymmetric tertiary amine with yields up to 98%.

N-Alkylation of amines with alcohols under UV irradiation at room temperature has been achieved by Shi's group using a TiO_2 (P25) supported Cu and Mo photocatalyst (Cu-Mo/TiO_2).[222] A variety of aromatic and aliphatic amines were selectively converted into the corresponding secondary amines or tertiary amines in moderate to excellent yields without the addition of any cocatalysts

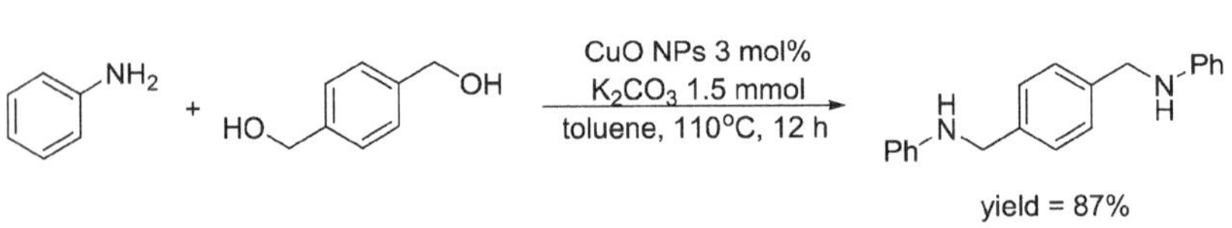

SCHEME 1.56 The *N*-alkylation of aniline with 1,4-phenylenedimethanol catalyzed by CuO NPs.

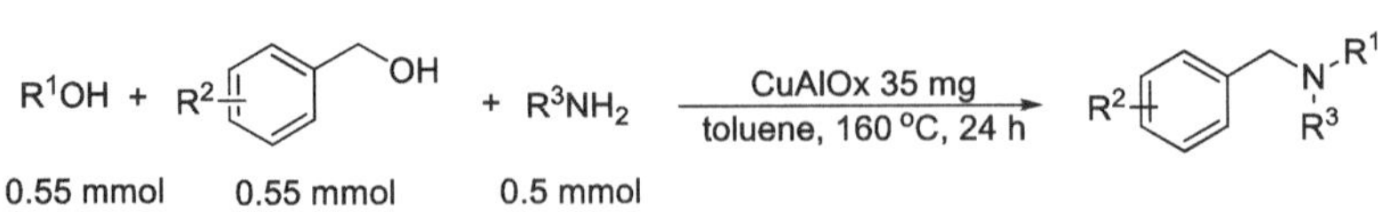

SCHEME 1.57 The synthesis of asymmetrical tertiary amines catalyzed by $CuAlO_x$.

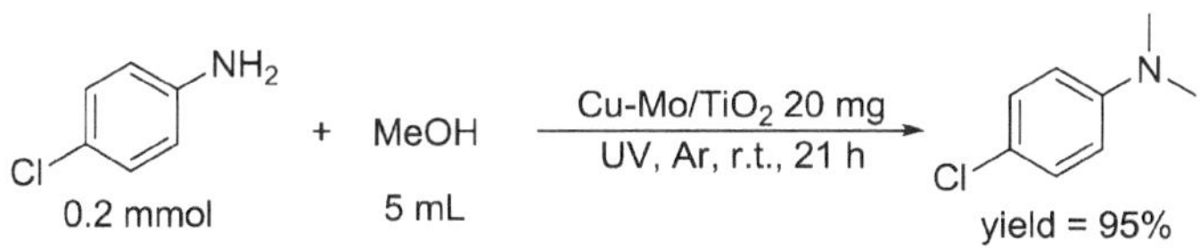

SCHEME 1.58 The photocatalyzed *N*-alkylation of p-Cl-aniline with methanol.

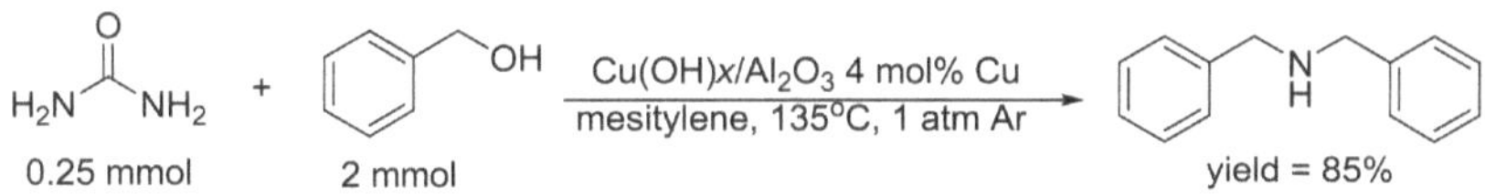

SCHEME 1.59 The *N*-alkylation of ammonia with benzyl alcohol catalyzed by supported copper hydroxide catalysts.

such as bases and organic ligands. Of note is the feasibility of this catalytic system in the alkylation of anilines containing halogen substituents with alcohols, yielding up to 95% of desired products (Scheme 1.58).

N-Alkylation of ammonia (in situ generated from urea or aqueous ammonia) with alcohols to secondary amines has been efficiently generated by supported copper hydroxide catalysts, i.e., $Cu(OH)_x/Al_2O_3$ and $Cu(OH)_x/TiO_2$.[223] For instance, the reaction of urea with benzyl alcohol gave the corresponding dibenzylamine with 85% yield (Scheme 1.59).

1.3.2 Ni Catalyst System

Different nickel catalysts have been historically used in the *N*-alkylation of amines using alcohols as electrophilic source. Thus, particles of nickel obtained by reduction of NiO at 300°C were effective catalysts in the alkylation of different aromatic amines, such as aniline and paratoluidine, with aliphatic alcohols, such as methanol, ethanol, and cyclohexanol. For example, the corresponding *N*-ethylaniline was obtained with a 30% yield by using 10% weight of nickel catalyst, 2/1 ratio of ethanol/aniline, at 180°C in an autoclave for 12 h.[224]

Better results were achieved by using nickel particles (100 mesh) as catalyst and potassium as base in alkylation of aniline with benzyl alcohol (200 mol%), as well as the corresponding 4-substituted derivatives, in xylene at 150°C.[225] Under optimized conditions, the reaction could also be performed using simple aliphatic primary alcohols, such as hexanol and decanol.

Other catalysts, such as nickel supported in SiO_2 or mixtures of SiO_2/Al_2O_3, were able to catalyze the reaction of 2-aminoethanol with ammonia to give ethylenediamine,[226] with the second supported catalyst being more effective than the first one. However, the yields are lower than 25%.

N-Alkyl substituted anilines were prepared by using Raney-nickel (*c*.60% weight) in a refluxing mixture of aniline and an excess of alcohol

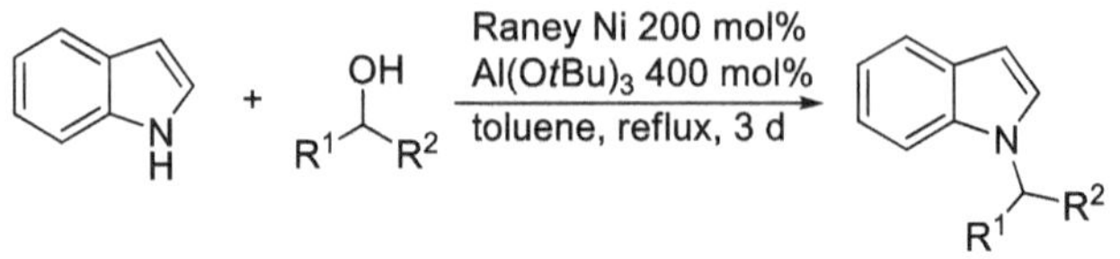

SCHEME 1.60 The *N*-alkylation of indole with secondary alcohols catalyzed by Raney-Ni.

(500 mol%) for 16 h. For straight-chain primary alcohols different from methanol, the yields were 78%–83%, while those derived from branched primary alcohols gave the corresponding *N*-alkyl anilines in only 41%–49% yields.[227]

Following the previous methods, indole could be successfully alkylated using an excess of a secondary alcohol (4000 mol%) in the presence of an excess of aluminum *tert*-butoxide, and catalyzed by Raney-nickel, after 3 days at toluene reflux (Scheme 1.60).[228]

Raney-nickel was effective not only in the monoalkylation of amines but also in double alkylation.[229] For example, the alkylation of 1,6-hexanediamine using equimolecular amounts of 1,4-butanediol gave the azepine derivative with a 15% yield. Several secondary and tertiary amines were prepared in 34%–80% yield using nickel catalyst (12 mol%) under a hydrogen atmosphere (100 atm pressure) at 200°C in a few hours, using different aliphatic and cyclic amines as nucleophiles, and ethanol, 1-butanol, and cyclohexanol (200 mol%) as electrophiles.[230]

Different nitrogen-containing heterocyclic compounds were prepared using amino alcohol derivatives as starting materials and Raney-nickel as catalyst in a hydrogen atmosphere. In one instance, 1-aminopropan-2-ol underwent a double hydrogen autotransfer process to form a mixture of *cis*/*trans*-piperazine and the related aromatic pyrazine (Scheme 1.61).[231] Under optimized reaction conditions, the aliphatic heterocycle could be obtained as the main product.

The *N*-alkylation of primary amines and *N*-sulfinyl derivatives in alcohols (1.5 mL) at room temperature has been achieved using Raney-Ni catalyst (1.2 g) (Scheme 1.62).[232] Aromatic primary amines and *N*-sulfinylamides were monoalkylated briefly (15–40 min) with good yields (77%–89%). When alcohols

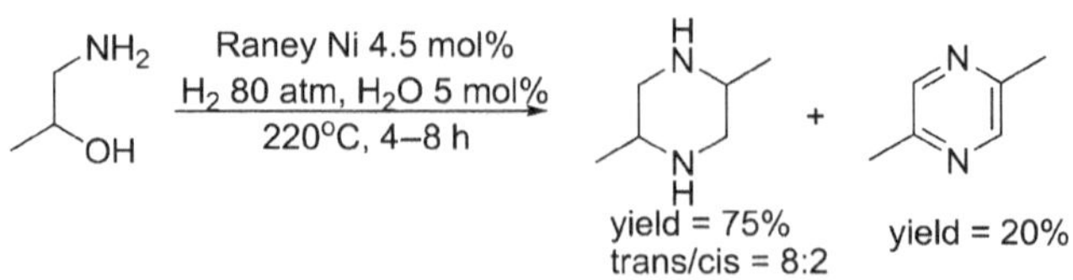

SCHEME 1.61 The double *N*-alkylation of 1-aminopropan-2-ol using Raney-Ni.

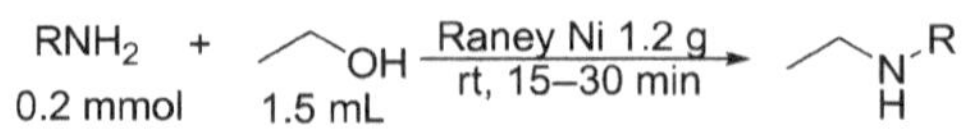

SCHEME 1.62 The *N*-alkylation of amines with alcohols at room temperature catalyzed by Raney-Ni.

with steric hindrance, such as *n*-BuOH and *i*-PrOH, were used as alkylating reagents, longer reaction time was required. Investigation of the reaction mechanism revealed that this process involved a radical mechanism in which the substrate is homolytically cleaved after absorption on the metal surface.

A γ-Al_2O_3 supported Ni and Cu bimetallic nanoparticles catalyst (45 wt.% Ni, Ni/Cu mass ratio = 4.5/1.0) prepared by electroless plating method has been used for the *N*-alkylation of amines with alcohols.[233] Although the catalyst allowed fast conversions, high selectivity for amines and alcohols with various structures in an Ar atmosphere in *o*-xylene, the base (NaOH) and Lewis acidic ($CaCl_2$) cocatalysts were required. Additionally, the catalyst still has a stable catalytic activity after two consecutive cycles regeneration.

A general nonnoble metal $NiCuFeO_x$ catalyst for the benign amination of alcohols with amines and ammonia has been developed by the Shi group.[234] This catalyst exhibited positive substrates generality comparable to previously reported catalysts including those noble metal catalysts. Typically, primary amines could be efficiently transformed into secondary amines and *N*-heterocyclic compounds, and secondary amines could be *N*-alkylated to synthesize tertiary amines. Note that primary and secondary amines could be produced through a one-pot reaction of ammonia and alcohols (Scheme 1.63). Moreover, the magnetic property of this catalyst makes it easily separable from the reaction mixture, and it could be recovered and reused for five runs without obvious deactivation.

Ni/θ-Al_2O_3 prepared by in situ H_2-reduction of NiO/θ-Al_2O_3 shows high activity for the alkylation of anilines and aliphatic amines with various alcohols (benzyl and aliphatic alcohols) under additive free conditions.[235] For the reaction of aniline with an aliphatic alcohol, the catalyst shows a higher turnover number (TON) than noble metal-based catalysts. The catalyst is also effective for synthesis of primary amines from alcohols and NH_3 (Scheme 1.64).[236] Various aliphatic alcohols are tolerated, and the turnover numbers were higher than

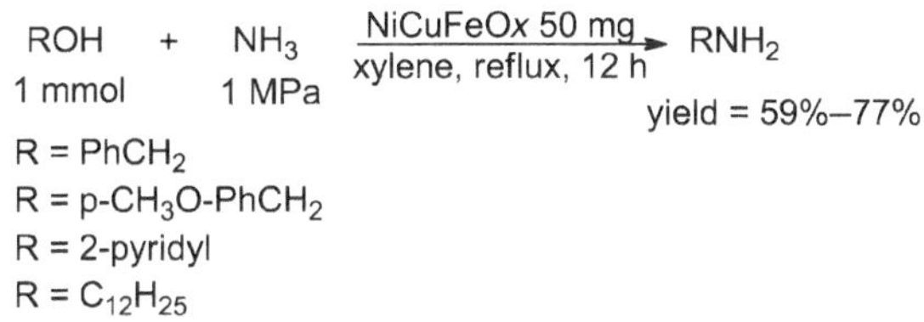

SCHEME 1.63 The synthesis of primary amine from alcohol and ammonia.

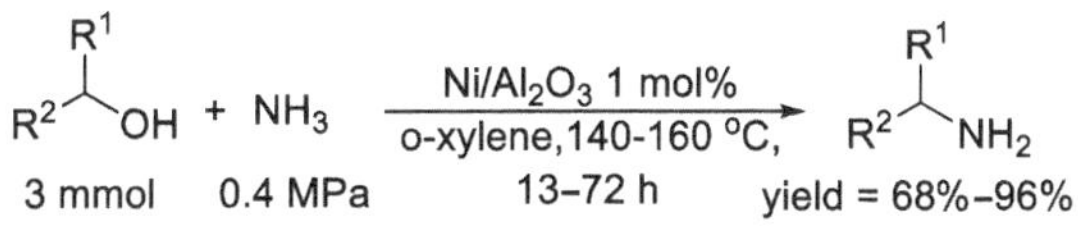

SCHEME 1.64 The synthesis of primary amines from alcohols and NH_3 catalyzed by Ni/Al_2O_3.

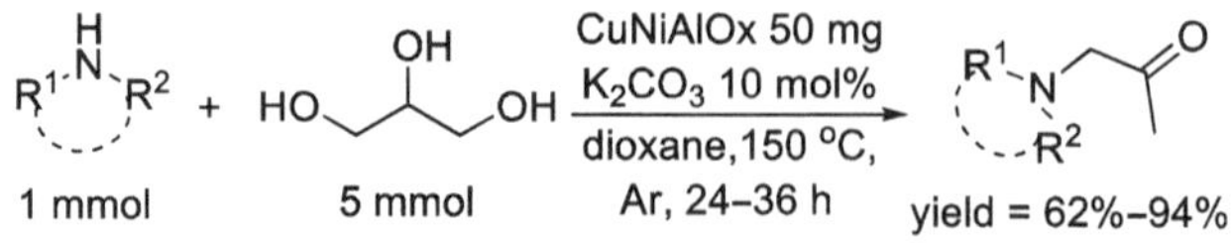

SCHEME 1.65 The amination of glycerol with amines using $CuNiAlO_x$.

those of Ru-based homogeneous catalysts. The catalyst was recoverable and was reused for two runs without obvious deactivation.

Recently the one-pot amination of glycerol with amines has been reported by Shi's group (Scheme 1.65).[237] Highly valuable prochiral aminoketones could be synthesized by selective activation and transformation of glycerol with amines over a simple heterogeneous $CuNiAlO_x$ catalyst. Good to excellent yields were obtained with aliphatic secondary amines. But in the case of aliphatic primary amines and aromatic amines, much lower yields were obtained. The catalyst is recyclable and can be reused three times without obvious deactivation.

1.3.3 Fe/Mn/Co Catalyst Systems

Unmodified commercial magnetite has been shown to be an active, stable, and highly selective catalyst for the selective monoalkylation of aromatic amines with benzylic alcohols (Scheme 1.66).[238] Even working with long reaction times and 2 equiv. KO*t*Bu, this is the first time that unmodified magnetite has been used as a recyclable catalyst in the *N*-alkylation of amines using alcohols as the initial source of electrophiles.

Iron oxide nanoparticles supported on a variety of biomaterials are catalytically active in the microwave-assisted *N*-alkylation of (substituted) amines with benzyl alcohols (Scheme 1.67).[239] Fe-HMS (hexagonal mesoporous silica) was found to be the most active catalyst in the reaction and able to catalyze the

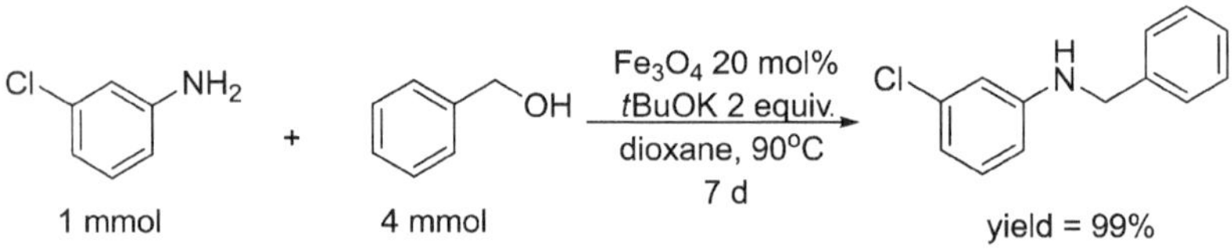

SCHEME 1.66 The Fe-catalyzed *N*-alkylation of 3-chloroaniline with benzyl alcohol.

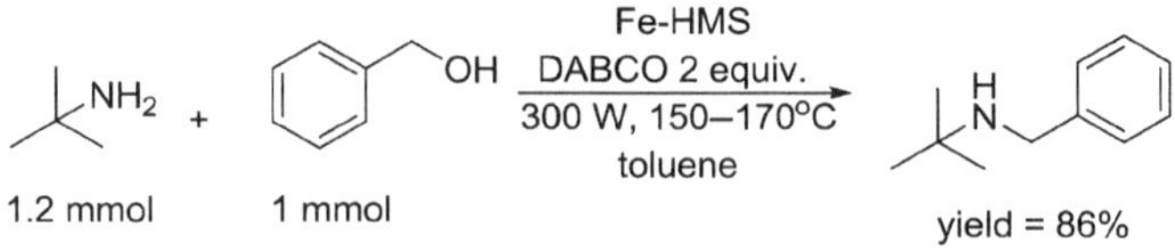

SCHEME 1.67 The *N*-alkylation of *tert*-butylamine with benzyl alcohol catalyzed by Fe-HMS.

N-alkylation of *tert*-butylamine with benzyl alcohol to afford the corresponding secondary amine with 86% yield, although 2 equiv. base DABCO was required.

$Fe_2(SO_4)_3{\cdot}xH_2O$ on silica as a reusable catalyst for the amination of secondary benzylic and allylic alcohols with primary amides and aromatic amines with electron-withdrawing groups in solvent-free conditions was reported by Zhu et al. in 2014.[240] Of note is the use of benzylic alcohols with electron-donating groups which allowed the reaction to perform at room temperature.

Manganese dioxide is employed as an in situ oxidant for one-pot conversion of alcohols into amines. In combination with $NaBH_4$ and CH_3OH solvent, a 89% yield of *N*-benzyl-isobutylamine was obtained by one-pot *N*-alkylation of *iso*-butylamine with benzyl alcohol.[241] This method has also been applied for the synthesis of the tertiary amines by replacing $NaBH_4$ and CH_3OH with polymer-supported cyanoborohydride (PSCBH) and AcOH.[242] However, the drawbacks of the method were obvious, i.e., the application of 10 equiv. MnO_2 as an oxidant and 2–5 equiv. $NaBH_4$ or PSCBH as a reductant.

In 2011, Xu et al. reported that *N*-alkylation of sulfonamides and amines with benzyl alcohols in air could be achieved in the presence of catalytic amounts of manganese dioxide catalyst and base (Scheme 1.68).[243] A yield of up to 99% secondary amines could be obtained and the aerobic condition is a more effective alcohol activation alternative than inert atomosphere.

In 1999, the bimetal Co-Fe alloy catalyst was used in the amination of biomass-based 1,3-propanediol with ammonia .[244] The reaction using supercritical ammonia, 95 wt% Co-5 wt% Fe catalyst and H_2 (3 mol%) at 195°C in a continuous fixed-bed reactor under pressure of 150 bar gave the expected 1,3-diaminopropane with the highest yield of 32% (Scheme 1.69).

1.3.4 Ru Catalyst System

In 2009, Shi and Beller et al. reported *N*-alkylation of sulfonamides with alcohols using a heterogeneous ruthenium catalyst.[245] A variety of coupling reactions of benzylic alcohols and sulfonamides including various heterocycles were

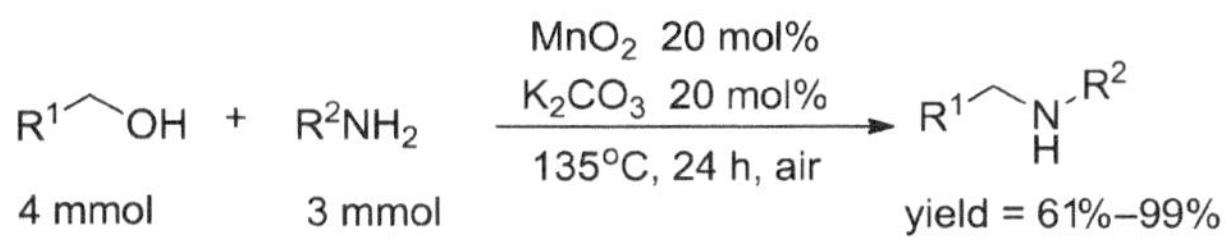

SCHEME 1.68 The Mn-catalyzed *N*-alkylation of amines with benzyl alcohols.

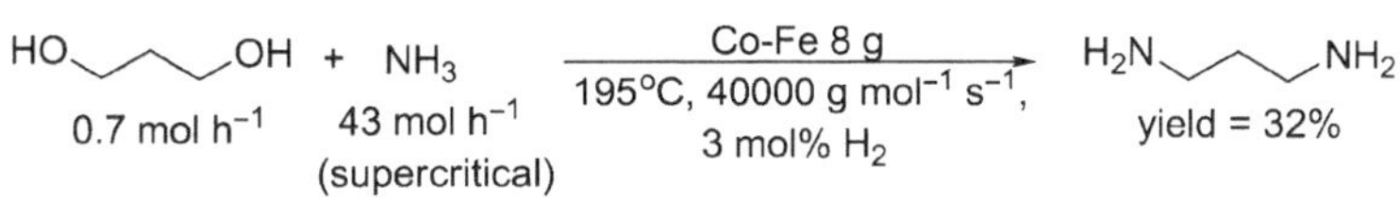

SCHEME 1.69 The amination of 1,3-propanediol with ammonia catalyzed by Co-Fe.

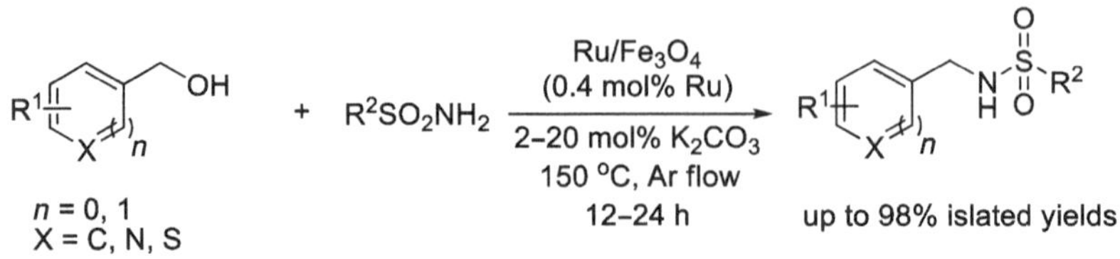

SCHEME 1.70 The *N*-alkylation of sulfonamides with alcohols catalyzed by heterogeneous ruthenium catalyst.

successfully realized in the presence of nano-Ru/Fe_3O_4 and K_2CO_3 at 150°C, often with an isolated yield greater than 80% (Scheme 1.70).

The heterogeneous ruthenium catalysts for the *N*-alkylation of primary and secondary amines with primary alcohols have been investigated by the Mizuno group. Applying the supported ruthenium hydroxide catalyst Ru(OH)*x*/Al_2O_3, a variety of aromatic and heteroaromatic amines were selectively converted into the corresponding secondary amines in moderate to excellent yields without any cocatalysts such as bases and stabilizing ligands.[246] In the case of Ru(OH)*x*/TiO_2, *N*-alkylation of ammonia (or its surrogates, such as urea, NH_4HCO_3, and $(NH_4)_2CO_3$) and amines with alcohols, including primary and secondary alcohols, was efficiently produced under anaerobic conditions (Scheme 1.71).[247] Various types of symmetrically and unsymmetrically substituted "tertiary" amines could be synthesized by the *N*-alkylation of ammonia (or its surrogates) and amines with "primary" alcohols. By contrast, *N*-alkylation of ammonia surrogates (i.e., urea and NH_4HCO_3) with "secondary" alcohols selectively produced corresponding symmetrically substituted "secondary" amines, even in the presence of excess amounts of alcohols, which is likely due to the steric hindrance of the secondary alcohols and/or secondary amines produced.

Ramón and coworkers reported impregnated ruthenium on magnetite to catalyze the *N*-alkylation of aromatic and heteroaromatic amines, sulfonamides and sulfonamides, in all cases alcohols giving secondary amines.[248] In the case of sulfinamides, this is the first time that these amino compounds have been alkylated following a hydrogen autotransfer process, allowing the use of chiral sulfinamides and secondary alcohols to give the alkylated compound with a diastereomeric ratio of 92:8. In these cases, after alkylation, a simple acid deprotection produced expected good yields of primary amines. The ruthenium catalyst is quite sensitive, and small modifications of the reaction medium can change the final product. Alkylation of amines using potassium hydroxide

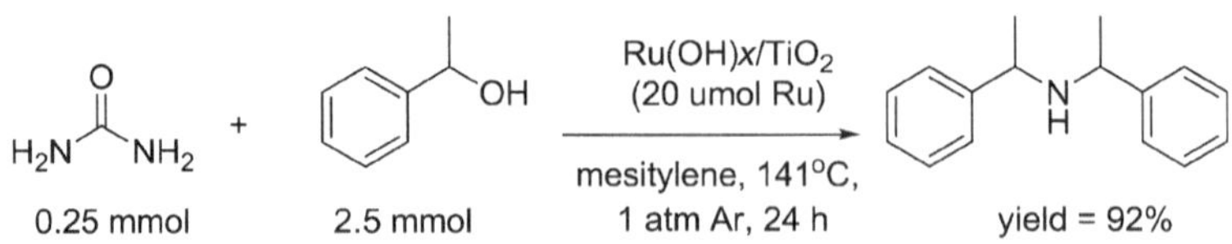

SCHEME 1.71 *N*-Alkylation of ammonia with secondary alcohol catalyzed by Ru(OH)*x*/TiO_2.

renders the N-monoalkylated amines, and the same process using sodium hydroxide yields the related amines. The catalyst can be easily recovered with a simple magnet and can be reused up to 10 times, showing the same activity. However, over stoichiometric amounts (130 mol%) of KOH were required.

In 2017, one-pot amination of glycerol with aqueous ammonia to alkyl amines was reported using supported Ru catalysts.[249] Ru/C catalyst exhibits a remarkable activity for conversion of glycerol and sorbitol to C_1—C_3 alkyl amines in the aqueous phase. C_1—C_3 amines including methylamine (MA), ethylamine (EA), and propylamines can be effectively obtained with 48.6% selectivity and 96.2% conversion of glycerol in a one pot process. In the same year, Rose and coworkers reported a similar process.[250] The amination of isomannide with ammonia were achieved in aqueous solutions using solid Ru-based catalysts. A mixture of mono- and diamines was obtained. Additionally, Murzin extended the application of Ru/C catalyst to the direct amination of dodecanol with ammonia to dodecylamine.[251] A dodecylamine yield of 83.8% was achieved with Ru/C catalyst at 150°C, 4 bar NH_3 and 2 bar H_2 after 24 hours of reaction. It was found that ammonia pressure did not influence the reaction, while extra hydrogen in the reaction mixture was beneficial for dodecanol conversion and dodecylamine selectivity.

1.3.5 Pd Catalyst System

As early as 1974, Murahashi et al. showed that palladium black was able to catalyze the reaction of arylmethanols or allyl alcohols with primary or secondary amines.[252] Under the optimized conditions (palladium black 0.2 g, 120°C), N-substituted pyrroles can be prepared by the reaction of but-2-ene-1,4-diol with excellent yields of primary amines.

A heterogeneous palladium catalyst, Pd/AlO(OH) composed of palladium nanoparticles entrapped in boehmite nanofibers, has shown high activity and excellent efficiency in the N-alkylation of aliphatic amines with benzyl alcohols without any additives (Scheme 1.72).[253] For example, the reaction of n-hexyl amine and benzyl alcohol gave the corresponding secondary amine a yield of 88%.

Pd/C and $Pd(OH)_2$/C have been reported to be effective for the N-alkylation of amino acids in C1—C4 aliphatic alcohols at room temperature, but 1 atm H_2 is necessary.[254] Both primary and secondary amino acids in methanol went smoothly and produced the corresponding tertiary amines in good to excellent yields (Scheme 1.73).

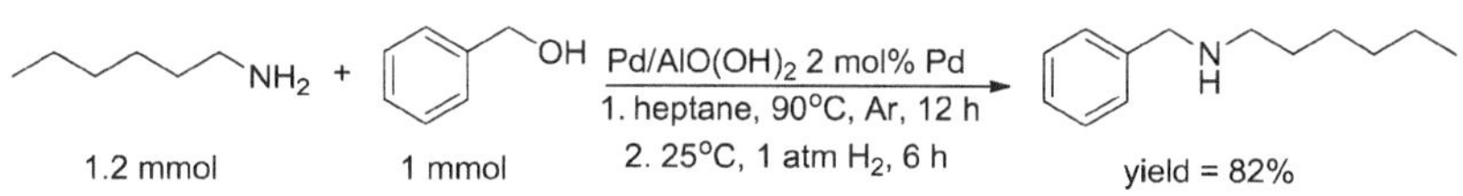

SCHEME 1.72 N-Alkylation of aliphatic amines with benzyl alcohols catalyzed by Pd/AlO(OH).

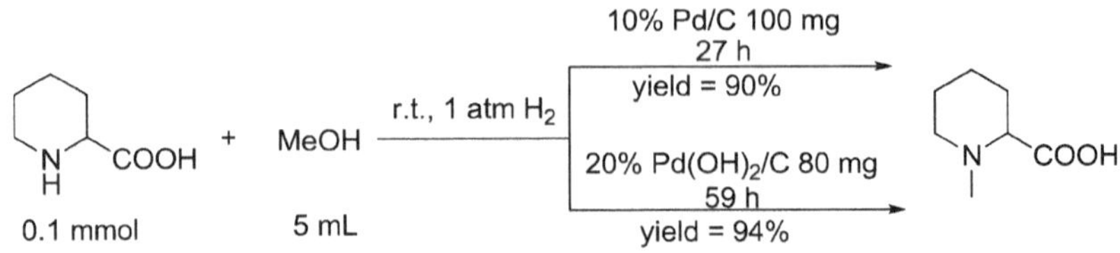

SCHEME 1.73 *N*-Alkylation of amino acid with methanol catalyzed by Pd/C and $Pd(OH)_2/C$.

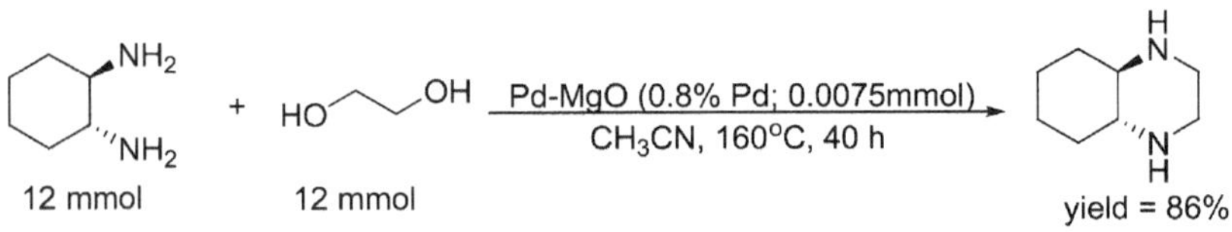

SCHEME 1.74 One-pot synthesis of piperazines from 1,2-diamines and ethylene glycol catalyzed by Pd/MgO.

A bifunctional metal Pd/MgO catalyst has been used in the selective mono-alkylation of amines with alcohols (Scheme 1.74).[255] The reaction is sensitive to the crystallite size of the Pd and the highest activity was found with a Pd crystallite size below 2.5 nm in diameter. The high efficiency of the catalytic system extends the process to the one-pot synthesis of piperazines.

In 2011, Shi's group reported a Pd/Fe_2O_3 catalyst that successfully performed *N*-alkylation reaction of anilines and primary aliphatic amines with primary benzylic and aliphatic alcohols such as octanol.[256] A difference of 20°C in temperature caused a notable shift in selectivity from secondary to tertiary amine formation (Scheme 1.75). By changing the support to TiO_2, *N,N*-dimethylation of different amines with methanol could be achieved with UV irradiation at room temperature.[257] A wide range of *N,N*-dimethyl amines were one-pot synthesized up to 98% by applying aliphatic amines and aromatic primary amines as starting materials. It is noteworthy that up to 90% yield of 4-chloro-*N,N*-dimethylaniline was obtained by adjusting the Pd loadings on the TiO_2 and the dehalogenation reaction was suppressed.

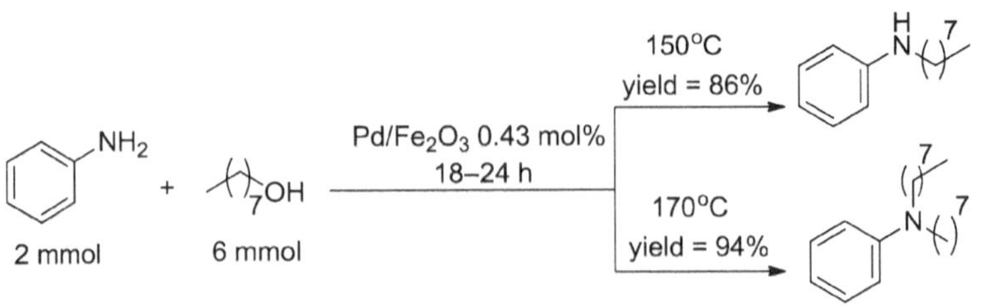

SCHEME 1.75 The *N*-alkylation of aniline with octanol catalyzed by Pd/Fe_2O_3.

Recently, Hermange and Astruc found that the commercial Pd/C was effective for the synthesis of secondary amines from amines and alcohols.[258] This catalyst could be recycled five times without a significant decrease in activity. Good yields were obtained with anilines containing electron-donating group and benzyl alcohols, but 3 equiv. amount of sodium formate is necessary.

1.3.6 Pt Catalyst System

As early as 1983, the Kagiya group has synthesized symmetrical secondary amines from primary amines using photocatalysis by platinized titanium dioxide (TiO_2/Pt) suspension in water.[259] By changing the reaction medium to alcohols, unsymmetrical secondary and tertiary amines could be obtained at room temperature.[260]

Vapor-phase amination of cyclohexnol with ammonia has been explored with various Pt-based catalysts such as Pt/C,[261] Pt/Al_2O_3,[262] and Pt/SiO_2.[263] A mixture of cyclohexylamine and aniline was usually obtained.

A heterogeneous bimetallic Pt-Sn/γ-Al_2O_3 catalyst (0.5 wt% Pt, Pt/Sn molar ratio = 1:3) has been used in the syntheses of secondary and tertiary amines by highly efficient direct *N*-alkylation of primary and secondary amines with alcohols (Scheme 1.76).[264] Additionally, the direct synthesis of diamines has been efficiently realized from the *N*-alkylation of amines with diols by the bimetallic Pt-Sn/y-Al_2O_3 catalyst.[265]

Biomass-based alcohol 2,5-tetrahydrofurandimethanol (THFDM) has been selectively transformed into 8-oxa-3-azabicyclo[3.2.1] octane (OABCO), a valuable building block for the synthesis of bioactive molecules, via one-pot aminocyclization with ammonia catalyzed by Pt/NiCuAlO_x.[266] Under optimized conditions (200°C, 11 h, 0.5 MPa hydrogen, 0.4 MPa ammonia), the OABCO yield reached 58% with 100% THFDM conversion (Scheme 1.77).

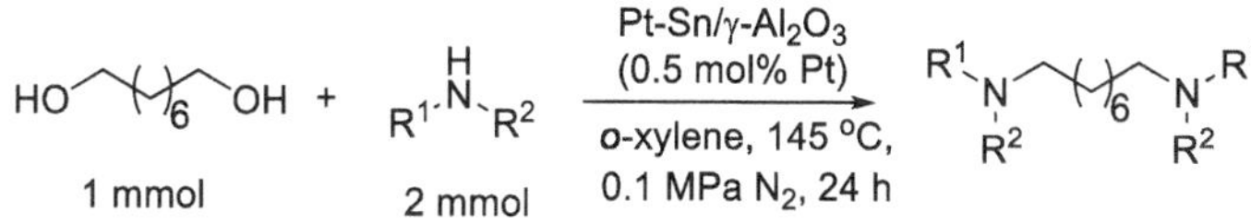

SCHEME 1.76 The *N*-alkylation of aniline with diols catalyzed by Pt-Sn/γ-Al_2O_3.

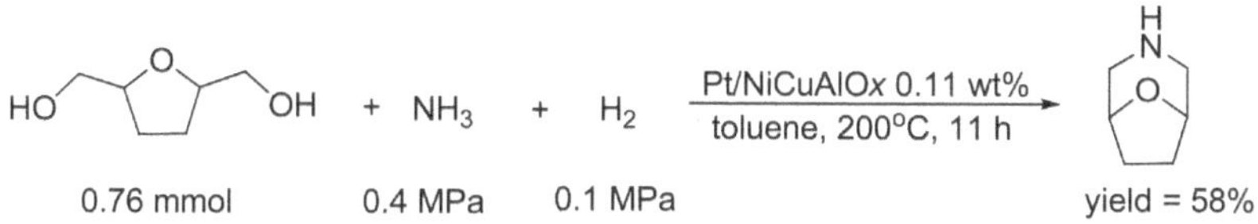

SCHEME 1.77 Amination of 2,5-tetrahydrofurandimethanol with ammonia catalyzed by Pt/NiCuAlO_x.

1.3.7 Au Catalyst System

Gold clusters supported on porous coordination polymers have been used as catalysts for one-pot *N*-alkylation of aniline with benzyl alcohol to produce the corresponding secondary amines under N_2 atmosphere without any hydrogen-acceptor in 2009.[267] Researchers obtained 48% yield of *N*-benzylaniline in the presence of Au/Al-MIL53 (1.6 nm) and 50 mol% Cs_2CO_3 at 130°C (Scheme 1.78). It was found that the catalyst was unstable and difficult to be recycled due to the aggregation of Au clusters to form larger NPs (>10 nm) after the one-pot reactions.

Au/TiO_2-VS catalyst (VS = very small particles, 1.8 nm) has been used in the *N*-alkylation of amines with alcohols (Scheme 1.79).[268] The reaction of aniline and primary as well as secondary aliphatic alcohols gave the corresponding secondary amines with high yields. When aliphatic amines were converted with these alcohols, only moderate yields were achieved since self-coupling of the amines occurred.

In 2012, Haruta et al. investigated the support effects on supported gold-catalyzed one-pot *N*-alkylation of aniline with benzyl alcohol to produce secondary amine under the equimolar amounts of substrates without additives.[269] It is found that selectivity to secondary amine was changed by the kinds of supports. Gold on ZrO_2 exhibited the highest catalytic performance and achieved 94% selectivity to secondary amine. Surface hydroxyl groups of ZrO_2 played important roles in the deprotonation of alcohol in the alcohol dehydrogenation step, as adsorption sites of aniline, and as a proton source in the hydrogen transfer step.

In 2013, Simakova et al. extended the application of Au/ZrO_2 to the amination of myrtenol[270] and then investigated the Kinetic behavior of the reaction,[271]

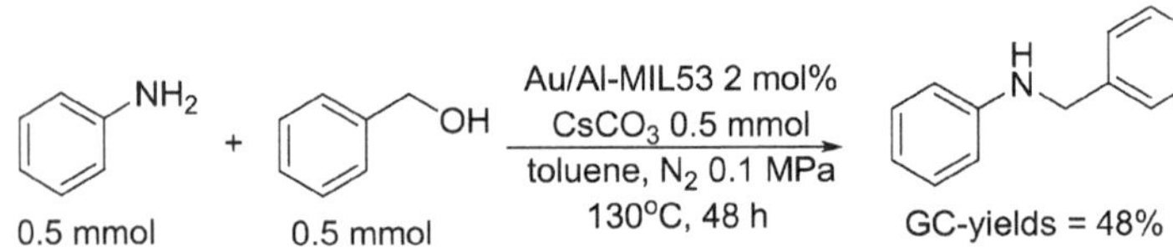

SCHEME 1.78 Au-catalyzed *N*-alkylation of aniline with benzyl alcohol.

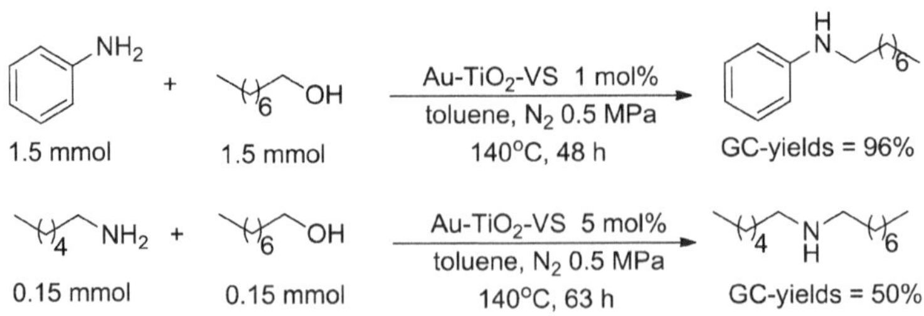

SCHEME 1.79 *N*-Alkylation of amines with alcohols catalyzed Au/TiO_2-VS.

the effect of hydrogen addition,[272] and extra hydrogen donors[273] on selectivity to the target amines, as well as the effect of catalyst preredox treatment on the catalytic activity.[274]

1.3.8 Ag Catalyst System

An alumina-supported silver cluster with a Lewis acidic cocatalyst (polyvalent metal salts, such as $FeCl_3{\cdot}6H_2O$) acts as a heterogeneous and recyclable catalyst for direct *N*-alkylation of anilines with benzyl alcohols (Scheme 1.80).[275] Systematic studies on the effects of metal (Pt, Pd, Au, and Ag), silver particle size, support oxides (CeO_2, ZrO_2, Al_2O_3, SiO_2), and Lewis acidic additives show four important factors required to achieve selective *N*-alkylation of anilines: (1) metal with a weak metal-hydrogen bond energy (i.e., Ag), (2) smaller size of the silver cluster, (3) the support having both acidic and basic sites (i.e., Al_2O_3), and (4) additives with high Lewis acidity (i.e., Fe^{III} salt).

A similar catalyst Ag/Al_2O_3 prepared by alumina-entrapped Ag was demonstrated to be highly active and selective in the *N*-alkylation of aniline with benzyl alcohols.[276] The highest activity was found with 2.4 wt% Ag/Al_2O_3. In the presence of a base, for example, Cs_2CO_3 or K_3PO_4 as promoter, excellent conversion and selectivity of *N*-benzylaniline were obtained. With secondary amines such as piperidine and pyrrolidine, the mixtures of amides and amines were obtained.

The hybrid material $Ag_6Mo_{10}O_{33}$ has been used in the *N*-alkylation reactions amines, carboxamides, and sulfonamides with benzylic alcohols. Up to 99% isolated yields were obtained under relatively mild conditions without additive organic ligand. This unusual catalyst was further shown to be active in the coupling of primary aliphatic alcohols, such as decanol with aniline (Scheme 1.81).[277]

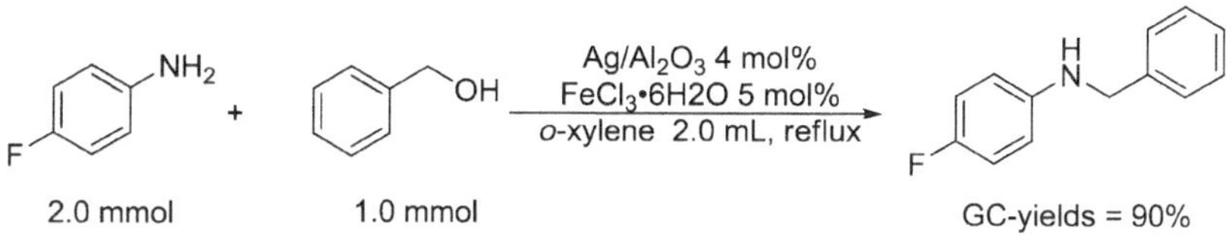

SCHEME 1.80 Ag-catalyzed *N*-alkylation of anilines with alcohols.

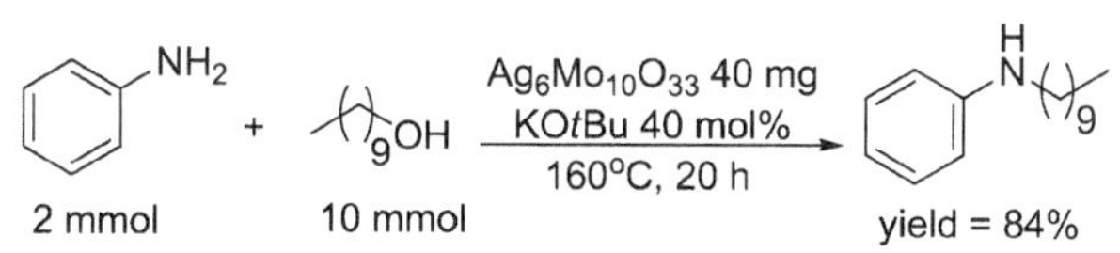

SCHEME 1.81 *N*-Alkylation of aniline with 1-decanol catalyzed by$Ag_6Mo_{10}O_{33}$.

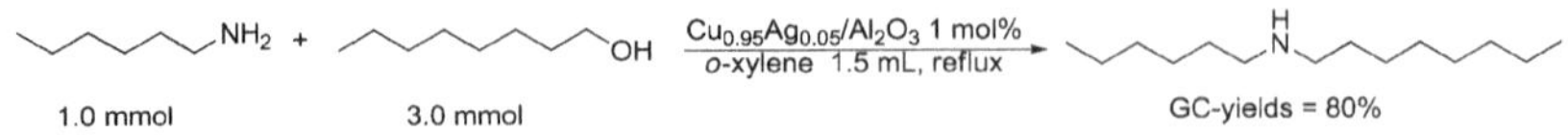

SCHEME 1.82 N-Alkylation of hexyl amines with 1-octanol catalyzed by $Cu_{0.95}Ag_{0.05}/Al_2O_3$.

A series of metal(M)-loaded Al_2O_3 catalysts (M/Al_2O_3) and bimetallic catalysts, Cu_xAg_{1-x}/Al_2O_3 with different compositions, $Cu_{0.95}Ag_{0.05}/MO_x$ with different supports (MO_x), and $Cu_{0.95}M'_{0.05}/Al_2O_3$ with different promoter (M′), were prepared by an impregnation method, followed by H_2-reduction at 600°C. For the *N*-alkylation of amines with alcohols, Al_2O_3-supported copper-silver bimetallic catalysts with a Cu/Ag molar ratio of 95/5 ($Cu_{0.95}Ag_{0.05}/Al_2O_3$) was found to be the most effective heterogeneous catalyst. *N*-alkylation of anilines and aliphatic amines with various alcohols (benzyl and aliphatic alcohols) was achieved with a small amount of the catalyst (1 mol%) (Scheme 1.82).[278] Structural studies indicate that small Ag nanoclusters are supported on Cu nanoparticles possibly through Ag—O—Cu bonds at the silver-copper boundary. This bimetallic structure can be crucial to effectively promote alcohol dehydrogenation and hydride transfer to the amines.

1.3.9 Transition Metal-Free Catalyst System

One of the oldest examples of *N*-alkylation by a hydrogen autotransfer process used silica gel as catalyst.[279] In this early work, simple aniline was alkylated with different aliphatic primary alcohols (100–200 mol%) at very high temperatures to give a mixture of substituted anilines with moderated yields.

Alumina (Al_2O_3) has been used as catalysts in the *N*-alkylation of aniline with methanol.[280] When alumina prepared from aluminum isopropoxide was used as a heterogeneous catalyst in a conventional flow reactor and using a vaporized mixture of aniline and methanol (15,000 mol%) at 400°C, the main product was the double alkylated compound. However, when the catalyst used was commercially available γ-Al_2O_3 under the same reaction conditions, the main product was *N*-methylaniline with a lower yield.

The alkylation of aniline with methanol over γ-Al_2O_3 was studied in a fixed-bed, integral-flow reactor at atmospheric pressure and 300–425°C.[281] *N*-Methylation is the predominant reaction over this temperature range. Increasing reaction temperature and methanol/aniline molar ratio enhances the selectivity of *N,N*-dimethylaniline.

A better result was obtained using gas-phase conditions and γ-Al_2O_3 at atmospheric pressure and a lower temperature (200°C), obtaining a yield of 41% *N*-methylaniline. Under similar conditions, different alcohols, such as 1- and 2-propanol (30,000 mol%), could be used as the initial source of the electrophile. For the propyl alcohol, yields up to 72% for *N*-propylaniline and 8% for *N,N*-dipropylaniline were reached. Other amines such as octylamine, benzylamine, and chiral *R*-methylbenzylamine could be alkylated under similar reaction conditions using methanol, 1-propanol, or benzylic alcohol.[282] Even ammonia could be alkylated using benzylic alcohol, although in this case the reaction was performed at 170°C under ammonia atmosphere, yielding only 48% of benzylamine as well as 23% of di- and trialkylated amines.

The selective alkylation of aniline with methanol was investigated over various metallosilicates synthesized in two different media: alkaline medium (NaOH) and fluorine medium (NH_4F).[283] Temperature programmed desorption (TPD) of NH_3 and aniline alkylation suggested that the selectivity for *N,N*-dimethylaniline (NNDMA) is significantly changed by the medium acid sites. The high selectivity was obtained when the metallosilicates have a large number of medium acid sites (567–673 K).

Direct *N*-alkylation of anilines with alcohols including benzylic, heteroaromatic, allylic, and aliphatic alcohols was successfully performed by using nanosized zeolite beta at 135°C under aerobic condition, with higher catalytic activity than other conventional zeolites.[284] Surprisingly, *N*-alkylation of 1,2,3,4-tetrahydroquinoline with benzyl alcohol gave the corresponding tertiary amine in 17% yield, whereas 1,2,3,4-tetrahydroisoquinoline yielded 72% yield of the tertiary amine. Of note, furfuryl alcohol was converted with aniline to afford the corresponding secondary amines with a 52% yield (Scheme 1.83). The catalyst was successfully recovered and reused three times without significant loss of activity.

Carbon catalysts without transition metal for the direct *N*-alkylation of amines with alcohols were achieved by Shi's group in 2015 (Scheme 1.84).[285] The detailed catalyst characterizations revealed that C=O group in the carbon catalysts may be a possible catalytically active site, and high surface area is also important for gaining high activity. The activity of the carbon catalyst remained unchanged after being reused five times. This was the first example in which a carbon-based material without transition metal is used as the catalyst for alcohol amination via the borrowing-hydrogen reaction mechanism.

Few-layers graphene oxide (GO) and carboxylic acid-functionalized GO ($GO\text{-}CO_2H$) have been found to promote the amination of (*E*)-1,3-

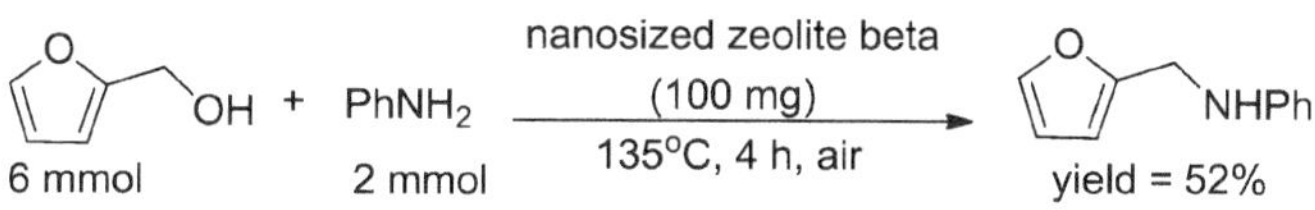

SCHEME 1.83 Amination of furfuryl alcohol with aniline catalyzed by nanosized zeolite beta.

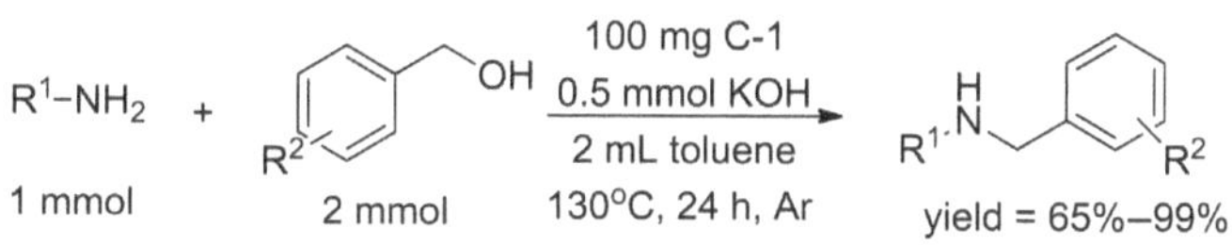

SCHEME 1.84 Carbon-catalyzed *N*-alkylation of amines with alcohols via the borrowing-hydrogen reaction mechanism.

diphenylprop-2-en-1-ol with 4-methylbenzenesulfonamide at 80°C in water.[286] Good yield was obtained in the presence of carbon catalyst (5 wt%) at 80°C in water. The reaction could also be performed in cooler conditions (50°C) only by increasing the catalyst loadings (50 wt%).

1.4 ACID CATALYSTS

In 2003, Ishii et al. showed the metal triflates ($La(OTf)_3$, $Yb(OTf)_3$, $Hf(OTf)_4$) to be active for the amination of secondary benzyl alcohols with amides.[287] Moderate to good yields were obtained in the presence of 1 mol% metal triflates in nitromethane. A similar system was reported by Zhou et al. in 2008.[288] Amination of secondary allylic and propargylic alcohols with primary amides and aromatic amines with strong electron-withdrawing groups were performed at room temperature, although higher catalyst amount (5 mol% $Yb(OTf)_3$) was necessary for good to excellent yields.

Following this work, some other metal triflates, such as $In(OTf)_3$,[289] $Ag(OTf)_3$,[290,291] $Bi(OTf)_3$,[292] $Al(OTf)_3$,[293,294] have also been used in the amination of alcohols and good results were obtained.

In 2006, Campagne and Prim et al. showed that $NaAuCl_4$ (5 mol%) could efficiently catalyze the amination of secondary benzylic alcohols with primary sulfonamides and 4-nitroaniline to produce good yields of the corresponding benzylic amine derivatives. Of note, the catalyst system allowed the reaction to perform at room temperature.

A similar catalytic performance was observed with $AuCl_3$ as a catalyst for the amination of allylic alcohols with primary sulfonamides and aromatic amines with electron-withdrawing groups.[295] But compared with $NaAuCl_4$ described by Campagne Prim et al., a lower catalyst loading (2 mol%) was required.

Campagne et al. investigated the amination of secondary benzylic, allylic, and propargylic alcohols with benzsulfamides in the presence of $NaAuCl_4{\cdot}2H_2O$ (5 mol%).[296] Benzylic and allylic alcohols gave the corresponding *N*-alkyl amines in good to excellent yields, but only moderate yield was achieved with propargylic alcohols.

Although various attempts have been done for the Au-catalyzed amination of allylic alcohols, the amines starting materials focused on the use of rather stable nitrogen compounds. More useful amines substrates containing sensitive groups were seldom explored. In 2011, Ohshima and Mashima et al. reported the direct amination of allylic and benzylic alcohols with synthetically useful, but acid-sensitive Boc, Bus, and Dios protected amines with various functionalities (OTBS, OTHP, etc.) using 1 mol% of $NaAuCl_4{\cdot}2H_2O$ under mild conditions.[297] Most of the substrates afforded good to excellent yields, but in the case of the substrates bearing more acid-sensitive functional groups, such as TBS and THP, only moderate yields were obtained.

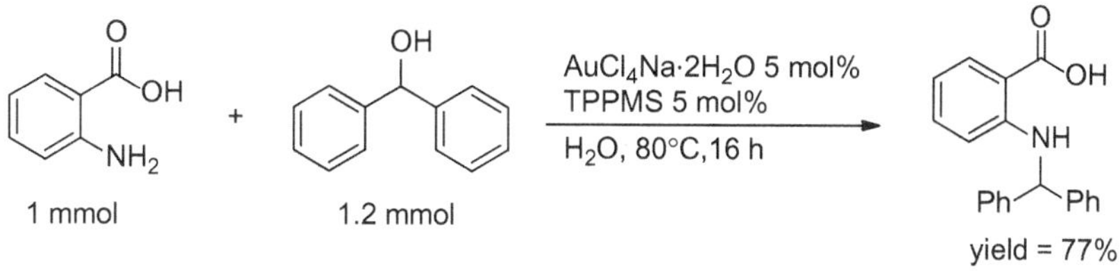

SCHEME 1.85 Au-catalyzed amination of benzhydryl with anthranilic acids

Similarly, in 2013, Hikawa and Yokoyama showed that the unprotected anthranilic acids could be used as the starting materials to react with benzhydryl alcohols to synthesize the corresponding *N*-alkyl amines in the presence of water-soluble Au(III)/TPPMS (Scheme 1.85).[298] Good to excellent yields were obtained and the sensitive carboxyl group was well tolerated. A similar catalytic performance was also observed when the combination of $Co(hfac)_2{\cdot}xH_2O$ (hfac: hexafluoroacetylacetone) and TPPMS was used.[299]

The Au/Ag double catalyst system was also investigated by Liu et al. in 2008.[300] The multisubstituted pyrroles were synthesized from the reactions of suitably substituted (*Z*)-enynols with amines or sulfonamides in the presence of 5 mol% $(p\text{-}MeOC_6H_4)_3P$-AuCl and 10 mol% $AgBF_4$ (Au/Ag). Moderate to good yields were obtained but in some cases the Lewis acid $BF_3{\cdot}Et_2O$ was required.

A similar protocol was developed by Widenhoefer et al.[301] The combination of $[P(t\text{-}Bu)_2\text{-}o\text{-}biphenyl]AuCl$ (5 mol%) and $AgSbF_6$ (5 mol%) could efficiently catalyze the intermolecular amination of allylic alcohols with cyclic ureas and sulfonamides to produce the corresponding allylic-substituted product in good to excellent yields. It is noteworthy that in the case of γ-unsubstituted or γ-methyl-substituted allylic alcohols, the desired product with high γ-regioselectivity and *syn*-stereoselectivity was obtained. Further, by increasing the reaction temperature to 100°C, the intramolecular amination of allylic alcohols with alkylamines could be also achieved to produce the substituted pyrrolidine and piperidine derivatives.[302]

A $BiCl_3$-catalyzed amination of secondary propargylic alcohols was developed by Zhan et al. in 2006.[303] Under optimized conditions, *p*-toluenesulfonamide and benzamide gave moderate to good yields, but no product was observed when acetamide, aniline and piperidine were used as the substrates. Further, Zhan et al. improved the process by changing the catalyst to simple $FeCl_3$ (5 mol%).[304]

Following the work of Zhan, a number of alcohol amination reactions including the amination of secondary benzylic and allylic alcohols with various amides,[305] the amination of Baylis-Hillman alcohol with sulfonamide,[306] and the intramolecular amination of *N*-protected amino allylic alcohol,[307] have been reported in the presence of $FeCl_3$ and $FeCl_3{\cdot}6H_2O$.

A comparative study of $FeCl_3{\cdot}6H_2O$ and TfOH as catalysts for allylic amination reaction was performed by Nájera et al. in 2012.[308] The results showed that similar or better yields are obtained in most of the cases, but the Brønsted acid TfOH allowed the reaction to perform with lower catalyst loadings and milder reaction conditions. Additionally, they demonstrated that $FeCl_3{\cdot}6H_2O$ was also effective for the amination of allylic alcohols in water.[309] However, the yields are lower compared with that in organic solvents, which can be ascribed to the lower solubility of organic compounds in aqueous media and/or to the formation of less Lewis acidic $[Fe(H_2O)_{6-n}(OH)_n]^{+3-n}Cl_{3-n}$ species under the aqueous conditions.

N-Alkylation of tosylhydrazones with benzyl alcohols in the presence of an acid catalyst was achieved by Reddy et al. in 2009.[310] Tris(pentafluorophenyl) borane was found to be effective for the reaction of primary and secondary benzylic alcohols and tosylhydrazones, and good to excellent yields were obtained.

In 2007, Cozzi et al. showed that the amination of optically active ferrocenyl alcohol with carbamates was efficiently catalyzed by a catalytic amount (5–10 mol%) of commercially available indium tribromide ($InBr_3$) (Scheme 1.86).[311] The desired products were generated with good to excellent yields and good retention of configuration.

Sulfonic acids, such as *p*-toluenesulfonic acid monohydrate (PTS) or polymer-bound *p*-toluenesulfonic acid, have been used to catalyze the amination of secondary allylic and propargylic alcohols.[312–314] Primary amides and aromatic amines with electron-withdrawing groups could be well tolerated and good to excellent yields of the desired products were obtained. When dodecylbenzenesulfonic acid (DBSA) was used as a catalyst in water, the reaction of benzhydrols and amines gave the desired *N*-alkyl products in moderate to good yields.[315]

HBF_4 (aqueous solution) showed a similar catalytic performance for the amination of secondary propargylic alcohols.[316]

Although several homogeneous acid catalysts have been used for the amination of alcohols with amines and amides and good results were obtained, these catalyst systems often are difficult to recycle, which was not desirable from the point view of industrial production.

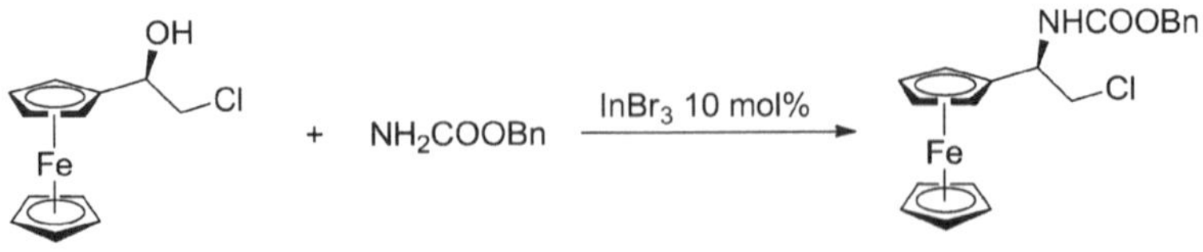

86% *ee*

yield = 86%, 86% *ee*

SCHEME 1.86 Amination of optically active ferrocenyl alcohol with carbamates catalyzed by $InBr_3$.

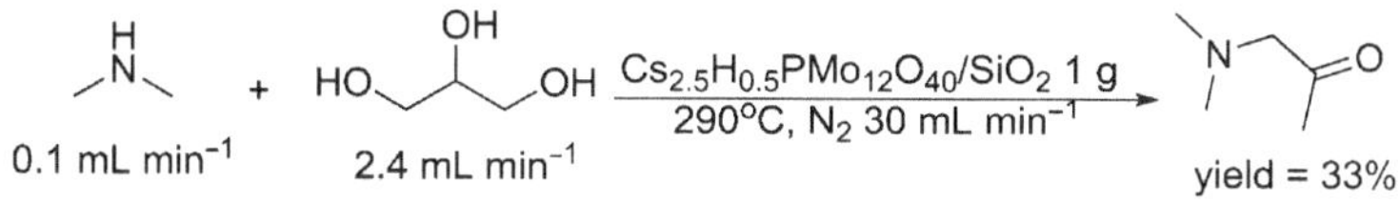

SCHEME 1.87 Amination of glycerol with dimethylamine catalyzed by $Cs_{2.5}H_{0.5}PMo_{12}O_{40}/SiO_2$.

In 2006, Kaneda et al. reported the proton-exchanged montmorillonites as a recyclable solid acid catalyst for the amination of secondary allylic and benzylic alcohols with anilines and amides.[317] The catalyst was recovered easily from the reaction mixture and was reusable at least five times without any loss of activity. Further, the reaction of indoles and secondary benzyl alcohols were also achieved with higher reaction temperature (100°C).[318]

Another reusable acid catalyst system was developed by Shimizu in 2008.[319] A water-soluble calix[4]resorcinarene sulfonic acid (10 mol%) was used as the catalyst for the amination of secondary allylic and benzylic alcohols. Three examples of sulfonamides were tested at 60°C in water and excellent yields were obtained. It is noteworthy that the aqueous phase containing catalyst could be reused at least five times without any loss of activity.

In addition to organic acids, some inorganic acids, such as phosphotungstic acid,[320] phosphomolybdic acid on silica,[321,322] $Cs_{2.5}H_{0.5}PMo_{12}O_{40}$ supported on a mesoporous silica (Scheme 1.87),[323] have also been explored for the synthesis of *N*-alkyl amines from the reaction of amines and alcohols.

1.5 CONCLUSIONS AND OUTLOOK

In summary, the main advancements in this field of the *N*-alkylation of amines with alcohols have been discussed in this chapter. With respect to alcohols, the selective conversions of carbohydrate alcohols, polyhydric alcohols, diols, aminoalcohols, as well as different functionalized alcohols containing nitro, cyano, double bond, and methoxycarbonyl groups have been described. With respect to amines, the application of ammonia and ammonia equivalents is noteworthy and both the homogeneously and heterogeneously catalyzed protocols for the synthesis of primary amines from ammonia and alcohols have been achieved. Some of the new catalyst systems show high reactivity resulting in low catalyst loadings and ambient conditions, although these examples still represent exceptions and require an equivalent amount of strong base (KO*t*Bu) or catalyst, or extra energy (light).

Nevertheless, some important goals remain to be solved. For example, no *N*-alkylation of urea itself with alcohol is reported so far. Hence, more active catalysts are still needed. A further challenge in this area is the development of less expensive catalysts, which are based on metals such as Fe, Cu, or Mn. With these catalysts in hand, we can anticipate the development of new research areas for amination involving biomass-based alcohol with market opportunities.

REFERENCES

1. Lawrence, S. A. *Amines: synthesis, properties and applications.* Cambridge University Press: Cambridge, 2004.
2. von Hofmann, A. *Eur. J. Org. Chem.* **1851**, *78*, 253.
3. Magano, J.; Dunetz, J. R. *Chem. Rev.* **2011**, *111*, 2177.
4. Aubin, Y.; Fischmeister, C.; Thomas, C. M.; Renaud, J.-L. *Chem. Soc. Rev.* **2010**, *39*, 4130.
5. Jiang, D.; Fu, H.; Jiang, Y.; Zhao, Y. *J. Org. Chem.* **2007**, *72*, 672.
6. Alex, K.; Tillack, A.; Schwarz, N.; Beller, M. *ChemSusChem* **2008**, *1*, 333.
7. Crozet, D.; Urrutigoïty, M.; Kalck, P. *ChemCatChem* **2011**, *3*, 1102.
8. Das, S.; Zhou, S.; Addis, D.; Enthaler, S.; Junge, K.; Beller, M. *Top. Catal.* **2010**, *53*, 979.
9. Junge, K.; Wendt, B.; Shaikh, N.; Beller, M. *Chem. Commun.* **2010**, *46*, 1769.
10. Steinhuebel, D.; Sun, Y.; Matsumura, K.; Sayo, N.; Saito, T. *J. Am. Chem. Soc.* **2009**, *131*, 11316.
11. Nef, J. *Eur. J. Org. Chem.* **1901**, *318*, 137.
12. Guillena, G.; Ramón, J. D.; Yus, M. *Chem. Rev.* **2009**, *110*, 1611.
13. Grigg, R.; Mitchell, T. R. B.; Sutthivaiyakit, S.; Tongpenyai, N. *J. Chem. Soc. Chem. Commun.* **1981**, 611.
14. Watanabe, Y.; Tsuji, Y.; Ohsugi, Y. *Tetrahedron Lett.* **1981**, *22*, 2667.
15. Trost, B. M.; Van Vranken, D. L. *Chem. Rev.* **1996**, *96*, 395.
16. Muzart, J. *Tetrahedron* **2005**, *61*, 4179.
17. Muzart, J. *Eur. J. Org. Chem.* **2007**, *2007*, 3077.
18. Muzart, J. *Tetrahedron* **2008**, *64*, 5815.
19. Detz, R. J.; Hiemstra, H.; van Maarseveen, J. H. *Eur. J. Org. Chem.* **2009**, *2009*, 6263.
20. Miyake, Y.; Uemura, S.; Nishibayashi, Y. *ChemCatChem* **2009**, *1*, 342.
21. Biannic, B.; Aponick, A. *Eur. J. Org. Chem.* **2011**, *2011*, 6605.
22. Emer, E.; Sinisi, R.; Capdevila, M. G.; Petruzziello, D.; De Vincentiis, F.; Cozzi, P. G. *Eur. J. Org. Chem.* **2011**, *2011*, 647.
23. Bauer, E. B. *Synthesis* **2012**, *44*, 1131.
24. Poli, G.; Prestat, G.; Liron, F.; Kammerer-Pentier, C. (2012). (). In: *Transition metal catalyzed enantioselective allylic substitution in organic synthesis;* Kazmaier, U., Ed.; Springer Berlin Heidelberg: Berlin, Heidelberg, 2012; p 1.
25. Sundararaju, B.; Achard, M.; Bruneau, C. *Chem. Soc. Rev.* **2012**, *41*, 4467.
26. Butt, N. A.; Zhang, W. *Chem. Soc. Rev.* **2015**, *44*, 7929.
27. Zhao, Y.; Foo, S. W.; Saito, S. *Angew. Chem. Int. Ed.* **2011**, *50*, 3006.
28. Hamid, M. H. S. A.; Slatford, P. A.; Williams, J. M. J. *Adv. Synth. Catal.* **2007**, *349*, 1555.
29. Lamb, G. W.; Williams, J. M. *Chim. Oggi* **2008**, *26*, 17.
30. Bähn, S.; Imm, S.; Neubert, L.; Zhang, M.; Neumann, H.; Beller, M. *ChemCatChem* **2011**, *3*, 1853.
31. Nixon, T. D.; Whittlesey, M. K.; Williams, J. M. J. *Dalton Trans.* **2009**, 753.
32. Gunanathan, C.; Milstein, D. *Science* **2013**, *341*, 1229712.
33. Shimizu, K.-i. *Catal. Sci. Technol.* **2015**, *5*, 1412.
34. Ma, X.; Xu, Q.; Ma, X.; Su, C. *Top. Curr. Chem.* **2016**, *374*, 27.
35. Qing, X.; Qiang, L. *Chin. J. Org. Chem.* **2012**, *33*, 18.
36. Guillena, G.; Ramón, D. J.; Yus, M. *Chem. Rev.* **2009**, *110*, 1611.
37. Dobereiner, G. E.; Crabtree, R. H. *Chem. Rev.* **2009**, *110*, 681.
38. Nef, J. U. *Justus Liebigs Ann. Chem.* **1901**, *318*, 137.
39. Arcelli, A.; Porzi, G. *J. Organomet. Chem.* **1982**, *235*, 93.

40. Ganguly, S.; Joslin, F. L.; Roundhill, D. M. *Inorg. Chem.* **1989**, *28*, 4562.
41. Ganguly, S.; Roundhill, D. M. *Polyhedron* **1990**, *9*, 2517.
42. Sakano, T.; Ishii, H.; Yamaguchi, I.; Osakada, K.; Yamamoto, T. *Inorg. Chim. Acta* **1999**, *296*, 176.
43. Marsella, J. A. *J. Org. Chem.* **1987**, *52*, 467.
44. Marsella, J. A. *J. Organomet. Chem.* **1991**, *407*, 97.
45. Tsuji, Y.; Huh, K. T.; Ohsugi, Y.; Watanabe, Y. *J. Org. Chem.* **1985**, *50*, 1365.
46. Felföldi, K.; Klyavlin, M.; Bartók, M. *J. Organomet. Chem.* **1989**, *362*, 193.
47. Tsuji, Y.; Yokoyama, Y.; Huh, K.-T.; Watanabe, Y. *Bull. Chem. Soc. Jpn.* **1987**, *60*, 3456.
48. Watanabe, Y.; Ohta, T.; Tsuji, Y. *Bull. Chem. Soc. Jpn.* **1983**, *56*, 2647.
49. Murahashi, S.-I.; Kondo, K.; Hakata, T. *Tetrahedron Lett.* **1982**, *23*, 229.
50. Watanabe, Y.; Morisaki, Y.; Kondo, T.; Mitsudo, T.-a. *J. Org. Chem.* **1996**, *61*, 4214.
51. Naskar, S.; Bhattacharjee, M. *Tetrahedron Lett.* **2007**, *48*, 3367.
52. Del Zotto, A.; Baratta, W.; Sandri, M.; Verardo, G.; Rigo, P. *Eur. J. Inorg. Chem.* **2004**, *2004*, 524.
53. Nishibayashi, Y.; Wakiji, I.; Hidai, M. *J. Am. Chem. Soc.* **2000**, *122*, 11019.
54. Milton, M. D.; Inada, Y.; Nishibayashi, Y.; Uemura, S. *Chem. Commun.* **2004**, 2712.
55. Nishibayashi, Y.; Milton, M. D.; Inada, Y.; et al. *Chem. Eur. J.* **2005**, *11*, 1433.
56. Nishibayashi, Y.; Imajima, H.; Onodera, G.; Hidai, M.; Uemura, S. *Organometallics* **2004**, *23*, 26.
57. Tillack, A.; Hollmann, D.; Michalik, D.; Beller, M. *Tetrahedron Lett.* **2006**, *47*, 8881.
58. Hollmann, D.; Tillack, A.; Michalik, D.; Jackstell, R.; Beller, M. *Chem. Asian J.* **2007**, *2*, 403.
59. Ruffolo, R. R., Jr.; Bondinell, W.; Hieble, J. P. *J. Med. Chem.* **1995**, *38*, 3681.
60. Duncton, M. A.; Roffey, J. R.; Hamlyn, R. J.; Adams, D. R. *Tetrahedron Lett.* **2006**, *47*, 2549.
61. Tanaka, N.; Hatanaka, M.; Watanabe, Y. *Chem. Lett.* **1992**, *21*, 575.
62. Hamid, M. H. S. A.; Williams, J. M. J. *Chem. Commun.* **2007**, 725.
63. Hamid, M. H. S. A.; Williams, J. M. J. *Tetrahedron Lett.* **2007**, *48*, 8263.
64. Hamid, M. H. S. A.; Allen, C. L.; Lamb, G. W.; et al. *J. Am. Chem. Soc.* **2009**, *131*, 1766.
65. Tillack, A.; Hollmann, D.; Mevius, K.; Michalik, D.; Baehn, S.; Beller, M. *Eur. J. Org. Chem.* **2008**, *2008*, 4745.
66. Luo, J.; Wu, M.; Xiao, F.; Deng, G. *Tetrahedron Lett.* **2011**, *52*, 2706.
67. Sundberg, R. J. *Indoles.* Academic Press: Cambridge, 1996.
68. Negwer, M.; Scharnow, H.-G. *Organic-chemical drugs and their synonyms: (an international survey).* Wiley-VCH: Weinheim, 2001.
69. Bähn, S.; Imm, S.; Mevius, K.; et al. *Chem. Eur. J.* **2010**, *16*, 3590.
70. Agrawal, S.; Lenormand, M.; Martin-Matute, B. *Org. Lett.* **2012**, *14*, 1456.
71. Pridmore, S. J.; Slatford, P. A.; Daniel, A.; Whittlesey, M. K.; Williams, J. M. J. *Tetrahedron Lett.* **2007**, *48*, 5115.
72. Monrad, R. N.; Madsen, R. *Org. Biomol. Chem.* **2011**, *9*, 610.
73. Enyong, A. B.; Moasser, B. *J. Org. Chem.* **2014**, *79*, 7553.
74. Jumde, V. R.; Gonsalvi, L.; Guerriero, A.; Peruzzini, M.; Taddei, M. *Eur. J. Org. Chem.* **2015**, *2015*, 1829.
75. Dang, T. T.; Ramalingam, B.; Seayad, A. M. *ACS Catal.* **2015**, *5*, 4082.
76. Marichev, K. O.; Takacs, J. M. *ACS Catal.* **2016**, *6*, 2205.
77. Hayes, K. *Appl. Catal. A Gen.* **2001**, *221*, 187.
78. Gunanathan, C.; Milstein, D. *Angew. Chem. Int. Ed.* **2008**, *120*, 8789.
79. Imm, S.; Bähn, S.; Neubert, L.; Neumann, H.; Beller, M. *Angew. Chem. Int. Ed.* **2010**, *49*, 8126.

80. Pingen, D.; Müller, C.; Vogt, D. *Angew. Chem. Int. Ed.* **2010**, *49*, 8130.
81. Nakagawa, N.; Derrah, E. J.; Schelwies, M.; Rominger, F.; Trapp, O.; Schaub, T. *Dalton Trans.* **2016**, *45*, 6856.
82. Pera-Titus, M.; Shi, F. *ChemSusChem* **2014**, *7*, 720.
83. Imm, S.; Bähn, S.; Zhang, M.; et al. *Angew. Chem. Int. Ed.* **2011**, *50*, 7599.
84. Pingen, D.; Diebolt, O.; Vogt, D. *ChemCatChem* **2013**, *5*, 2905.
85. Zhang, M.; Imm, S.; Bähn, S.; Neumann, H.; Beller, M. *Angew. Chem. Int. Ed.* **2011**, *50*, 11197.
86. Fujita, K.-i.; Yamamoto, K.; Yamaguchi, R. *Org. Lett.* **2002**, *4*, 2691.
87. Fujita, K.-i.; Yamaguchi, R. *Synlett* **2005**, *2005*, 560.
88. Fujita, K.-i.; Fujii, T.; Yamaguchi, R. *Org. Lett.* **2004**, *6*, 3525.
89. Fujita, K.-i.; Li, Z.; Ozeki, N.; Yamaguchi, R. *Tetrahedron Lett.* **2003**, *44*, 2687.
90. Fujita, K.-i.; Enoki, Y.; Yamaguchi, R. *Tetrahedron* **2008**, *64*, 1943.
91. Balcells, D.; Nova, A.; Clot, E.; Gnanamgari, D.; Crabtree, R. H.; Eisenstein, O. *Organometallics* **2008**, *27*, 2529.
92. Saidi, O.; Blacker, A. J.; Farah, M. M.; Marsden, S. P.; Williams, J. M. J. *Chem. Commun.* **2010**, *46*, 1541.
93. Saidi, O.; Blacker, A. J.; Lamb, G. W.; Marsden, S. P.; Taylor, J. E.; Williams, J. M. J. *Org. Process Res. Dev.* **2010**, *14*, 1046.
94. Kawahara, R.; Fujita, K.-i.; Yamaguchi, R. *J. Am. Chem. Soc.* **2010**, *132*, 15108.
95. Kawahara, R.; Fujita, K.-i.; Yamaguchi, R. *Adv. Synth. Catal.* **2011**, *353*, 1161.
96. Prades, A.; Corberan, R.; Poyatos, M.; Peris, E. *Chem. Eur. J.* **2008**, *14*, 11474.
97. Bartoszewicz, A.; Marcos, R.; Sahoo, S.; Inge, A. K.; Zou, X.; Martín-Matute, B. *Chem. Eur. J.* **2012**, *18*, 14510.
98. Aramoto, H.; Obora, Y.; Ishii, Y. *J. Org. Chem.* **2009**, *74*, 628.
99. Gnanamgari, D.; Sauer, E. L. O.; Schley, N. D.; Butler, C.; Incarvito, C. D.; Crabtree, R. H. *Organometallics* **2009**, *28*, 321.
100. Blank, B.; Madalska, M.; Kempe, R. *Adv. Synth. Catal.* **2008**, *350*, 749.
101. Blank, B.; Michlik, S.; Kempe, R. *Adv. Synth. Catal.* **2009**, *351*, 2903.
102. Michlik, S.; Hille, T.; Kempe, R. *Adv. Synth. Catal.* **2012**, *354*, 847.
103. Blank, B.; Michlik, S.; Kempe, R. *Chem. Eur. J.* **2009**, *15*, 3790.
104. Michlik, S.; Kempe, R. *Chem. Eur. J.* **2010**, *16*, 13193.
105. Andrushko, N.; Andrushko, V.; Roose, P.; Moonen, K.; Börner, A. *ChemCatChem* **2010**, *2*, 640.
106. Li, F.; Xie, J.; Shan, H.; Sun, C.; Chen, L. *RSC Adv.* **2012**, *2*, 8645.
107. Yuan, K.; Jiang, F.; Sahli, Z.; Achard, M.; Roisnel, T.; Bruneau, C. *Angew. Chem. Int. Ed.* **2012**, *124*, 9006.
108. Li, J.-Q.; Andersson, P. G. *Chem. Commun.* **2013**, *49*, 6131.
109. Michlik, S.; Kempe, R. *Angew. Chem. Int. Ed.* **2013**, *52*, 6326.
110. Ruch, S.; Irrgang, T.; Kempe, R. *Chem. Eur. J.* **2014**, *20*, 13279.
111. Wetzel, A.; Wöckel, S.; Schelwies, M.; et al. *Org. Lett.* **2013**, *15*, 266.
112. Campos, J.; Sharninghausen, L. S.; Manas, M. G.; Crabtree, R. H. *Inorg. Chem.* **2015**, *54*, 5079.
113. Liu, S.; Rebros, M.; Stephens, G.; Marr, A. C. *Chem. Commun.* **2009**, 2308.
114. Cumpstey, I.; Agrawal, S.; Borbas, K. E.; Martín-Matute, B. *Chem. Commun.* **2011**, *47*, 7827.
115. Yamashita, Y.; Gopalarathnam, A.; Hartwig, J. F. *J. Am. Chem. Soc.* **2007**, *129*, 7508.
116. Roggen, M.; Carreira, E. M. *J. Am. Chem. Soc.* **2010**, *132*, 11917.
117. Bandini, M. *Angew. Chem. Int. Ed.* **2011**, *50*, 994.

118. Lafrance, M.; Roggen, M.; Carreira, E. M. *Angew. Chem. Int. Ed.* **2012**, *51*, 3470.
119. Bandini, M.; Cera, G.; Chiarucci, M. *Synthesis* **2012**, *44*, 504.
120. Defieber, C.; Ariger, M. A.; Moriel, P.; Carreira, E. M. *Angew. Chem. Int. Ed.* **2007**, *46*, 3139.
121. Zhang, Y.; Lim, C. S.; Sim, D. S. B.; Pan, H. J.; Zhao, Y. *Angew. Chem. Int. Ed.* **2014**, *126*, 1423.
122. Rong, Z.-Q.; Zhang, Y.; Chua, R. H. B.; Pan, H.-J.; Zhao, Y. *J. Am. Chem. Soc.* **2015**, *137*, 4944.
123. Atkins, K.; Walker, W.; Manyik, R. *Tetrahedron Lett.* **1970**, *11*, 3821.
124. Yang, S.-C.; Hsu, Y.-C.; Gan, K.-H. *Tetrahedron* **2006**, *62*, 3949.
125. Hirai, Y.; Nagatsu, M. *Chem. Lett.* **1994**, *23*, 21.
126. Hirai, Y.; Shibuya, K.; Fukuda, Y.; Yokoyama, H.; Yamaguchi, S. *Chem. Lett.* **1997**, *26*, 221.
127. Masuyama, Y.; Kagawa, M.; Kurusu, Y. *Chem. Lett.* **1995**, *24*, 1121.
128. Kimura, M.; Futamata, M.; Shibata, K.; Tamaru, Y. *Chem. Commun.* **2003**, 234.
129. Sakamoto, M.; Shimizu, I.; Yamamoto, A. *Bull. Chem. Soc. Jpn.* **1996**, *69*, 1065.
130. Kayaki, Y.; Koda, T.; Ikariya, T. *J. Org. Chem.* **2004**, *69*, 2595.
131. Yang, S.-C.; Chung, W.-H. *Tetrahedron Lett.* **1999**, *40*, 953.
132. Yang, S.-C.; Hung, C.-W. *J. Org. Chem.* **1999**, *64*, 5000.
133. Yang, S.-C.; Hung, C.-W. *Synthesis* **1999**, *1999*, 1747.
134. Yang, S.-C.; Tsai, Y.-C. *Organometallics* **2001**, *20*, 763.
135. Shue, Y.-J.; Yang, S.-C.; Lai, H.-C. *Tetrahedron Lett.* **2003**, *44*, 1481.
136. Hikawa, H.; Yokoyama, Y. *J. Org. Chem.* **2011**, *76*, 8433.
137. Hikawa, H.; Yokoyama, Y. *Org. Biomol. Chem.* **2011**, *9*, 4044.
138. Wagh, Y. S.; Sawant, D. N.; Dhake, K. P.; Bhanage, B. M. *Catal. Sci. Technol.* **2012**, *2*, 835.
139. Banerjee, D.; Jagadeesh, R. V.; Junge, K.; Junge, H.; Beller, M. *Angew. Chem. Int. Ed.* **2012**, *51*, 11556.
140. Banerjee, D.; Jagadeesh, R. V.; Junge, K.; Junge, H.; Beller, M. *ChemSusChem* **2012**, *5*, 2039.
141. Ozawa, F.; Okamoto, H.; Kawagishi, S.; Yamamoto, S.; Minami, T.; Yoshifuji, M. *J. Am. Chem. Soc.* **2002**, *124*, 10968.
142. Ozawa, F.; Yoshifuji, M. *Dalton Trans.* **2006**, 4987.
143. Ozawa, F.; Yoshifuji, M. *C. R. Chim.* **2004**, *7*, 747.
144. Ozawa, F.; Ishiyama, T.; Yamamoto, S.; Kawagishi, S.; Murakami, H.; Yoshifuji, M. *Organometallics* **2004**, *23*, 1698.
145. Kinoshita, H.; Shinokubo, H.; Oshima, K. *Org. Lett.* **2004**, *6*, 4085.
146. Piechaczyk, O.; Doux, M.; Ricard, L.; le Floch, P. *Organometallics* **2005**, *24*, 1204.
147. Nishikata, T.; Lipshutz, B. H. *Org. Lett.* **2009**, *11*, 2377.
148. Ghosh, R.; Sarkar, A. *J. Org. Chem.* **2011**, *76*, 8508.
149. Jing, J.; Huo, X.; Shen, J.; Fu, J.; Meng, Q.; Zhang, W. *Chem. Commun.* **2017**, *53*, 5151.
150. Piechaczyk, O.; Thoumazet, C.; Jean, Y.; le Floch, P. *J. Am. Chem. Soc.* **2006**, *128*, 14306.
151. Mora, G.; Deschamps, B.; Van Zutphen, S.; Le Goff, X. F.; Ricard, L.; Le Floch, P. *Organometallics* **2007**, *26*, 1846.
152. Usui, I.; Schmidt, S.; Keller, M.; Breit, B. *Org. Lett.* **2008**, *10*, 1207.
153. Tao, Y.; Zhou, Y.; Qu, J.; Hidai, M. *Tetrahedron Lett.* **2010**, *51*, 1982.
154. Wang, M.; Xie, Y.; Li, J.; Huang, H. *Synlett* **2014**, *25*, 2781.
155. Banerjee, D.; Junge, K.; Beller, M. *Angew. Chem. Int. Ed.* **2014**, *53*, 13049.
156. Feng, S. L.; Liu, C. Z.; Li, Q.; Yu, X. C.; Xu, Q. *Chin. Chem. Lett.* **2011**, *22*, 1021.
157. Bertoli, M.; Choualeb, A.; Lough, A. J.; Moore, B.; Spasyuk, D.; Gusev, D. G. *Organometallics* **2011**, *30*, 3479.
158. Zhu, Z.; Espenson, J. H. *J. Org. Chem.* **1996**, *61*, 324.

159. Ohri, R. V.; Radosevich, A. T.; Hrovat, K. J.; et al. *Org. Lett.* **2005**, *7*, 2501.
160. Abdukader, A.; Jin, H.; Cheng, Y.; Zhu, C. *Tetrahedron Lett.* **2014**, *55*, 4172.
161. Utsunomiya, M.; Miyamoto, Y.; Ipposhi, J.; Ohshima, T.; Mashima, K. *Org. Lett.* **2007**, *9*, 3371.
162. Das, K.; Shibuya, R.; Nakahara, Y.; Germain, N.; Ohshima, T.; Mashima, K. *Angew. Chem. Int. Ed.* **2012**, *124*, 154.
163. Mora, G.; Piechaczyk, O.; Houdard, R.; Mézailles, N.; Le Goff, X.-F.; le Floch, P. *Chem. Eur. J.* **2008**, *14*, 10047.
164. Ohshima, T.; Miyamoto, Y.; Ipposhi, J.; Nakahara, Y.; Utsunomiya, M.; Mashima, K. *J. Am. Chem. Soc.* **2009**, *131*, 14317.
165. Reddy, C. R.; Madhavi, P. P.; Reddy, A. S. *Tetrahedron Lett.* **2007**, *48*, 7169.
166. Yang, H.; Fang, L.; Zhang, M.; Zhu, C. *Eur. J. Org. Chem.* **2009**, *2009*, 666.
167. Zhang, M.; Yang, H.; Cheng, Y.; Zhu, Y.; Zhu, C. *Tetrahedron Lett.* **2010**, *51*, 1176.
168. Rösler, S.; Ertl, M.; Irrgang, T.; Kempe, R. *Angew. Chem. Int. Ed.* **2015**, *54*, 15046.
169. Mastalir, M.; Tomsu, G.; Pittenauer, E.; Allmaier, G.n.; Kirchner, K. *Org. Lett.* **2016**, *18*, 3462.
170. Zhang, G.; Yin, Z.; Zheng, S. *Org. Lett.* **2015**, *18*, 300.
171. Shi, F.; Tse, M. K.; Cui, X.; et al. *Angew. Chem. Int. Ed.* **2009**, *121*, 6026.
172. Cui, X.; Shi, F.; Tse, M. K.; et al. *Adv. Synth. Catal.* **2009**, *351*, 2949.
173. Martínez-Asencio, A.; Ramón, D. J.; Yus, M. *Tetrahedron Lett.* **2010**, *51*, 325.
174. Martinez-Asencio, A.; Ramon, D. J.; Yus, M. *Tetrahedron* **2011**, *67*, 3140.
175. Li, F.; Shan, H.; Kang, Q.; Chen, L. *Chem. Commun.* **2011**, *47*, 5058.
176. Cui, X.; Shi, F.; Zhang, Y.; Deng, Y. *Tetrahedron Lett.* **2010**, *51*, 2048.
177. Bala, M.; Verma, P. K.; Sharma, U.; Kumar, N.; Singh, B. *Green Chem.* **2013**, *15*, 1687.
178. Yan, T.; Feringa, B. L.; Barta, K. *Nat. Commun.* **2014**, *5*, 5602.
179. Pan, H.-J.; Ng, T. W.; Zhao, Y. *Chem. Commun.* **2015**, *51*, 11907.
180. Rawlings, A. J.; Diorazio, L. J.; Wills, M. *Org. Lett.* **2015**, *17*, 1086.
181. Yan, T.; Feringa, B. L.; Barta, K. *ACS Catal.* **2015**, *6*, 381.
182. Emayavaramban, B.; Roy, M.; Sundararaju, B. *Chem. Eur. J.* **2016**, *22*, 3952.
183. Mastalir, M.; Glatz, M.; Gorgas, N.; et al. *Chem. Eur. J.* **2016**, *22*, 12316.
184. Mastalir, M.; Stöger, B.; Pittenauer, E.; Puchberger, M.; Allmaier, G.; Kirchner, K. *Adv. Synth. Catal.* **2016**, *358*, 3824.
185. Furukawa, J.; Kiji, J.; Yamamoto, K.; Tojo, T. *Tetrahedron* **1973**, *29*, 3149.
186. Bricout, H.; Carpentier, J.-F.; Mortreux, A. J. *Mol. Catal. A Chem.* **1998**, *136*, 243–251.
187. Elangovan, S.; Neumann, J.; Sortais, J.-B.; Junge, K.; Darcel, C.; Beller, M. *Nat. Commun.* **2016**, *7*, 12641.
188. Neumann, J.; Elangovan, S.; Spannenberg, A.; Junge, K.; Beller, M. *Chem. Eur. J.* **2017**, *23*, 5410.
189. Dai, X.; Cui, X.; Deng, Y.; Shi, F. *RSC Adv.* **2015**, *5*, 43589.
190. Xu, Q.; Li, Q.; Zhu, X.; Chen, J. *Adv. Synth. Catal.* **2013**, *355*, 73.
191. Guérin, C.; Bellosta, V.; Guillamot, G.; Cossy, J. *Org. Lett.* **2011**, *13*, 3534.
192. Martinez-Montero, L.; Gotor, V.; Gotor-Fernandez, V.; Lavandera, I. *Green Chem.* **2017**, *19*, 474.
193. Tauber, K.; Fuchs, M.; Sattler, J. H.; et al. *Chem. Eur. J.* **2013**, *19*, 4030.
194. Khan, I. A.; Saxena, A. K. *J. Org. Chem.* **2013**, *78*, 11656.
195. Trillo, P.; Baeza, A.; Nájera, C. *J. Org. Chem.* **2012**, *77*, 7344.
196. Du, Y.; Oishi, S.; Saito, S. *Chem. Eur. J.* **2011**, *17*, 12262.
197. Wu, W.; Rao, W.; Er, Y. Q.; Loh, J. K.; Poh, C. Y.; Chan, P. W. H. *Tetrahedron Lett.* **2008**, *49*, 2620.

198. García, A.; Castedo, L.; Domínguez, D. *Synlett* **1993**, *1993*, 271.
199. Zhu, A.; Li, L.; Wang, J.; Zhuo, K. *Green Chem.* **2011**, *13*, 1244.
200. Mitsunobu, O. *Synthesis* **1981**, *1981*, 1.
201. Aston, J.; Peterson, T.; Holowchak, J. *J. Am. Chem. Soc.* **1934**, *56*, 153.
202. Schwoegler, E. J.; Adkins, H. *J. Am. Chem. Soc.* **1939**, *61*, 3499.
203. Schneider, H. J.; Adkins, H.; McElvain, S. *J. Am. Chem. Soc.* **1952**, *74*, 4287.
204. Becker, J.; Niederer, J. P.; Keller, M.; Hölderich, W. F. *Appl. Catal. A Gen.* **2000**, *197*, 229.
205. Barrault, J.; Essayem, N.; Guimon, C. *Appl. Catal. A Gen.* **1993**, *102*, 151.
206. Baiker, A.; Richarz, W. *Tetrahedron Lett.* **1977**, *18*, 1937.
207. Baiker, A.; Richarz, W. *Synth. Commun.* **1978**, *8*, 27.
208. Runeberg, J.; Baiker, A.; Kijenski, J. *Appl. Catal.* **1985**, *17*, 309.
209. Hammerschmidt, W.; Baiker, A.; Wokaun, A.; Fluhr, W. *Appl. Catal.* **1986**, *20*, 305.
210. Kijeński, J.; Niedzielski, P.; Baiker, A. *Appl. Catal.* **1989**, *53*, 107.
211. Göbölös, S.; Hegedüs, M.; Kolosova, I.; Maciejewski, M.; Margitfalvi, J. *Appl. Catal. A Gen.* **1998**, *169*, 201.
212. Yamakawa, T.; Tsuchiya, I.; Mitsuzuka, D.; Ogawa, T. *Catal. Commun.* **2004**, *5*, 291.
213. Abe, H.; Yokota, Y.; Okabe, K. *Appl. Catal.* **1989**, *52*, 171.
214. Kimura, H.; Taniguchi, H. *Appl. Catal. A Gen.* **2005**, *287*, 191.
215. Kimura, H.; Tsutsumi, S.-I.; Tsukada, K. *Appl. Catal. A Gen.* **2005**, *292*, 281.
216. Kimura, H.; Taniguchi, H. *Catal. Lett.* **1996**, *40*, 123.
217. Likhar, P. R.; Arundhathi, R.; Kantam, M. L.; Prathima, P. S. *Eur. J. Org. Chem.* **2009**, *2009*, 5383.
218. Chen, X.; Luo, H.; Qian, C.; He, C. *React. Kinet. Mech. Catal.* **2011**, *104*, 163.
219. Mittapelly, N.; Reguri, B. R.; Mukkanti, K. *Der Pharma Chem.* **2011**, *3*, 180.
220. Santoro, F.; Psaro, R.; Ravasio, N.; Zaccheria, F. *RSC Adv.* **2014**, *4*, 2596.
221. Pang, S.; Deng, Y.; Shi, F. *Chem. Commun.* **2015**, *51*, 9471.
222. Zhang, L.; Zhang, Y.; Deng, Y.; Shi, F. *Catal. Sci. Technol.* **2015**, *5*, 3226.
223. He, J.; Yamaguchi, K.; Mizuno, N. *Chem. Lett.* **2010**, *39*, 1182.
224. Guyot, A.; Fournier, M. *Bull. Soc. Chim. Fr.* **1930**, *47*, 203.
225. Pratt, E. F.; Frazza, E. J. *J. Am. Chem. Soc.* **1954**, *76*, 6174.
226. Barnes, C. M.; Rase, H. F. *Ind. Eng. Chem. Prod. Res. Dev.* **1981**, *20*, 399.
227. Rice, R. G.; Kohn, E. J. *J. Am. Chem. Soc.* **1955**, *77*, 4052.
228. De Angelis, F.; Grasso, M.; Nicoletti, R. *Synthesis* **1977**, *1977*, 335.
229. Rice, R. G.; Kohn, E. J.; Daasch, L. W. *J. Org. Chem.* **1958**, *23*, 1352.
230. Winans, C. F.; Adkins, H. *J. Am. Chem. Soc.* **1932**, *54*, 306.
231. Langdon, W.; Levis, W., Jr.; Jackson, D. *Ind. Eng. Chem. Process Des. Dev.* **1962**, *1*, 153.
232. Garcia Ruano, J. L.; Parra, A.; Aleman, J.; Yuste, F.; Mastranzo, V. M. *Chem. Commun.* **2009**, 404.
233. Sun, J.; Jin, X.; Zhang, F.; Hu, W.; Liu, J.; Li, R. *Catal. Commun.* **2012**, *24*, 30.
234. Cui, X.; Dai, X.; Deng, Y.; Shi, F. *Chem. Eur. J.* **2013**, *19*, 3665.
235. Shimizu, K.; Imaiida, N.; Kon, K.; Siddiki, S.; Satsuma, A. *ACS Catal.* **2013**, *3*, 998.
236. Shimizu, K.-i.; Kon, K.; Onodera, W.; Yamazaki, H.; Kondo, J. N. *ACS Catal.* **2013**, *3*, 112.
237. Dai, X.; Rabeah, J.; Yuan, H.; Brückner, A.; Cui, X.; Shi, F. *ChemSusChem* **2016**, *9*, 3133.
238. Martinez, R.; Ramon, D. J.; Yus, M. *Org. Biomol. Chem.* **2009**, *7*, 2176.
239. Gonzalez-Arellano, C.; Yoshida, K.; Luque, R.; Gai, P. L. *Green Chem.* **2010**, *12*, 1281.
240. Li, L.; Zhu, A.; Zhang, Y.; Fan, X.; Zhang, G. *RSC Adv.* **2014**, *4*, 4286.
241. Kanno, H.; Taylor, R. J. *Tetrahedron Lett.* **2002**, *43*, 7337.
242. Blackburn, L.; Taylor, R. J. *Org. Lett.* **2001**, *3*, 1637.

243. Yu, X.; Liu, C.; Jiang, L.; Xu, Q. *Org. Lett.* **2011**, *13*, 6184.
244. Fischer, A.; Maciejewski, M.; Buergi, T.; Mallat, T.; Baiker, A. *J. Catal.* **1999**, *183*, 373.
245. Shi, F.; Tse, M. K.; Zhou, S.; et al. *J. Am. Chem. Soc.* **2009**, *131*, 1775.
246. Kim, J. W.; Yamaguchi, K.; Mizuno, N. *J. Catal.* **2009**, *263*, 205.
247. Yamaguchi, K.; He, J.; Oishi, T.; Mizuno, N. *Chem. Eur. J.* **2010**, *16*, 7199.
248. Cano, R.; Ramón*, D. J.; Yus, M. *J. Org. Chem.* **2011**, *76*, 5547.
249. Du, F.; Jin, X.; Yan, W.; Zhao, M.; Thapa, P. S.; Chaudhari, R. V. *Catal. Today* **2017**, https://doi.org/10.1016/j.cattod.2017.03.028.
250. Niemeier, J.; Engel, R. V.; Rose, M. *Green Chem.* **2017**, *19*, 2839.
251. Ruiz, D.; Aho, A.; Saloranta, T.; et al. *Chem. Eng. J.* **2017**, *307*, 739.
252. Murahashi, S.-I.; Shimamura, T.; Moritani, I. *J. Chem. Soc. Chem. Commun.* **1974**, 931.
253. Kwon, M. S.; Kim, S.; Park, S.; Bosco, W.; Chidrala, R. K.; Park, J. *J. Org. Chem.* **2009**, *74*, 2877.
254. Xu, C.-P.; Xiao, Z.-H.; Zhuo, B.-Q.; Wang, Y.-H.; Huang, P.-Q. *Chem. Commun.* **2010**, *46*, 7834.
255. Corma, A.; Ródenas, T.; Sabater, M. J. *Chem. Eur. J.* **2010**, *16*, 254.
256. Zhang, Y.; Qi, X.; Cui, X.; Shi, F.; Deng, Y. *Tetrahedron Lett.* **2011**, *52*, 1334.
257. Zhang, L.; Zhang, Y.; Deng, Y.; Shi, F. *RSC Adv.* **2015**, *5*, 14514.
258. Liu, X.; Hermange, P.; Ruiz, J.; Astruc, D. *ChemCatChem* **2016**, *8*, 1043.
259. Nishimoto, S.; Ohtani, B.; Yoshikawa, T.; Kagiya, T. *J. Am. Chem. Soc.* **1983**, *105*, 7180.
260. Ohtani, B.; Osaki, H.; Nishimoto, S.; Kagiya, T. *J. Am. Chem. Soc.* **1986**, *108*, 308.
261. Baker, R. S. *US Patent 3442950.* 1960.
262. N. Goetz, L. Hupfer, W. Hoffmann, Baumann M., US Patent 4,355,180, 1982.
263. Hideaki, H.; Yasushi, K.; Takahiro, S.; Katsuhiko, W. *Bull. Chem. Soc. Jpn.* **1987**, *60*, 55.
264. He, W.; Wang, L.; Sun, C.; et al. *Chem. Eur. J.* **2011**, *17*, 13308.
265. Wang, L.; He, W.; Wu, K.; He, S.; Sun, C.; Yu, Z. *Tetrahedron Lett.* **2011**, *52*, 7103.
266. Cui, X.; Yuan, H.; Li, J.-P.; et al. *Catal. Commun.* **2015**, *58*, 195.
267. Ishida, T.; Kawakita, N.; Akita, T.; Haruta, M. *Gold Bull.* **2009**, *42*, 267.
268. He, L.; Lou, X. B.; Ni, J.; et al. *Chem. Eur. J.* **2010**, *16*, 13965.
269. Ishida, T.; Takamura, R.; Takei, T.; Akita, T.; Haruta, M. *Appl. Catal. A Gen.* **2012**, *413–414*, 261.
270. Demidova, Y. S.; Simakova, I. L.; Estrada, M.; et al. *Appl. Catal. A Gen.* **2013**, *464–465*, 348.
271. Demidova, Y. S.; Simakova, I. L.; Wärnå, J.; Simakov, A.; Murzin, D. Y. *Chem. Eng. J.* **2014**, *238*, 164.
272. Demidova, Y. S.; Suslov, E. V.; Simakova, I. L.; et al. *J. Mol. Catal. A Chem.* **2017**, *426*(Part A), 60.
273. Demidova, Y. S.; Suslov, E. V.; Simakova, I. L.; et al. *Mol. Catal.* **2017**, *433*, 414.
274. Simakova, I. L.; Demidova, Y. S.; Estrada, M.; et al. *Catal. Today* **2017**, *279*(Part 1), 63.
275. Shimizu, K.; Nishimura, M.; Satsuma, A. *ChemCatChem* **2009**, *1*, 497.
276. Liu, H.; Chuah, G.-K.; Jaenicke, S. *J. Catal.* **2012**, *292*, 130.
277. Cui, X.; Zhang, Y.; Shi, F.; Deng, Y. *Chem. Eur. J.* **2011**, *17*, 1021.
278. Shimizu, K.-i.; Shimura, K.; Nishimura, M.; Satsuma, A. *RSC Adv.* **2011**, *1*, 1310.
279. Brown, A. B.; Reid, E. E. *J. Am. Chem. Soc.* **1924**, *46*, 1836.
280. Matsuhashi, H.; Arata, K. *Bull. Chem. Soc. Jpn.* **1991**, *64*, 2605.
281. Ko, A.-N.; Yang, C.-L.; Lin, H.-e. *Appl. Catal. A Gen.* **1996**, *134*, 53.
282. Valot, F.; Fache, F.; Jacquot, R.; Spagnol, M.; Lemaire, M. *Tetrahedron Lett.* **1999**, *40*, 3689.

283. Park, Y.; Park, K.; Woo, S. *Catal. Lett.* **1994**, *26*, 169.
284. Reddy, M. M.; Kumar, M. A.; Swamy, P.; et al. *Green Chem.* **2013**, *15*, 3474.
285. Yang, H.; Cui, X.; Dai, X.; Deng, Y.; Shi, F. *Nat. Commun.* **2015**, *6*, 6478.
286. Gómez-Martínez, M.; Baeza, A.; Alonso, D. A. *ChemCatChem* **2017**, *9*, 1032.
287. Noji, M.; Ohno, T.; Fuji, K.; Futaba, N.; Tajima, H.; Ishii, K. *J. Org. Chem.* **2003**, *68*, 9340.
288. Huang, W.; Shen, Q. S.; Wang, J. L.; Zhou, X. G. *Chin. J. Chem.* **2008**, *26*, 729.
289. Liu, Y.-L.; Liu, L.; Wang, D.; Chen, Y.-J. *Tetrahedron* **2009**, *65*, 3473.
290. Sreedhar, B.; Reddy, P. S.; Reddy, M. A.; Neelima, B.; Arundhathi, R. *Tetrahedron Lett.* **2007**, *48*, 8174.
291. Giner, X.; Trillo, P.; Nájera, C. *J. Organomet. Chem.* **2011**, *696*, 357.
292. Qin, H.; Yamagiwa, N.; Matsunaga, S.; Shibasaki, M. *Angew. Chem. Int. Ed.* **2007**, *119*, 413.
293. Ohshima, T.; Ipposhi, J.; Nakahara, Y.; Shibuya, R.; Mashima, K. *Adv. Synth. Catal.* **2012**, *354*, 2447.
294. Gohain, M.; Marais, C.; Bezuidenhoudt, B. C. *Tetrahedron Lett.* **2012**, *53*, 1048.
295. Guo, S.; Song, F.; Liu, Y. *Synlett* **2007**, *2007*, 0964.
296. Georgy, M.; Boucard, V.; Debleds, O.; Zotto, C. D.; Campagne, J.-M. *Tetrahedron* **2009**, *65*, 1758.
297. Ohshima, T.; Nakahara, Y.; Ipposhi, J.; Miyamoto, Y.; Mashima, K. *Chem. Commun.* **2011**, *47*, 8322.
298. Hikawa, H.; Suzuki, H.; Yokoyama, Y.; Azumaya, I. *J. Org. Chem.* **2013**, *78*, 6714.
299. Hikawa, H.; Ijichi, Y.; Kikkawa, S.; Azumaya, I. *Eur. J. Org. Chem.* **2017**, *2017*, 465.
300. Lu, Y.; Fu, X.; Chen, H.; Du, X.; Jia, X.; Liu, Y. *Adv. Synth. Catal.* **2009**, *351*, 129.
301. Mukherjee, P.; Widenhoefer, R. A. *Org. Lett.* **2010**, *12*, 1184.
302. Mukherjee, P.; Widenhoefer, R. A. *Org. Lett.* **2011**, *13*, 1334.
303. Zhan, Z.-p.; Yang, W.-z.; Yang, R.-f.; Yu, J.-I.; Li, J.-p.; Liu, H.-j. *Chem. Commun.* **2006**, 3352.
304. Zhan, Z.-p.; Yu, J.-l.; Liu, H.-j.; et al. *J. Org. Chem.* **2006**, *71*, 8298.
305. Jana, U.; Maiti, S.; Biswas, S. *Tetrahedron Lett.* **2008**, *49*, 858.
306. Lee, K. Y.; Lee, H. S.; Kim, J. N. *Bull. Kor. Chem. Soc.* **2008**, *29*, 1099.
307. Guérinot, A.; Serra-Muns, A.; Gnamm, C.; Bensoussan, C.; Reymond, S.; Cossy, J. *Org. Lett.* **2010**, *12*, 1808.
308. Trillo, P.; Baeza, A.; Nájera, C. *Eur. J. Org. Chem.* **2012**, *2012*, 2929.
309. Trillo, P.; Baeza, A.; Nájera, C. *ChemCatChem* **2013**, *5*, 1538.
310. Reddy, C. R.; Jithender, E. *Tetrahedron Lett.* **2009**, *50*, 5633.
311. Vicennati, P.; Cozzi, P. G. *Eur. J. Org. Chem.* **2007**, *2007*, 2248.
312. Sanz, R.; Martínez, A.; Álvarez-Gutiérrez, J. M.; Rodríguez, F. *Eur. J. Org. Chem.* **2006**, *2006*, 1383.
313. Sanz, R.; Martínez, A.; Miguel, D.; Álvarez-Gutiérrez, J. M.; Rodríguez, F. *Adv. Synth. Catal.* **2006**, *348*, 1841.
314. Pan, Y.-m.; Zheng, F.-j.; Lin, H.-x.; Zhan, Z.-p. *J. Org. Chem.* **2009**, *74*, 3148.
315. Shirakawa, S.; Kobayashi, S. *Org. Lett.* **2007**, *9*, 311.
316. Barreiro, E.; Sanz-Vidal, A.; Tan, E.; Lau, S. H.; Sheppard, T. D.; Díez-González, S. *Eur. J. Org. Chem.* **2015**, *2015*, 7544.
317. Motokura, K.; Nakagiri, N.; Mori, K.; et al. *Org. Lett.* **2006**, *8*, 4617.
318. Motokura, K.; Nakagiri, N.; Mizugaki, T.; Ebitani, K.; Kaneda, K. *J. Org. Chem.* **2007**, *72*, 6006.

319. Shirakawa, S.; Shimizu, S. *Synlett* **2008**, *2008*, 1539.
320. Wang, G. W.; Shen, Y. B.; Wu, X. L. *Eur. J. Org. Chem.* **2008**, *2008*, 4367.
321. Yadav, J.; Reddy, B. S.; Rao, T. S.; Krishna, B. B. M.; Kumar, G. N. *Chem. Lett.* **2007**, *36*, 1472.
322. Srihari, P.; Shyam Sunder Reddy, J.; Bhunia, D. C.; Mandal, S.; Yadav, J. *Synth. Commun.* **2008**, *38*, 1448.
323. Safariamin, M.; Paul, S.; Moonen, K.; Ulrichts, D.; Dumeignil, F.; Katryniok, B. *Catal. Sci. Technol.* **2016**, *6*, 2129.

Chapter 2

N-Alkyl Amine Synthesis by Catalytic Coupling of Amines

2.1 INTRODUCTION

Except the methodologies discussed in the previous chapters, the *N*-alkyl amines can be prepared by the cross coupling of amine with other amines as the alkylation agent. It is clear that this process was carried out via hydrogen autotransfer methodologies, where the aminal intermediate was generated by the reaction of in situ formed imine and another amine (Scheme 2.1).[1] First, the alkyl amine was dehydrogenated to an imine. This was followed by the addition of other nucleophilic amine to the in situ formed imine, to generate an aminal intermediate with the elimination of ammonium molecule. Finally, the formed imine is hydrogenated to the final product with in situ generated [M-H] species.

2.2 HOMOGENEOUS CATALYST SYSTEM

2.2.1 Ir Catalyst System

In 2009, Williams and coworkers reported the first examples of selective amine alkylation even when both of the amines were able to undergo oxidation to an imine. Ir complex showed high activity and selectivity on the alkylating of

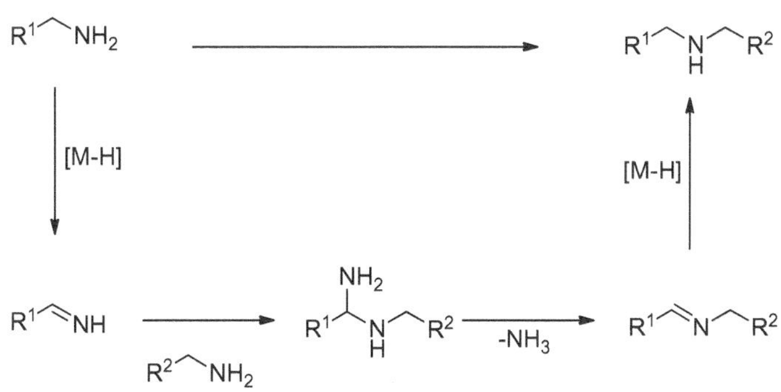

SCHEME 2.1 The proposed reaction pathway of the amine alkylation with another amine.

Catalytic Amination for *N*-Alkyl Amine Synthesis. https://doi.org/10.1016/B978-0-12-812284-6.00002-0

aniline with ethyl, cyclohexyl, and isopropyl amines and the corresponding products were obtained in good to excellent yields in optimal conditions (1 mol% Ir, xylene, refluxing for 3 h) (Scheme 2.2). Interestingly, the furan-containing substrate was also successfully tolerated, although some decomposition occurred, which lead to a lower product yield.

The high selectivity for the desired alkyl amines in the presence of two amines might be to be favored by the formation of secondary amines containing one branched and one unbranched substituent, thereby relieving the known steric strain within amines containing more than one branched substituent.[2] The diisopropylamine and other branched amines can be transferred to the corresponding product successfully because the assumed imine intermediates derived from them are much more stable compared with imines formed from oxidation of a primary amine.

In 2014, Wang and coworkers synthesized several new bisbenzoxazolyl iridium (III) complexes for the *N*-alkylation of amine with another amine as electrophile. Under optimal conditions (2 mol% of Ir/$AgNTf_2$ catalyst, 1 equiv., of Cs_2CO_3, xylene, and 155°C), it produces the corresponding products in yields of 57%–85% (Scheme 2.3).[3] The catalytic activity of these complexes is greatly enhanced by noncoordinating anions, while the experiments have excluded the possibility of a "silver effect" (bimetallic catalysis or silver-assisted metal catalysis) from the experiments.

In 2015, Zou and coworkers reported an efficient catalyst system for the *N*-alkylation of amines with amines under mild conditions, using cyclometallated iridium complexes as catalysts. Almost all substituted anilines reacted with diisopropylamine successfully to produce the desired products in excellent yields.[4–5] However, corresponding transformed *para*-acetyl aniline product only yielded 19% because of the easily reducible carbonyl compounds which decreased the selectivity of the reaction. Interestingly, anilines substituted at the *meta-* or *ortho* positions are also viable substrates generating the desired products in positive amounts, demonstrating the high steric hindrance has effect on the activity but did not suppress the reaction. When two amino groups were present, both amine groups were alkylated producing excellent yields.

Very recently, an efficient method for secondary amines synthesis has been developed through alanine triazole iridium-catalyzed cross-coupling of an aromatic amine and an alkyl amine, borrowing hydrogen being the basis of this strategy.[6] Moderate to good yields (53%–79%) were obtained with xylene as the solvent and sodium hydroxide as the base.

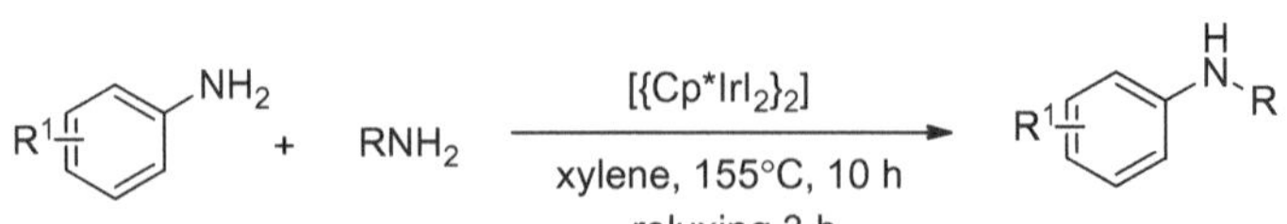

SCHEME 2.2 *N*-Alkylation of amine catalyzed by Ir complex.

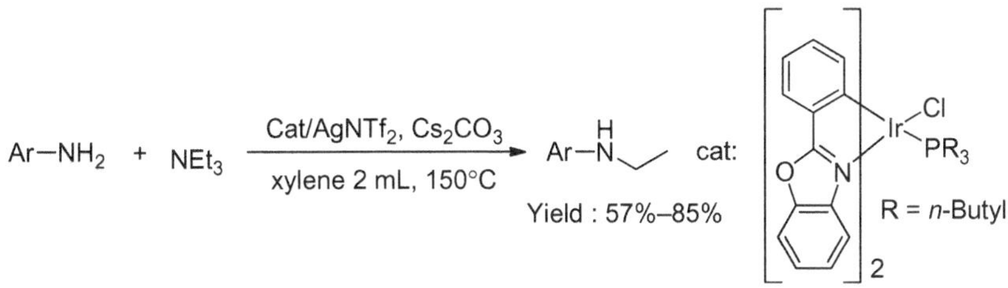

SCHEME 2.3 Alkylation of amines with tertiary amines catalyzed by bisbenzoxazolyl iridium (III) complexes.

2.2.2 Ru Catalyst System

Till today, ruthenium complexes have been proved to be the most successful and widely applied catalysts not only in the *N*-alkylation of amines using alcohols as electrophiles but also for a similar transformation using amines as an electrophilic source. The coupling reaction of primary amines was first reported in 1981 with $RuCl_2(PPh_3)_3$ as catalyst at 180°C for 5h producing the corresponding symmetrical secondary amines with excellent yields in the absence of a solvent (Scheme 2.4).[7] However, in the case of a long chain amine (*n*-dodecylamine) and α-substituted amines (cyclohexyl and isopropylamine), yields were poor. Yields could be improved by the use of either a larger amount of catalyst or tetrahydrofuran as solvent (THF/amine volume ratio 1/1).

$RuCl_2(PPh_3)_3$ was also used in the isomerization of *N,N*-dimethylalkylamines using a similar procedure, and the starting materials were converted to the corresponding *N*-methyldialkylkamines with good yields (Scheme 2.5).[8] Again, a larger amount of catalyst or tetrahydrofuran played the key role in this transformation. For example, in the case of *N,N*-dimethyloctylamine, both di- and tri-octylamine yields of 15% and 30%, were obtained, respectively. After 12h with 2.5% of catalyst, 25% yields were obtained and 30% yields were obtained after 7h with 7.5% of catalyst. However, both byproducts are almost absent if the reaction is performed in THF as solvent. Compared to the coupling reaction of primary amines, the process formally involved the dehydrogenation of the initial amine with the formation of the methylene derivative. After pyrolysis of the methylene derivative, the corresponding *N*-methylalkylamine was generated which acted as nucleophilic partner of the reaction, and attached to the alkylene derivative (electrophilic partner of the reaction). With the elimination of the methylamine, the nucleophilic addition ended, producing the corresponding aminal intermediate. Finally, the desired products were obtained by the in-situ hydrogenation. Similar products

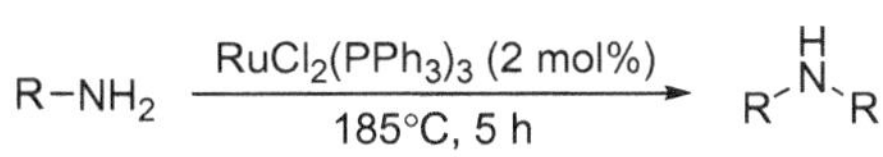

SCHEME 2.4 Ru-catalyzed coupling of primary amines.

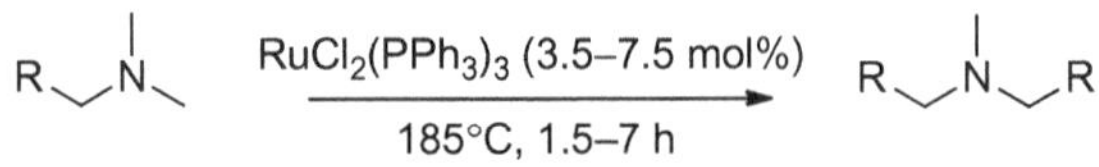

SCHEME 2.5 Catalytic isomerization of tertiary amines.

could be obtained when the *N*-methylalkylamines were used as starting materials, as expected.

Systemic investigation revealed that this reaction was likely a first-order dependence on the concentration of both the catalyst and amine where the rate-determining step was the catalytic dehydrogenation of amine.[9] During the reaction, no imine intermediate was observed in the reaction media, which indicated that the reaction of imine with the starting amine and the elimination of NH_3 were accelerated. Moreover, the desired secondary amine was observed and good yield was obtained finally, demonstrating that the dissociation of final product from the ruthenium intermediate complex was easy.

In 1982, some heterocyclic amines were synthetized from aliphatic diamines in the presence $RuCl_2(PPh_3)_3$. The best yields were obtained by heating the diamines in a sealed glass tube at 180°C for 5 h with diphenyl ether as solvent. As shown in Scheme 2.6, the highest yield was obtained for the six-membered ring product.[10] As expected, the combination of $RuCl_3$ and suitable amount of PPh_3 ligand exhibited similar catalytic activity.

Interestingly, the transalkylation of amines was successfully used for the preparation of quinolones. For example, a variety of different quinolone derivatives could be prepared by cross coupling tetraalkylammonium halide with different anilines. When aniline was reacted with allylammonium chlorides under the optimal reaction conditions (aniline/allylammonium chlorides = 6, $RuCl_2(PPh_3)_3$ (5 mol%), and $SnCl_2{\cdot}2H_2O$ (1 mmol) in H_2O/dioxane at 180°C for 20 h), the corresponding product obtained a yield of 93% (Scheme 2.7).[11] When the aniline and ammonium salt were replaced by nitrobenzene and tripropylamine, the same quinolines could be also produced. In contrast to previously described methods, similar yield was obtained.[12]

The preparation of quinolines was also achieved by the reaction of aniline derivatives with triallylamine in the presence of the in situ generated ruthenium complexes which was obtained by reaction of $RuCl_3{\cdot}nH_2O$ (5 mol%) and PPh_3 (15 mol%) in the presence of $SnCl_2{\cdot}2H_2O$ (100 mol%).[13] When the reaction was carried out at 180°C in dioxane, the corresponding quinolines were obtained with yields of 24%–61%. Compared to electron withdrawing substituted starting regent, better results were obtained when anilines bearing electron donating groups were used. In 2000, Cho and coworkers reported that slightly better yields (21%–86%) could be achieved when (bis(diphenylphosphane)methane) was used as the ligand and 1-hexene was utilized as a hydrogen scavenger.[14]

In 1985, $Ru_3(CO)_{12}$ was reported to be an effective catalyst for the *N*-alkylation of amines using different amines as alkylation reagents.[15]

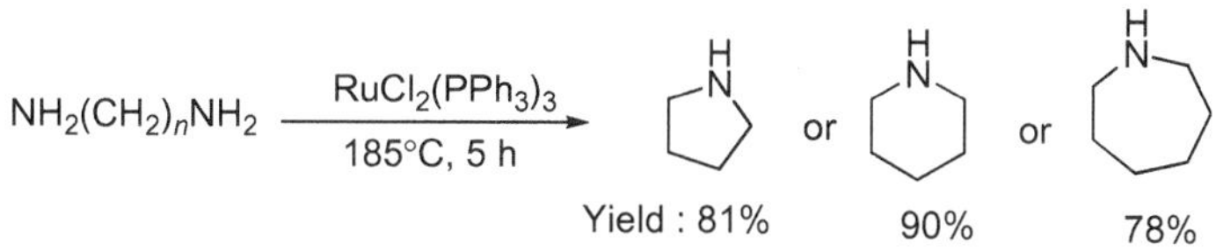

SCHEME 2.6 Heterocyclic amines synthesis from aliphatic diamines.

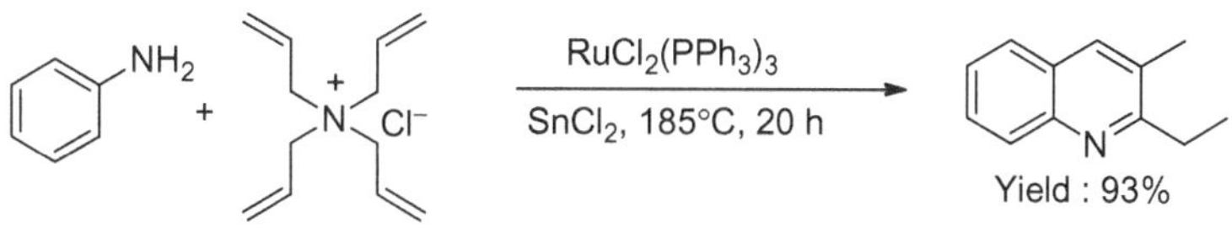

SCHEME 2.7 Synthesis of quinolines from the reaction of aniline and ammonium.

For instance, transalkyaltion reaction between triethylamine and tripropylamine could be performed successfully under the optimized conditions (Ru: 0.42 mol, 100°C, N_2: 7 bar) giving diethylpropylamine and ethyldipropylamine yielding 25% and 27%, respectively (Scheme 2.8). Mechanistic and kinetic studies for this reaction showed a very complex rate dependence.

In 2007, the Beller group reported bisruthenium complex-catalyzed alkylation of aromatic amines with alkyl amines as well as obtaining high activity and selectivity (Scheme 2.9).[16] Under the optimized conditions, different aromatic amines could be alkylated with primary amines to produce the desired secondary amines in good to excellent yields. The activity and selectivity of the alkylating products was not influenced by the position of the substitutes at the aromatic ring. Both the electron-donating and electron-withdrawing groups did not have any impact on the results. However, in the case of nitroaniline, low yield was obtained (20%) due to the possible reduction of the nitro group.

The aforementioned method was also employed in the reaction of aniline with *n*-hexylamine, di-*n*-hexylamine and tri-*n*-hexylamine.[17] Surprisingly, all the different hexylamines are converted to the same *N*-hexylaniline in high yields (75%–87%). It should be pointed out that the primary amine produced the highest yield while tertiary amine gave the lowest. Notably, a mixture of mono-, di-, and tri-*n*-hexylamine is also highly selectively converted to give *N*-hexylaniline under the optimized condition (1 mol% Shvo catalyst, aniline/hexamine = 2, *tert*-amyl alcohol as solvent, 150°C).

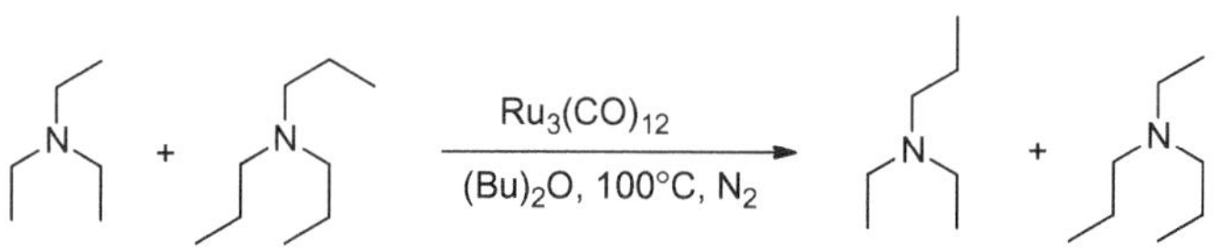

SCHEME 2.8 Cross coupling of triethylamine and tripropylamine.

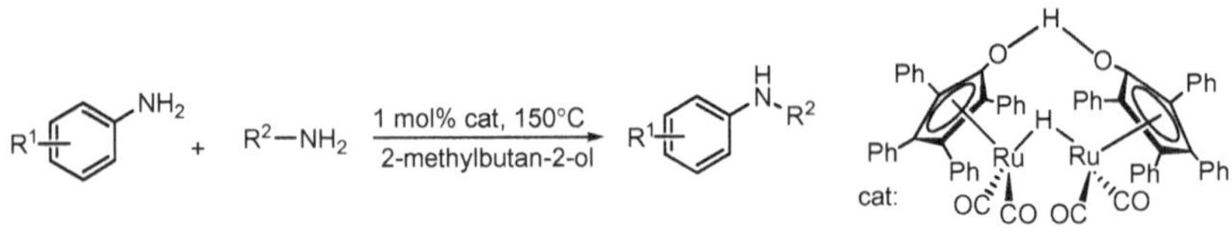

SCHEME 2.9 *N*-Alkylation of aniline derivatives with alkyl amines.

However, it is a challenge using cyclic amines as an electrophile source because the dehydrogenation of these compounds is difficult. Interestingly, when the reaction of aniline with pyrrolidine was performed in the presence of 1 mol% of Shvo catalyst, the desired products could be obtained in moderate yields (Scheme 2.10).[18] Further study revealed that the product yield depended on the electron density of the aromatic ring and the nucleophilicity of the amino group. It appears that the nucleophilic attack of the aryl amine is involved in the rate-determined step. As expected, the electron-rich aryl amines produced *N*-arylpyrrolidines in yields of 39%–67%, but 4-trifluoromethylaniline showed no reaction, even at higher temperatures.

The alkylation of aliphatic amines with other amines as electrophile occurred in similar circumstances with presence of Shvo catalyst.[19] It should be pointed out that *tert*-butyl amine derivatives such as 2,4,4-trimethylpentan-2-amine could not be converted to the corresponding imine which lead to difficulty in the electrophilic reaction. Interestingly, 2,4,4-trimethylpentan-2-amine was alkylated with benzylamine successfully, giving corresponding secondary amines selectively (Scheme 2.11). In addition to primary amine, symmetrical secondary amines (e.g., dibenzylamine) could be also used as electrophiles, producing desired products with a 76% yield. However, when tertiary amine (e.g., tribenzylamine) was used as nucleophilic amine, the alkylation reaction could not be carried out.

2.2.3 Pt Catalyst System

In 1983, the coupling reaction of primary amines was realized with $PtCl_2(PPh_3)_2$ complex (0.5 mol%) as catalyst. The best results could be obtained in the presence of $SnCl_2{\cdot}2H_2O$ (25 mol%) (Scheme 2.12).[20] In optimal conditions, the primary amines were successfully converted to the corresponding secondary ones with moderate to good yields (27%–81%), independent of the substitution on the starting amine. However, cyclization of 1,4-butanediamine occurred using this protocol resulting in low yields of the expected pyrrolidine.

2.2.4 Co Catalyst System

Recently, Yin and coworkers reported the first sample of a cobalt complex-catalyzed selective alkylation of amines with amines to synthesize a large

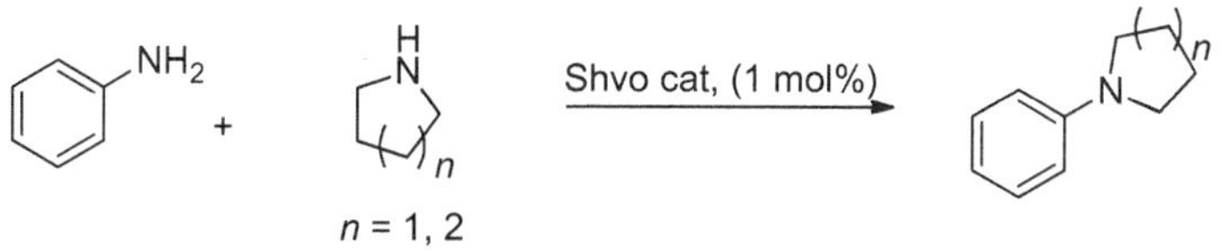

SCHEME 2.10 *N*-Alkylation of aniline derivatives with cyclic amines.

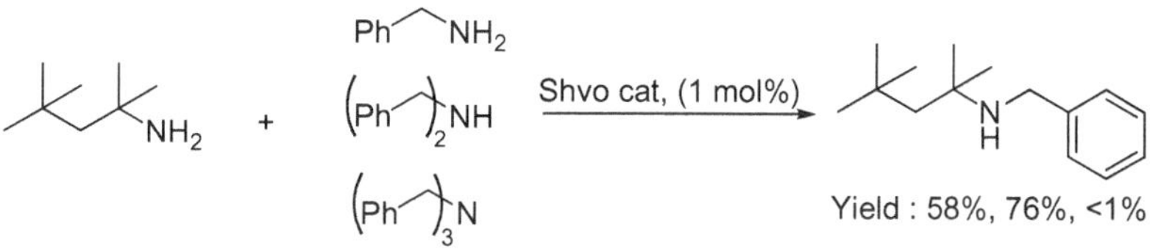

SCHEME 2.11 Catalytic *N*-alkylation of *tert*-alkylamines.

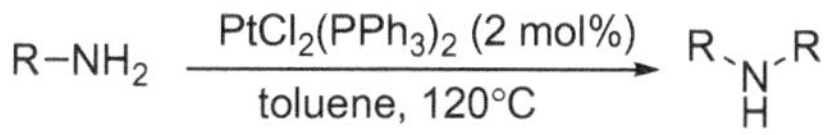

SCHEME 2.12 $PtCl_2(PPh_3)_2$-catalyzed amine coupling.

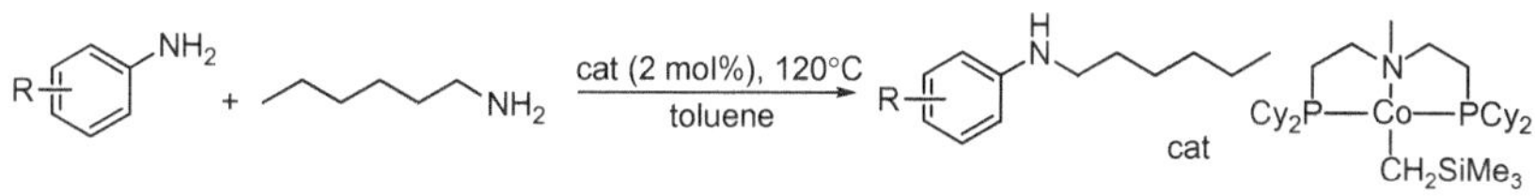

SCHEME 2.13 Co-catalyzed coupling reaction of amines.

variety of secondary amines.[21] A range of amine substrates have been converted to the corresponding products through hetero or homo coupling between amines. Cyclic sec-amines are also achieved from diamine precursors, however occurrences are rare. (Scheme 2.13).

2.3 HETEROGENEOUS CATALYST SYSTEM

2.3.1 Pd Catalyst System

Since the beginning of the last century, heterogeneous palladium catalysts immobilized on different supports have been extensively studied and applied in the coupling of amines. In one case, in the presence of hydrogen, benzylamine was converted to the corresponding dibenzylamine in 90% yield with Pd/Ba_2SO_4 (2.5 wt%, 2.5 mol%) as the catalyst.[22]

In 1931, Kindler reported that when the coupling of benzylamine was performed in the presence of palladium black (18 mol%) in refluxing ethanol for 3 h, the desired dibenzylamine was prepared successfully quantitatively.[23] However, this catalytic system suffered from limited substrates. For example, when 2-phenylethylamine was used to instead of benzylamine, only low yields (10%) of the corresponding symmetrical secondary amine were obtained.

In 1966, Anderson and coworker prepared a variety of catalysts wherein different metals, such as Pt, Pd, Ni, W, and Co, were supported on carbon. The activity of these prepared catalysts was evaluated in the reaction of methylamine and dimethylamine.[24] When the reaction was tested at 227°C under a hydrogen atmosphere, only palladium on carbon demonstrated activity, although the mixtures of dimethylamine and trimethylamine were observed as main products.

Based on the earlier studies of the activity of palladium for this reaction, more detailed investigations were performed. For instance, the desired dibenzylamine was formed in moderate yield when benzylamine was used as starting material in the presence of 5 mol% palladium black at 80°C where the main byproduct *N*-benzylideneaniline was formed yielding 41%.[25] Interestingly, allylamine could be transformed to the corresponding *N*-alkyl amine with high yields (95%). Surprisingly, secondary amines, featuring challenging substrates in such reactions due to the high steric hindrance, were transformed into the corresponding tertiary ones by simply increasing the reaction temperatures, though the desired product yields are low (%3–60%). It is interesting that secondary amines including piperidine and pyrrolidine were alkylated with other secondary amines resulting in corresponding tertiary amines yields of 5%–87%.[26] Furthermore, 4-di(pyrrolidin-1-yl)butane can be prepared via the *N*-alkylation of pyrrolidine under similar circumstances by heating the reaction at 200 °C for 5 h.

In general, it is a significant challenge to produce the asymmetric *N*-alkyl amine selectively using two different tertiary amines as starting materials due to the formation of the mixture of all statistically possible tertiary amines. For example, Murahashi reported that a mixture of tributylamine, trihexylamine, dibutylhexylamine, and butyldihexylamine was obtained when equimolecular amounts of tributylamine and trihexylamine were used.[27]

Subsequently, the application of the above methodology was applied in the selective synthesis of diamino-based compounds.[28] In one case, *N,N'*-bis(3-aminopropyl)-1,3-propanediamine was obtained with a 75% yield with reaction of triamine and 1,3-propanediamine in the presence of palladium. Meanwhile, polyamines were obtained in moderate to good yields with the cross-coupling triamines with azetidine (Scheme 2.14).

In 2004, Sajiki showed that the *N*-alkylation of different substituted anilines can be performed successfully with nitriles, which can be transformed in-situ into primary amine, in the presence of palladium on carbon catalyst in a hydrogen atmosphere.[29] Interestingly, similar compounds could be synthesized using nitroarenes instead of anilines (Scheme 2.15).

In 2005, Miyazawa and coworkers demonstrated the coupling of amines using a catalytic amount of Pd/C in water under microwave irradiation conditions.[30] In this case, *sec*-butylamine was converted to corresponding symmetric

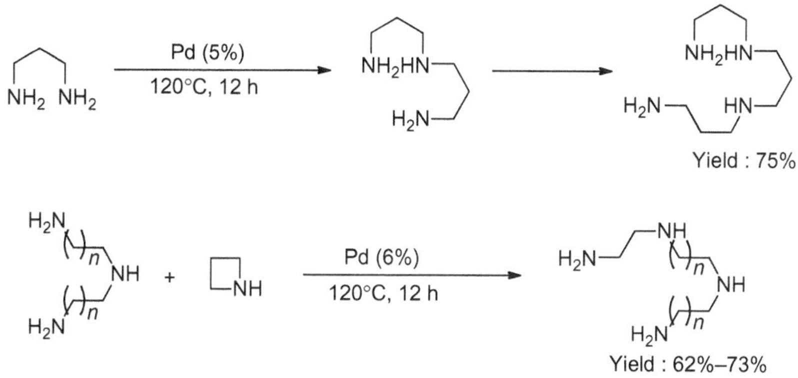

SCHEME 2.14 Selective *N*-alkylation of primary amines with azetidine using palladium black.

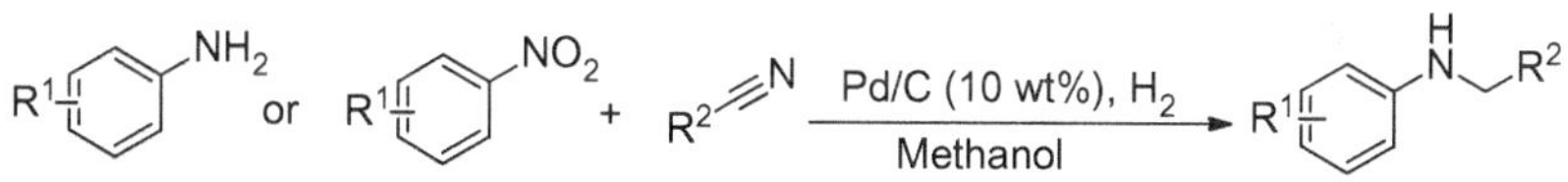

SCHEME 2.15 Palladium-catalyzed cross-coupling of amines or nitrobenzene with nitrile.

secondary amine with a 50% yield. Apart from the alkyl amine, the corresponding imine and aldehyde or ketone were also observed as by-products. The *N*-alkylation of aniline with amines could result as well, with Pd/C, ionic liquid [Bmim]PF_6 or THF as solvent.[31]

In 2011, Lubinu et al. reported the synthesis of *N*-alkylation of anilines with tertiary amines using a palladium carbon catalyst via transfer-hydrogenation methodology.[32] The reactions proceed efficiently and without hindrance, resulting on good to excellent yields of corresponding *N*-alkylanilines. For example, in the presence of palladium carbon catalyst, it is found that the amination of aniline with a sterically hindered substituent at the *ortho* position can be performed effectively, though a lower yield (66%) was observed. Interestingly, they explored the alkylation of aniline with the unsymmetrical tertiary amines and found that the reaction of transalkylation gave rise to a mixture of two *N*-alkylanilines (Scheme 2.16).

Shimizu demonstrated that Al_2O_3 supported Pd nanoclusters were very effective for the cross-coupling of benzylic and heterocyclic secondary amines.[33] Both the palladium particle size and support remarkably affect the

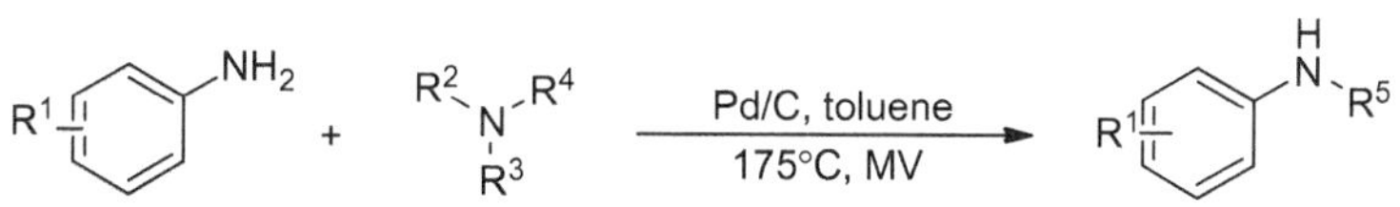

SCHEME 2.16 *N*-Alkylation of aniline with tertiary amines.

catalytic performance. Typically, the palladium nanoclusters supported on the amphoteric oxide (Al_2O_3) with an average particle size of 1.8 nm give the highest activity (Scheme 2.17).

The mechanism of the palladium-catalyzed transalkylation of amines was investigated by Murahashi[34] and it was revealed that the catalytic dehydrogenation of the starting electrophilic amine in the first step was rapid and without difficulty to form the corresponding imine. Second, the formed imine was coordinated with Pd rapidly giving the enamine-palladium complex, which is the important intermediate to react with nucleophilic amine producing an aminal. Finally, the hydrogenation of the aminal occurred with the generation of amine. Furthermore, the hydrolysis of tertiary amines can be realized by palladium black with the formation of the corresponding secondary ones and aldehyde in acidic aqueous solution.

2.3.2 Pt Catalyst System

In 1983, the first heterogeneous platinum catalyst, i.e., Pt/TiO_2, (5 wt%, 22 mol%), catalyzed coupling of amines under photoirradiation conditions.[35] The primary alkylamines could be converted into corresponding symmetrical secondary ones at room temperature, though only yields of 24%–33% were obtained. For example, the butan-1-amine was successfully transformed to the dibutylamine with 24% yield and 55% selectivity where the alcohol and aldehyde were observed as the main side products. Interestingly, the corresponding cyclic aliphatic amines can be generated from diamines yielding 20%–67% (Scheme 2.18).

In 2006, Miyazawa and coworkers investigated the coupling of primary amines using Pt/C catalyst under microwave irradiation, and the corresponding secondary ones were obtained in moderated yields.[36] It should be pointed out that the corresponding ketone was observed as the main side product when the reaction was carried out in the presence of aqueous solution.

In 2009, Largeron et al. developed the electro-catalytic synthesis of benzylic secondary amines.[37] The alkylation of substituted primary amines was performed in the presence of arylmethyl amines as an initial electrophilic source, resulting in the expected amines with moderate results (Scheme 2.19).

In 2011, He and coauthors displayed a heterogeneous bimetallic Pt-Sn/γ-Al_2O_3 (0.5 wt% Pt, Pt/Sn molar ratio = 1: 3) catalyzed self-coupling of primary amines through a borrowing-hydrogen strategy (Scheme 2.20).[38]

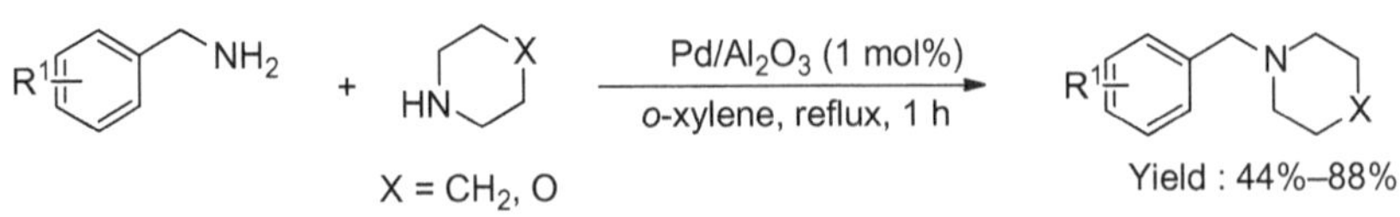

SCHEME 2.17 Selective alkylation of benzyl amines with heterocyclic secondary amines.

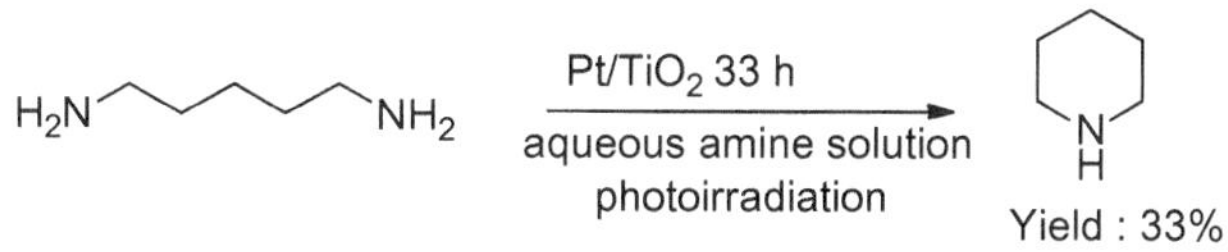

SCHEME 2.18 Cyclic aliphatic amines synthesis from diamine.

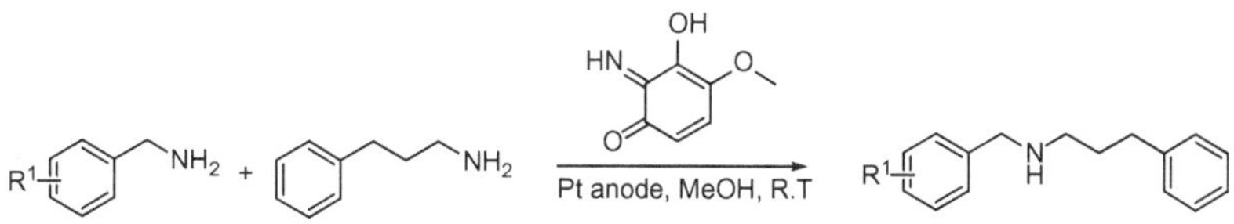

SCHEME 2.19 Electrocatalyzed coupling reaction of amines.

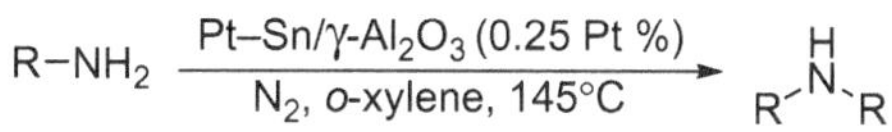

SCHEME 2.20 Pt-Sn/γ-Al_2O_3 catalyzed coupling reaction of amine.

A variety of different substituted benzyl amines were converted to corresponding secondary amines with excellent yields (87%–95%). It should be pointed out that the sterically hindered amines were also successfully transformed to the desired products with excellent yields. The heterogeneous Pt catalyst also showed good activity in the coupling of aliphatic amines.

Notably, the reaction of benzyl amine with aniline was carried out under similar conditions and formed a mixture of two secondary amines due to easy deaminative self-coupling of aliphatic primary amines, and the unsymmetrical secondary amine was formed as the main product with yields of 57%.

Shimizu et al. reported that a simple Pt/Al_2O_3 could be an active catalyst for the cross-coupling of different amines.[39–40] The prepared Pt/Al_2O_3 catalysts showed to be structure-sensitive for the mono-*N*-alkylation of amines with different amines in the size range of 0.8–24 nm and the highest activity was obtained when the average particle size of Pt was 0.8 nm. Among various supports, the moderate basicity Al_2O_3 gives higher activity compared to strongly acidic and basic supports because suitable basicity property of the support is essential.

Sulfur-loaded Pt/SiO_2 (Pt = 5 wt%) catalysts prepared by adding ammonium sulfate on Pt/SiO_2 followed by H_2-reduction at 500°C, were recently reported to be an active catalyst for mono-*N*-alkylation of aniline with primary and secondary amines. Moderate to excellent yields were obtained, although the materials with the high steric hindrance were used as substrates (Scheme 2.21).[41] Further study revealed that when the average size of Pt particle was varied from 3.3 to 6.8 nm, the catalyst with S/Pt ratio of 0.13 showed more than five times higher

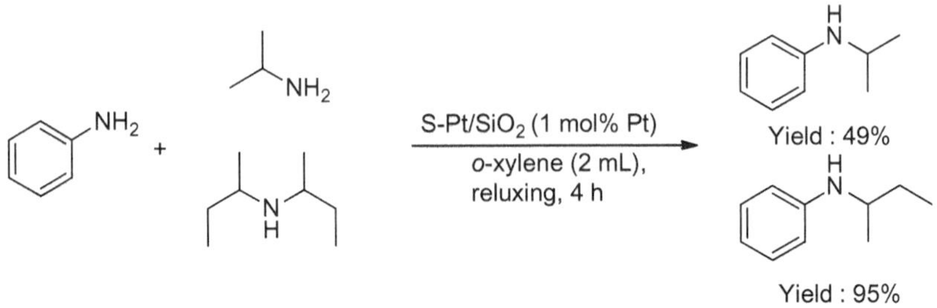

SCHEME 2.21 *N*-Alkylation of aniline with primary amine and secondary amine catalyzed by S-Pt/SiO_2.

TOF than unmodified catalysts. Combined with kinetic results and characterizations, XAFS (X-ray absorption fine structure), TEM, and CO adsorption IR, Spectroscopic studies showed that sulfur species did not change the structure and electronic states of Pt.

2.3.3 Ni Catalyst System

In 1923, the coupling reaction of amine was observed in Raney-nickel-catalyzed nitrile derivatives reduction. Excluding the formation of primary amine, symmetrical secondary amines were observed when the reaction was performed under optimal conditions (20 atm H_2 and 110–130°C, in tetralin or decalin). However, the yields of secondary amines were only 2%–47%.[42] In this case, the intermediate imine was produced either by the reduction of the starting nitrile in the first step or by the catalytic dehydrogenation of the produced primary amines. The final hydrogenation of this imine led to the formation of the secondary amine. The formation of secondary amine could be inhibited if an adequate amount of ammonia was added.[43]

Moderate yields of the corresponding secondary amines were obtained with 45% and 66% yields if either octanenitrile or nonanenitrile was used as the starting material.[44] Using Raney-nickel as catalyst, the formation of the alkyl amines was also observed in the reduction of other nitrogen-containing derivatives, such as amides and oximes. For example, in the reduction of benzaldoxime, benzamide, and other compounds, the corresponding secondary amines were observed in the presence of Raney-nickel (8–15 mol%) in low yields with the appropriate solvent. However, up to 99% yields were obtained in the catalytic alkylation of cyclohexylamine and 3-methylbutan-1-amine under similar reaction conditions.[45] At the same time, it was found that when the hydrogenation of thioamides was carried out in the presence of Raney-nickel at reflux of ethanol or dioxane, the coupling of amines was also observed as a side reaction and the alkyl amine was generated in moderate 40%–45% yields.[46]

The coupling of primary amines has been performed by the use of Raney-nickel under optimal conditions yielding the corresponding alkyl secondary amines with yields of 70%–86% in the absence of hydrogen.[47] Interestingly, cyclic secondary amines were also successfully converted to corresponding alkylating amines using primary alkyl amines as the source of electrophile (Scheme 2.22). The desired tertiary amines can be obtained in moderate yields, with the catalytic liberation of ammonia. However, a large excess of Raney-nickel (3–4 equiv. compared to starting materials) was required in above reactions and it is necessary to reduce the amount of Ni for industrial application.[48] Indeed, the amount of Ni catalysts in the coupling of primary amines could be reduced to produce corresponding symmetrical secondary amines in good to excellent yields (67%–88%).

2.3.4 Cu Catalyst System

In 1971, Pommersheim and a colleague reported the synthesis of diethylamine from ethylamine by using porous copper based catalysts. However, this reaction was carried out under harsh conditions (256°C and 1 bar H_2), and only 14% yield was obtained.[49] The kinetic study revealed that this transformation was performed with a zero-order equation rate, and the desorption of ammonia and ethylamine was the rate-determining step. Furthermore, both the reactivity of the catalyst and product distribution of the reaction were strongly influenced by the amount of ammonia and hydrogen presented on and over the catalyst's surface.

In 1984, Kijenski and coworkers displayed that Cu/Al_2O_3 was able to catalyze the autoalkylation and transalkylation of benzylamine and benzylmethylamine respectively in a flow system under 1 bar hydrogen pressure at 227–257°C.[50] For example, dibenzylamine was produced with yields ranging from 42% to 50% in the presence of benzylamine. Corresponding imine (*N*-benzylidene-1-phenylmethanamine) was also detectable as the major side product. Interestingly, when benzylmethylamine was used as the initial reagent, the coupling product was given only in 14%–17% yields and methylamine was observed as side product.

In these processes catalyzed by copper and other metals, the challenge was to prevent the formation of bulk copper nitride during the reaction which could decrease the activity of the Cu-based catalyst and make the catalyst deactivated finally. Detailed studies showed that the presence of hydrogen was necessary to avoid this shortcoming.[51] According to the different active metals involved in the amine coupling reaction, different deactivation processes had to be considered, namely, the formation of metal nitride, of metal carbide, and carbonaceous

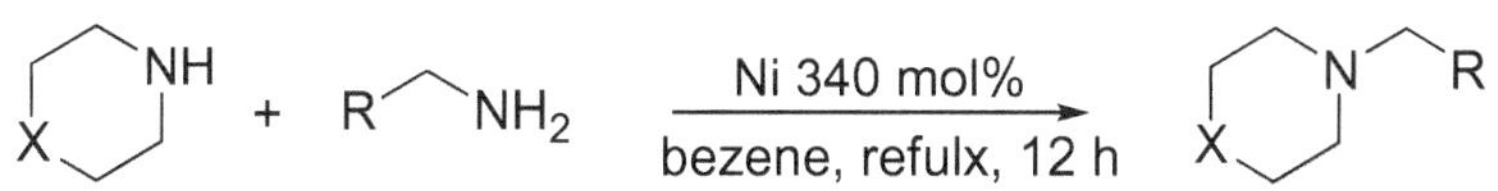

SCHEME 2.22 *N*-Alkylation of cyclic amines with primary amines catalyzed by Raney-nickel.

species. In the case of Cu-based catalyst, temperature-programmed desorption (TPD) and surface reaction (TPSR) investigation indicated that only copper nitride formation occurred with a copper catalyst. In contrast, all of the three deactivation processes were observed if other metal catalysts such as nickel and cobalt were employed.

The amine coupling reaction was also observed in the copper catalyzed amide reduction. In 1952, Schneider demonstrated that the symmetrical secondary amine dilaurylamine could be obtained by the reduction of lauramide using copper-chromite catalyst ($CuCr_2O_4$-$BaCr_2O_4$, 93 mol%) at 220°C and under 340–365 atm of hydrogen pressure. The corresponding didodecylamine was obtained with a moderate yield (52%).[52] In 1993, Barrault studied the reduction of dodecylnitrile using a copper-chromium catalyst supported on alumina. When the reaction was set out in a stainless-steel tubular flow reactor at 150°C and 30 atm of hydrogen pressure, a modest yield (34%) of didodecylamine was produced.[53]

2.3.5 Other Catalysts

The heterogeneous rhodium or ruthenium catalysts have also been demonstrated to be active for the alkylation of amine with other amine as electrophilic source.[54] For example, the hydrogenation of aniline was performed in the presence of heterogeneous rhodium or ruthenium catalysts (0.05 mol% Rh, 145°C, 54–68 bar H_2) and it was found that this reaction gave not only the hydrogenation of aniline to cyclohexanamine, but also its alkylation, to form dicyclohexylamine (47 h, 25% yield).

Rhodium on carbon (5 mol%) has also been used in the alkylation of different aliphatic primary amines and good yields (71%–96%) were obtained.[29] When nitriles were used as the starting materials, moderate to good yields were obtained under hydrogen atmosphere.

The coupling of amine has also been carried out by using montmorillonite as catalyst (20 wt%) at 220°C.[55] Primary amines resulted in the corresponding secondary ones with liberating ammonia. When cyclic secondary amines are used, however, interesting products are obtained. For instance, pyrrolidine was reacted at 200°C to yield 1,4-(l-pyrrolidyl)butanamine as the principal product, with increasing yields of 1,4-(l-pyrrolidyl)butanamine (40%) and 1,4-di-(l-pyrrol-idy1)butane (10%) as the reaction performed after 74 h (Scheme 2.23).

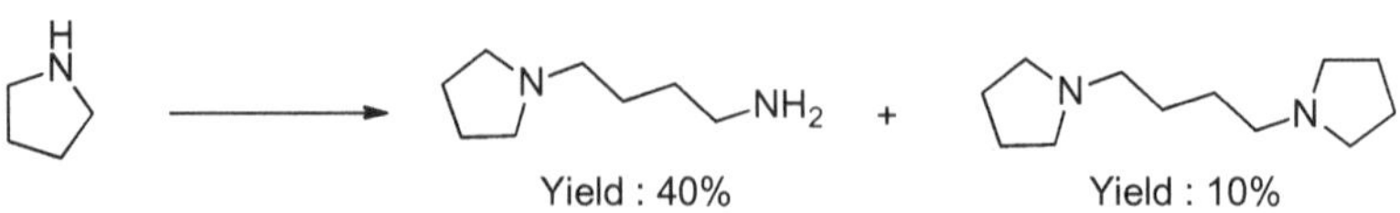

SCHEME 2.23 Montmorillonite-catalyzed coupling reaction of pyrrolidine.

2.4 CONCLUSIONS AND OUTLOOK

As reviewed in this chapter, recent progress in the area of alkylation of amines by amines to *N*-alkylamine through a hydrogen-borrowing mechanism has been summarized. It is apparent that homogeneous catalysts based on Ru and Ir complexes are highly active for the cross-coupling of different amines. This said, the use of reusable heterogeneous catalysts is preferable from environmental and economic viewpoints. Nevertheless, heterogeneous systems for amine coupling suffer from low selectivity, limited scope, low TON, or need of additives or special reaction methods (e.g., microwave heating, electrocatalysis, or photocatalysis). For these reasons, future efforts will focus on developing heterogeneous catalyst systems with high activity and good substrate scope. A further challenge in this area is the development of less expensive catalysts based on metals such as Ni, Cu, Co or other nonnoble metals, which will lead to highly desirable lower production cost for industrial scale production.

REFERENCES

1. Snyder, H. R.; Carnahan, R. E.; Lovejoy, E. R. *J. Am. Chem. Soc.* **1954**, *76*, 1301.
2. Saidi, O.; Blacker, A. J.; Farah, M. M.; Marsden, S. P.; Williams, J. M. J. *Angew. Chem. Int. Ed.* **2009**, *48*, 7375.
3. Wang, D. W.; Zhao, K. Y.; Xu, C. Y.; Miao, H. Y.; Ding, Y. Q. *ACS Catal.* **2014**, *4*, 3910.
4. Zou, Q. Z.; Wang, C.; Smith, J.; Xue, D.; Xiao, J. L. *Chem. Eur. J.* **2015**, *21*, 9656.
5. Michon, C.; MacIntyre, K.; Corre, Y.; Agbossou-Niedercorn, F. *ChemCatChem* **2016**, *8*, 1755.
6. Yu, X. L.; Zhao, R. R.; Wan, H. D.; Yang, Y. C.; Wang, D. W. *Tetrahedron Lett.* **2016**, *57*, 4588.
7. Khai, B. T.; Concilio, C.; Porzi, G. *J. Organomet. Chem.* **1981**, *208*, 249.
8. Arcelli, A.; Khai, B. T. *J. Organomet. Chem.* **1982**, *231*, C31.
9. Jung, C. W.; Fellmann, J. D.; Garrou, P. E. *Organometallics* **1983**, *2*, 1042.
10. Buithekhai, C. C.; Porzi, G. *J. Org. Chem.* **1981**, *46*, 1759.
11. Cho, C. S.; Kim, J. S.; Oh, B. H.; Kim, T. J.; Shim, S. C.; Yoon, N. S. *Tetrahedron* **2000**, *56*, 7747.
12. Cho, C. S.; Kim, T. K.; Kim, B. T.; Kim, T. J.; Shim, S. C. *J. Organomet. Chem.* **2002**, *650*, 65.
13. Cho, C. S.; Oh, B. H.; Shim, S. C. *Tetrahedron Lett.* **1999**, *40*, 1499.
14. Cho, C. S.; Oh, B. H.; Kim, J. S.; Kim, T. J.; Shim, S. C. *Chem. Commun.* **2000**, 1885.
15. Wilson, R. B.; Laine, R. M. *J. Am. Chem. Soc.* **1985**, *107*, 361.
16. Hollmann, D.; Bahn, S.; Tillack, A.; Beller, M. *Angew. Chem. Int. Ed.* **2007**, *46*, 8291.
17. Hollmann, D.; Bahn, S.; Tillack, A.; Beller, M. *Chem. Commun.* **2008**, *0*, 3199.
18. Hollmann, D.; Bahn, S.; Tillack, A.; Parton, R.; Altink, R.; Beller, M. *Tetrahedron Lett.* **2008**, *49*, 5742.
19. Bahn, S.; Hollmann, D.; Tillack, A.; Beller, M. *Adv. Synth. Catal.* **2008**, *350*, 2099.
20. Tsuji, Y.; Shida, J.; Takeuchi, R.; Watanabe, Y. *Chem. Lett.* **1984**, *13*, 889.
21. Yin, Z. W.; Zeng, H. S.; Wu, J.; Zheng, S. P.; Zhang, G. Q. *ACS Catal.* **2016**, *6*, 6546.
22. Rosenmund, K. W.; Jordan, G. *Ber. Dtsch. Chem. Ges.* **1925**, *58b*, 51.
23. Kindler, K. *Liebigs Annal. Chem.* **1931**, *485*, 113.
24. Anderson, J. R.; Clark, N. J. *J. Catal.* **1966**, *5*, 250.
25. Yoshimura, N.; Moritani, I.; Shimamura, T.; Murahashi, S.-I. *J. Am. Chem. Soc.* **1973**, *95*, 3038.

26. Yoshimura, N.; Moritani, I.; Shimamura, T.; Murahashi, S.-I. *J. Chem. Soc. Chem. Commun.* **1973**, (9), 307.
27. Murahashi, S.-I.; Hirano, T. Y. T. *J. Am. Chem. Soc.* **1978**, *100*, 348.
28. Murahashi, S. I.; Yoshimura, N.; Tsumiyama, T.; Kojima, T. *J. Am. Chem. Soc.* **1983**, *105*, 5002.
29. Sajiki, H.; Ikawa, T.; Hirota, K. *Org. Lett.* **2004**, *6*, 4977.
30. Miyazawa, A.; Tanaka, K.; Sakakura, T.; et al. *Chem. Commun.* **2005**, 2104.
31. Linciano, P.; Pizzetti, M.; Porcheddu, A.; Taddei, M. *Synlett* **2013**, *24*, 2249.
32. Lubinu, M. C.; De Luca, L.; Giacomelli, G.; Porcheddu, A. *Chem. Eur. J.* **2011**, *17*, 82.
33. Shimizu, K.; Shimura, K.; Ohshima, K.; Tamura, M.; Satsuma, A. *Green Chem.* **2011**, *13*, 3096.
34. Murahashi, S.-I.; Watanabe, T. *J. Am. Chem. Soc.* **1979**, *101*, 7429.
35. Nishimoto, S. I.; Ohtani, B.; Yoshikawa, T.; Kagiya, T. *J. Am. Chem. Soc.* **1983**, *105*, 7180.
36. Miyazawa, A.; Saitou, K.; Tanaka, K.; et al. *Tetrahedron Lett.* **2006**, *47*, 1437.
37. Largeron, M.; Fleury, M. B. *Org. Lett.* **2009**, *11*, 883.
38. He, W.; Wang, L. D.; Sun, C. L.; et al. *Chem. Eur. J.* **2011**, *17*.
39. Shimizu, K.; Ohshima, K.; Tai, Y.; Tamura, M.; Satsuma, A. *Catal. Sci. Technol.* **2012**, *2*, 730.
40. Shimizu, K.; Shimura, K.; Kato, K.; Tamagawa, N.; Tamura, M.; Satsuma, A. *J. Mol. Catal. A Chem.* **2012**, *353*, 171.
41. Shimizu, K.; Shimura, K.; Tamagawa, N.; Tamura, M.; Satsuma, A. *Appl. Catal. A Gen.* **2012**, *417*, 37.
42. von Braun J., Blessing G., Zobel F. Ber. Dtsch. Chem. Ges. 1923;56:1988.
43. Schwoegler, E. J.; Adkins, H. *J. Am. Chem. Soc.* **1939**, *61*, 3499.
44. Borrows, E. T.; Hargreaves, B. M. C.; Page, J. E.; Resuggan, J. C. L.; Robinson, F. A. *J. Chem. Soc.* **1947**, *0*, 197.
45. Winans, C. F.; Adkins, H. *J. Am. Chem. Soc.* **1932**, *54*, 306.
46. Kornfeld, E. C. *J. Org. Chem.* **1951**, *16*, 131.
47. Kindler, K.; Melamed, G.; Matthies, D. *Justus Liebigs Ann. Chem.* **1961**, *644*, 23.
48. De Angelis, F.; Grgurina, I.; Nicoletti, R. *Synthesis* **1970**, 70.
49. Pommersheim, J. M.; Coull, J. *AIChE J.* **1971**, *17*, 1075.
50. Kijenski, J.; Burger, J.; Baiker, A. *Appl. Catal.* **1984**, *11*, 295.
51. Baiker, A.; Monti, D.; Fan, Y. S. *J. Catal.* **1984**, *88*, 81.
52. Schneider, H.; Adkins, H.; McElvain, S. M. *J. Am. Chem. Soc.* **1952**, *74*, 4287.
53. Barrault, J.; Essayem, N.; Guimon, C. *Appl. Catal. A Gen.* **1993**, *102*, 151.
54. Greenfield, H. J. *J. Org. Chem.* **1964**, *29*, 3082.
55. Ballantine, J. A.; Purnell, H.; Rayanakorn, M.; Thomas, J. M.; Williams, K. J. *J. Chem. Soc. Chem. Commun.* **1981**, *0*, 9.

Chapter 3

N-Alkyl Amine Synthesis by Hydroamination of Alkene and Diene

3.1 INTRODUCTION

Hydroamination, which is the addition of amine N—H functionalities to unsaturated carbon—carbon bonds either in an inter- or intramolecular fashion (Scheme 3.1), provides amines in a waste-free manner starting from readily

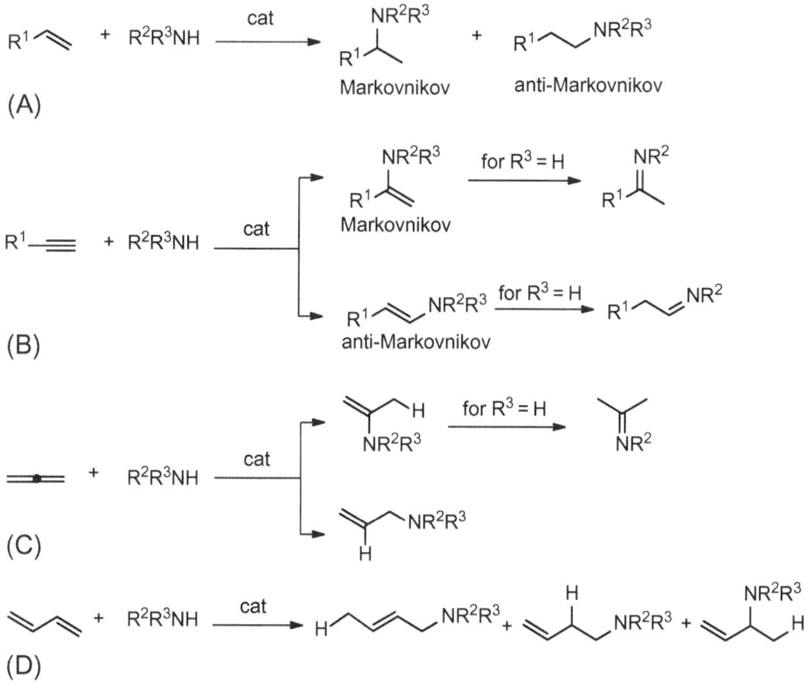

SCHEME 3.1 The catalytic hydroamination of (A) alkenes, (B) alkynes, (C) allenes, and (D) dienes leads to amines, imines, and enamines.

Catalytic Amination for *N*-Alkyl Amine Synthesis. https://doi.org/10.1016/B978-0-12-812284-6.00003-2

available precursors. Therefore, it is a facile and highly atom-economical way to access industrially relevant nitrogen containing basic and fine chemicals as well as naturally occurring alkaloid skeletons.[1–23]

In the hydroamination, unsaturated molecules including alkenes, alkynes, allenes, and dienes are normally involved. Alkenes can be divided into "unactivated" alkenes, ring-strained olefins, and conjugated alkenes (vinylarenes and 1,3-dienes). Strictly speaking, "unactivated" alkenes refer to unfunctionalized C=C bonds, i.e., cyclic or acyclic alkenes with H or alkyl substituents (Scheme 3.2). In general, the addition of amines to unactivated alkenes represents the greatest challenge, while reactions involving alkynes, allenes, strained alkenes, and conjugated alkenes proceed more readily.

The direct addition of amines to alkenes is thermodynamically feasible ($\Delta G^{\theta} = 14.7$ kJ mol^{-1} for the addition of ammonia to ethylene)[24] with a slightly exothermic to thermoneutral reaction enthalpy. Moreover, with the exception of aza-Michael reactions, this seemingly simple reaction is hampered by a high activation barrier caused by an electrostatic repulsion between the nitrogen lone pair of the approaching amine and the π-bond of the electron rich olefin. However, the activation barrier cannot be overcome by increasing the reaction temperature because the negative reaction entropy shifts the equilibrium toward the starting materials. Therefore, the unfavorable kinetic barriers can be overcome only with the use of catalysts.

Significant efforts have been made in the development of efficient catalyst systems for hydroamination reactions. Many catalytic systems based on transition metals (groups 3–5 and 8–10, as well as lanthanides and actinides) and main group metals (alkali and alkaline earth metals), as well as Brønsted acids, have been developed in recent decades.

This chapter focuses on the progress in the catalytic hydroamination of alkenes and dienes for *N*-alkyl amine synthesis. For the scope of this overview, reactions such as Aza-Michael[25,26] and Cope-type hydroamination,[27,28] will not be considered owing to the use of special substrates in the hydroamination. Furthermore, hydroamination of alkynes[6,13] and allenes [1,3,4,8,14,17,21–23] will not be discussed as these reactions mainly generate imines and enamines, discussion of which would stray from the subject of preparation of alkyl amines.

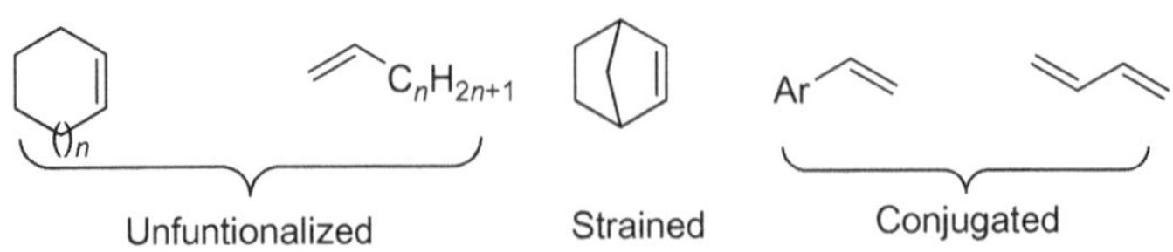

SCHEME 3.2 The types of alkene substrates.

3.2 INTERMOLECULAR HYDROAMINATION OF ALKENES

3.2.1 Alkali and Alkaline Earth Metal-Based Catalysts

In 1954, the first base-catalyzed hydroamination of ammonia and amines with alkenes in the presence of metallic sodium or their hydrides was documented by Whitman and coworkers. The reaction of ethylene and ammonia gave the three ethylamines in 42% conversion and a 70% yield under harsh conditions (175–200°C and 800–1000 tm, Scheme 3.3).[29]

Later, the reaction of diethylamine with ethylene to yield triethylamine in the presence of lithium diethylamide and *N,N,N′,N′*-tetramethylethylenediamine was studied. Because the TMEDA was used as a ligand, the reaction condition became mild (70–90°C and 6–60 atm).[30,31] The hydroamination of vinylarenes with primary and secondary amines was studied with 2 mol % of $LiN(SiMe_3)_2$/TMEDA at 120°C in the absence of solvent to give anti-Markovnikov addition (Scheme 3.4).[32]

The addition of aniline to ethylene has been studied using sodium anilide as a catalyst by Kolka in 1957.[33] In 1972, Falk et al. described the addition of amines to vinyl aromatics and 1,3 dienes in the presence of lithium amide, which was generated in situ by adding the desired amount of butyllithium to amine.[34] Since then, considerable advancements have been achieved in the hydroamination of amines, salts of amines, and olefins with *n*-butyllithium.[35–38] In addition, KOtBu and cesium hydroxide were also served as catalysts for hydroamination of aromatic olefins with anilines.[36,39]

In 2009, researchers reported hydroamination of activated alkenes with aliphatic amines by applying homoleptic complexes $[\{Ae[N\text{-}(SiMe_3)_2]_2\}_2]$ (Ae = Mg, Ca, Sr, Ba) and $[Ae\{CH(SiMe_3)_2\}_2(THF)_2]$ (Ae = Ca, Sr) as catalysts. The results demonstrated that the Sr complex was superior to that of Mg, Ca, or Ba. DFT and kinetic analysis demonstrated that this might be a consequence of the increased radius and accessibility of this cation, rather than a reflection of a reduced barrier for rate-determining alkene insertion.[40,41]

In 2012, several complete families of stable complexes of the large Ae metals were prepared and used as catalysts in the intermolecular hydroamination of activated alkenes. Contrary to expectations based on previous computations, the barium complexes displayed superior efficiency in comparison to other Ae metals complexes as barium with the largest atomic radius.[42]

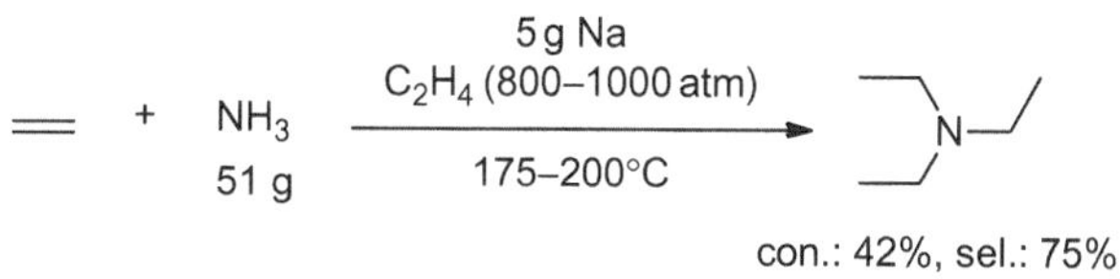

SCHEME 3.3 Sodium-catalyzed hydroamination of ammonia with ethylene.

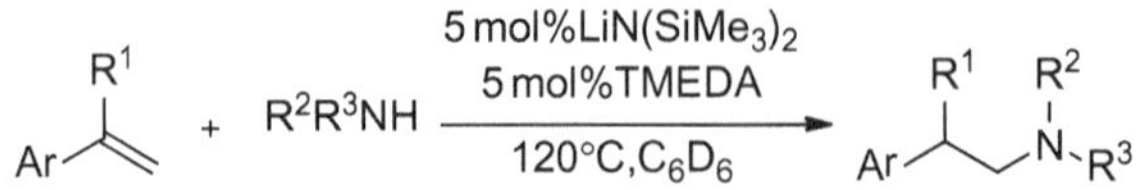

SCHEME 3.4 The hydroamination of vinylarenes with primary and secondary amines with $LiN(SiMe_3)_2$/TMEDA.

3.2.2 Rare Earth Metal-Based Catalysts

The rare-earth-metal-based catalyst system for intermolecular hydroamination nonfunctionalized alkene and diene with primary amines was achieved in 1996 by Marks with $Me_2SiCp'_2$-Nd-$CH(SiMe_3)_2$ complexes as catalyst (Scheme 3.5).[43] Nonfunctionalized alkene and diene reacted with *n*-propylamines to give high yields to the desired products. It is worth noting that the lanthanide-mediated intermolecular hydroamination regioselectively generates the anti-Markovnikov addition products when R = $SiMe_3$ or CH_2=CH. However, Markovnikov addition product was observed with R = $CH_3CH_2CH_2$.

This chemistry has been extended to a number of vinylarenes to afford anti-Markovnikov addition products utilizing $Cp'_2LnCH(SiMe_3)_2$ (Ln:La, Nd, Sm, Lu) as catalyst.[44] These regiochemical preferences are attributed to aryl-directing interactions of the weakly coordinating arene system and the electrophilic lanthanide center.

The mechanism might proceed through a metal-amido species, formed upon protonolysis of a metal-amido, metal-alkyl, or metal-aryl bond in the precatalyst, followed by insertion of the olefin into the metal-amido bond. The roughly thermoneutral insertion step is usually rate-determining. The resulting metal-alkyl species undergoes rapid protonolysis with a second amine molecule, regenerating the metal-amido species and releasing the hydroamination product.[44]

Recently, an yttrium-based catalyst system using a dianionic ene-diamido ligand displayed good activity in the anti-Markovnikov addition of pyrrolidine to styrene with 100% conversion.[45]

3.2.3 Late Transition Metal-Based Catalysts

3.2.3.1 Rh Catalyst System

In 1971, Coulson introduced the first example of a homogeneous transition metal-catalyzed hydroamination, whereby $RhCl_3 \cdot 3H_2O$ and $IrCl_3 \cdot 3H_2O$ were able to catalyze the addition of secondary alkylamines to ethylene, albeit at high reaction temperatures and pressures.[46] The less basic anilines also react with ethylene under the analogous catalyst system, and the TON is 30.[47] Upon further optimization of this reaction, it is observed that addition of the *n*-Bu_4PI (65 equiv. pyrrolidine/Rh) and iodine (2 equiv./Rh) produces the most active

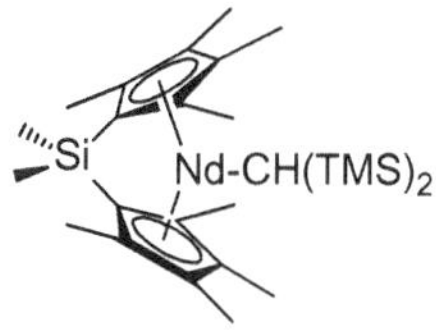

SCHEME 3.5 Me_2SiCp_2' $NdCH(SiMe_3)_2$ catalyst.

catalytic system reported to date. A TON number of 260 is achieved in the hydroamination of ethylene with aniline.[48]

[(DPEphos)Rh(COD)]BF_4 catalyzed hydroamination of *N*-allyl imines with cyclic amines has been presented to synthesis 1,2-diamines (Scheme 3.6).[49] The similar catalyst system also facilitates the anti-Markovnikov hydroamination of homoallylic amines. The proximal Lewis basic amine serves to promote the reactivity and enforce the regioselectivity through the formation of the favored metallacycle.[50]

In 1992, the combination of $[Rh(PEt_3)_2Cl]_2$ and LiNHPh displayed positive activity for the hydroamination/hydroarylation of norbornene with anilines, but the yield of hydroamination product was only 15%.[51] It should be noted that aniline and *m*- and *p*-toluidines afford a mixture of hydroamination and hydroarylation products whereas *o*-toluidine and diphenylamine only generate the hydroarylation product.[52] Further optimization has shown that ionic liquid [BMI][Br] as solvent with $RhCl_3$ catalyst not only improves TON from 0 to 8.3 but also increases the selectivity for hydroamination (84%) over hydroarylation.[53]

Under the similar reaction conditions mentioned above, the addition of aniline to styrene has been reported to generate a mixture of the hydroamination product (30%) and oxidative amination (65%) products with TON of 21.[54] Subsequent investigation determined $[Rh(cod)_2]BF_4$ and PPh_3 as the catalyst system for the hydroamination of the styrenes with secondary aliphatic amines, providing the anti-Markovnikov hydroamination products in 2%–26% yields.[55] Notably, the hydroamination of 2- and 4-vinylpyridine with secondary amines in some cases provided exclusive access to hydroamination products in excellent yield.[56] In marked contrast, [Rh(COD)(DPEphos)]BF_4 possesses

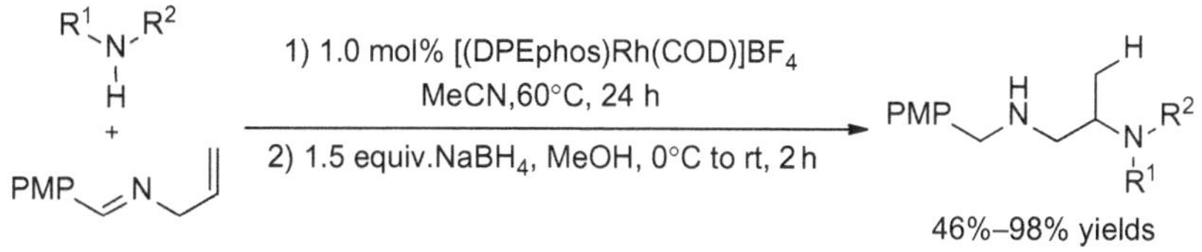

SCHEME 3.6 Rh-catalyzed intermolecular hydroamination of allyl imines for the synthesis of 1, 2-diamines.

significantly improved selectivities of anti-Markovnikov hydroamination products in all reactions.[57]

3.2.3.2 Pd Catalyst System

In 1974, Hegedus and colleagues introduced $Pd(PhCN)_2Cl_2$-promoted amination of unactivated terminal olefins with alkyl amines to give amines up to a 90% yield.[58,59] Recently, the $Pd(OAc)_2$-catalyzed anti-Markovnikov hydroamination of inert olefins using 8-aminoquinoline directing group, which is used to control the regiochemistry and prevent β-hydride elimination of the nucleopalladated intermediate, was reported (Scheme 3.7).[60] Unfortunately, this catalyst system seems to be limited to imides, amides, and the addition of stoichiometric amounts of Brønsted acid.

In 1986, it was reported that conjugated dienes react with amines in the presence of $Pd(PPh_3)_4$ and hydroiodide salts to afford hydroamination products with yields of up to 66%.[61] Since then, the breakthrough for Pd-catalyzed hydroamination of conjugated dienes with anilines has been achieved with the coaddition of $Pd(PPh_3)_4$/TFA and cationic palladium catalysts (Scheme 3.8).[62,63] Finally, the scope of amine substrates has been expanded to alkylamines for synthesis of 1,4-adducts product with a 92% yield.[64] The unsaturated heterocycles, such as 2,3-dihydrofuran and pyrans, can also react with amines in the presence of $K_2Pd(SCN)_4$ catalyst with up to a 98% yield.[65,66]

In 2000, the hydroamination of vinylarenes with anilines was achieved in the presence of $Pd(TFA)_2$, DPPF, and TfOH to generate Markovnikov addition products in high yields.[67–69] Additionally, a series of cationic (diphosphine)-palladium(II) complexes also displays good activity in the hydroamination of vinylarenes with anilines with 39%–100% yields.[70]

3.2.3.3 Ir Catalyst System

In 1988, Ir-catalyzed hydroamination of the norbornene with aniline was documented with $Ir(PEt_3)_2(C_2H_4)_2Cl_4$ and $ZnCl_2$ as catalyst to afford exo-2-(phenylamino) norbornane. By investigating the individual steps of the catalytic reaction, a previously unreported mechanism through N—H bond oxidative addition pathway has been established (Scheme 3.9).[71]

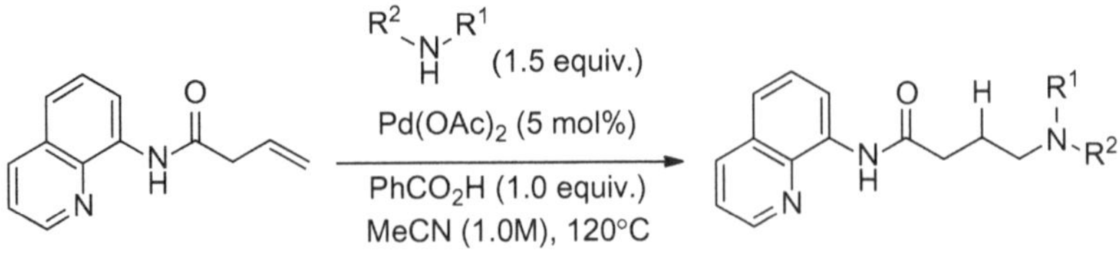

SCHEME 3.7 Pd-catalyzed the anti-Markovnikov hydroamination of inert olefins with imides and amides.

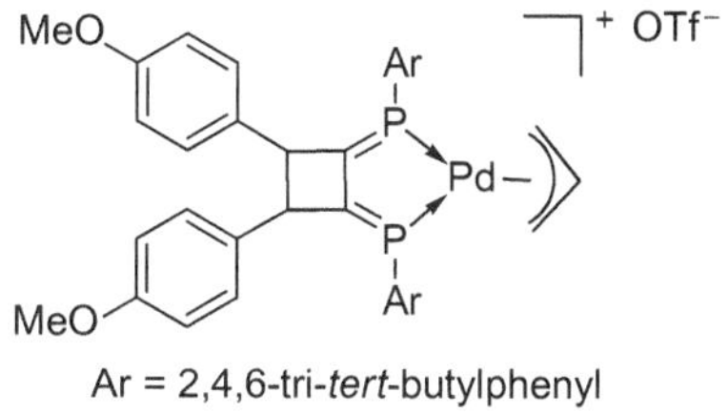

SCHEME 3.8 Cationic palladium catalyst.

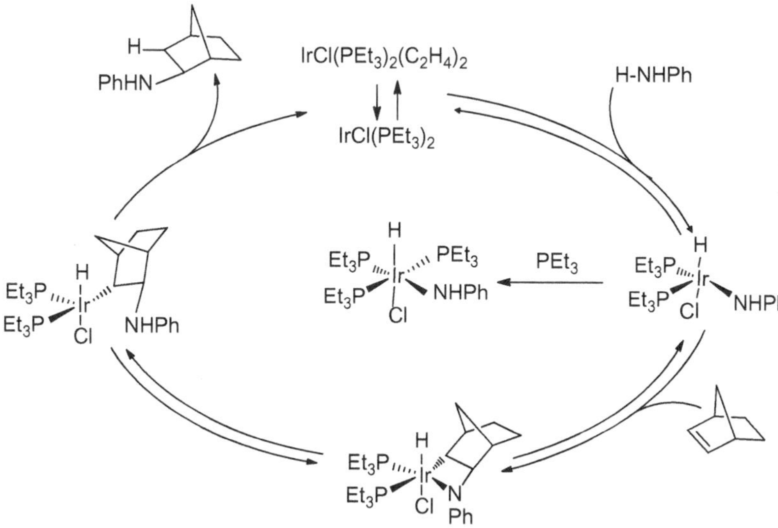

SCHEME 3.9 The mechanism of Ir-catalyzed hydroamination of the norbornene with aniline.

Two decades later, the combination of $[Ir(coe)_2Cl]_2$ with DTBM-Segphos was identified as an effective catalyst system for amination of unactivated olefins with arylamides and sulfonamides, providing the desired products in 65%–92% yields.[72] Subsequently, this catalyst system has been extended to facilitate the hydroamination of unactivated alkenes with indoles.[73]

3.2.3.4 Ru Catalyst System

In 2001, stoichiometric intermolecular hydroamination of secondary aliphatic amines was reported by Keim and Angelici, respectively.[74,75] Subsequently, the reaction of secondary aliphatic amines and vinylarenes was achieved to allow anti-Markovnikov hydroamination products in 40%–96% yields catalyzed by $Ru(cod)(2\text{-methylallyl})_2$, DPPPent, and TfOH.[76] The proposed mechanism for this process involves the generation of an η^6-styrene complex, followed by a nucleophilic attack of the amine and exchange of η-arene ligands (Scheme 3.10).[77]

Soon after, the cationic ruthenium complex $[(PCy_3)_2(CO)(Cl)\text{-}Ru{=}CHCH{=}C(CH_3)_2]BF_4$ is identified as an effective catalyst for the reaction of aniline and ethylene to form a 1:1 ratio of *N*-ethylaniline and 2-methylquinoline products. The analogous reaction with 1,3-dienes produces 1,2-addition products in moderate yields.[78]

3.2.3.5 Ni Catalyst System

In 2002, a simple colorimetric assay of various transition-metal catalysts showed that the combination of DPPF, $Ni(COD)_2$, and TFA exhibited positive activity in the hydroamination of dienes with alkylamines to form allylic amines with 38%–94% yields (Scheme 3.11).[79] The mechanism involves the generation of a nickel hydride, insertion of the diene to give a nickel allyl, and external attack of the amine on the coordinated allyl to afford the product and release the catalyst.

3.2.3.6 Pt Catalyst System

In 2004, the platinum-catalyzed hydroamination of ethylene with aniline was reported.[80] By applying $PtBr_2$ as a catalyst precursor in *n*-Bu_4PBr, aniline can react with ethylene to afford *N*-ethylaniline with 80 turnovers (Scheme 3.12). The TON is increased by decreasing the basicity of arylamine. The TON number reaches 250 under the same conditions if weakly basic 2-chloroaniline is employed. The activating role of *n*-Bu_4PBr can be rationalized by understanding that the presence of bromide ions increases the basicity of the platinum center in a zwitterionic intermediate. Since then, this chemistry has been extended to norbornene and 1-hexene utilizing the same catalyst system.[81–83] The proposed mechanism for this process involves ethylene activation with a platinum catalyst, followed by nucleophilic addition of aniline to

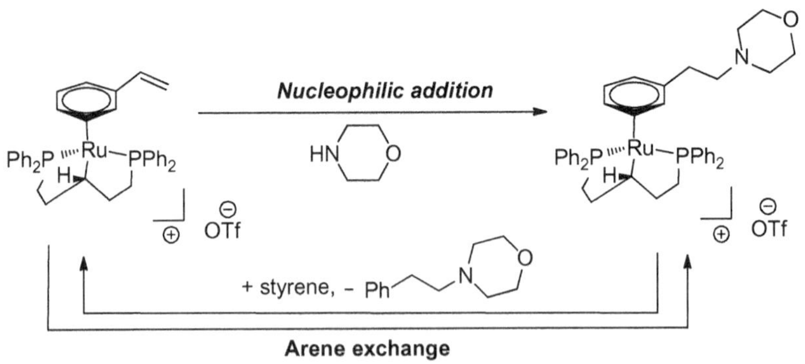

SCHEME 3.10 The mechanism for Ru-catalyzed anti-Markovnikov hydroamination of vinylarenes with secondary aliphatic amines.

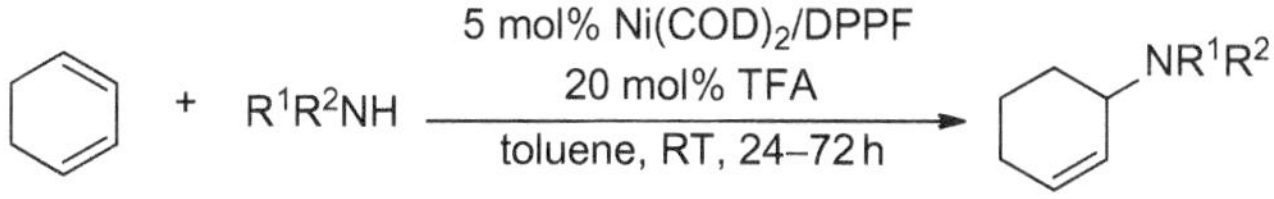

SCHEME 3.11 Ni-catalyzed hydroamination of dienes with alkylamines.

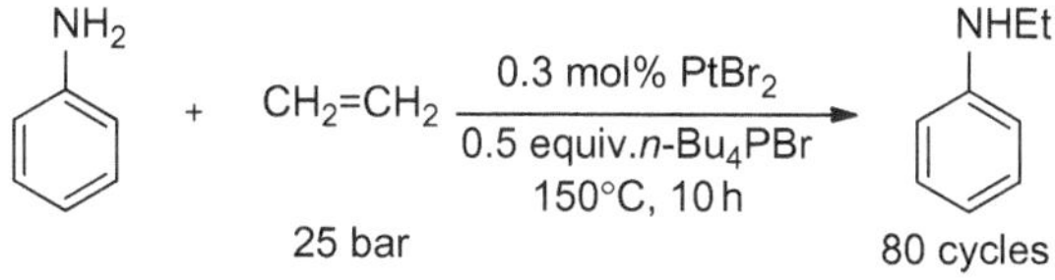

SCHEME 3.12 Pt-catalyzed hydroamination of ethylene with aniline.

the coordinated ethylene, intramolecular transfer of the ammonium proton to the metal center to generate a 5-coordinate (16-electron) Pt_{IV}-H intermediate, and final reductive elimination completes the catalytic cycle and produces the hydroamination product.[84–86]

Additionally, a variety of platinum-based catalyst systems have been studied with regard to hydroamination of ethylene with carboxamides and sulfonamides.[87–89] Among them, cationic $(COD)Pt(OTf)_2$ or $[PtCl_2(C_2H_4)]_2/AgBF_4$ catalyst system displays good activity up to a 95% yield.[88] A protocol for the hydroamination of vinyl arenes with carboxamides has been achieved when $[PtCl_2(H_2C{=}CH_2)]_2$ and $P(4\text{-}C_6H_4CF_3)_3$ are employed as well.[90]

3.2.3.7 *Au Catalyst System*

In 2006, He described the first example of gold-catalyzed addition of sulfonamides and carbamates to alkenes[91] and 1,3-dienes[92] under mild reaction conditions with a cationic gold(I) complex generated from 5 mol% of Ph_3PAuCl/AgOTf (Scheme 3.13).

Subsequent investigation indicates that $[(PhO)_3P]AuCl$/AgOTf possesses significantly improved reactivity compared to Ph_3PAuCl/AgOTf and the catalyst loading could be decreased to 0.05 mol%.[93] The mechanism of the

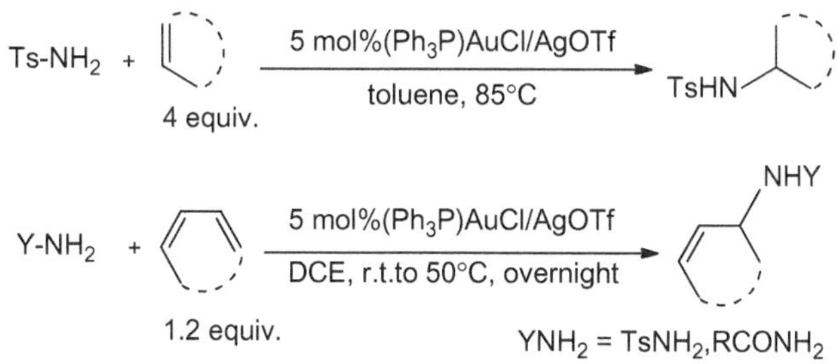

SCHEME 3.13 Gold-catalyzed addition of sulfonamides and carbamates to alkenes.

gold-catalyzed hydroamination process is considered similar to Pt-catalyzed processes, although with minor difference in proton transfer step.[94] Furthermore, comparative studies about the hydroamination of unactivated alkenes and dienes catalyzed by either cationic gold(I) triphenyl phosphite complexes or silver salts have been studied using sulfonamides, anilines and carbamates as nucleophiles. The gold catalyst shows a wider scope than silver salt, which is unable to catalyze the hydroamination of terminal alkenes.[95]

3.2.4 Acid Catalysts

In 1987, Deeba described the first example of zeolite-catalyzed amination of ethene with ammonia.[96] H-clinoptilolite displays the highest activity to form ethylamine with 14% conversion and 93% selectivity, albeit at high reaction temperature (370°C) and pressure (55 atm, Scheme 3.14). This methodology has been extended to the hydroamination of isobutene with ammonia and with proton-exchanged zeolite catalysts.[97–99] The number and strength of Brønsted acid sites are factors controlling the amination of 2-methylpropene over proton-exchanged zeolite catalysts.

A variety of acid catalysts, such as H-mont,[100] $InBr_3$,[101] $FeCl_3$,[102] $BiCl_3$,[103,104] $TiCl_4$,[105,106] $Cu(OTf)_2$,[107,108] TfOH,[109,110] and TfOH/silica gel,[111] have been studied for hydroamination of alkenes with amine, sulfonamides, carboxamides, and sulfamates. Subsequently, it has been demonstrated that $ZnCl_2$/silica is an active catalyst for the reaction of hexylamine with cyclohexene.[112]

In 2005, $PhNH_3B(C_6F_5)_4$ was identified as an active catalyst for the addition of several activated alkenes with anilines, affording a mixture of hydroamination products and hydroarylation products.[113] It is noteworthy that the acid counter anion has an obvious impact on efficiency of reaction, and thus ionic liquids as solvents can provide higher selectivity and yields than organic solvents.[114]

Furthermore, carbamates, sulfonamides, and carboxamides can react with 1,3-dienes to afford allylic amines in good yields catalyzed by $Bi(OTf)_3$/$Cu(CH_3CN)_4PF_6$/DPPE.[115] In addition, $H_3SiW_{12}O_{40}nH_2O$[116] and Amberlyst-15[117] could also facilitate the hydroamination of activated alkenes with these less basic amines to generate hydroamination products in moderate to good yields with ionic liquid [Bmim][BF_4] as a solvent. The proposed mechanism suggested Lewis acid-base interaction between the catalyst and amine substrate,

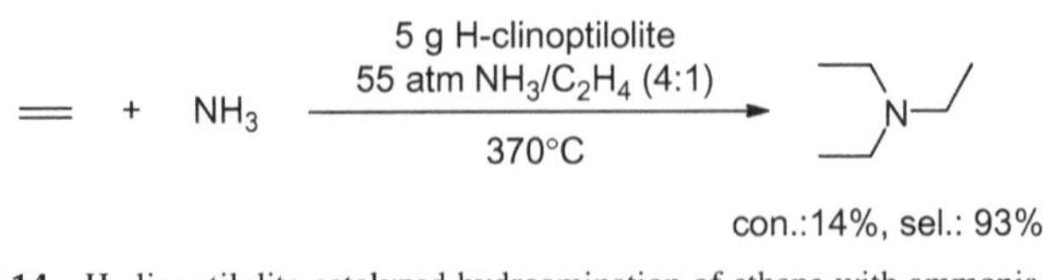

SCHEME 3.14 H-clinoptilolite-catalyzed hydroamination of ethene with ammonia.

and protonation of alkenes to form a cationic intermediate, followed by a nucleophilic attack of the amine, and finally deprotonation to give the hydroamination product.[104,118]

3.2.5 Other Catalysts

Excluding metal-based catalysts, hydroamination utilizing organic catalysts was also explored. For example, NBS was shown to be active for the hydroamination of activated styrenes with tosylamides or carbamates to allow amino derivatives in 60%–83% yields.[119] I_2 also shows appreciable catalytic activity for the addition of sulfonamides to styrenes in 75%–92% yields.[120] In 2014, an anti-Markovnikov intermolecular hydroamination reaction employing a new organic photoredox catalyst system was achieved by Nicewicz.[121] Tri-substituted aliphatic alkenes and styrene derivative can react with sulfonamide or heterocyclic amines in the presence of catalytic amounts of mesityl acridinium, PhSSPh and 2,6-lutidine to afford the anti-Markovnikov hydroamination products in 35%–85% yields (Scheme 3.15).

3.3 ASYMMETRIC INTERMOLECULAR HYDROAMINATION OF ALKENES

3.3.1 Rare Earth Metal Catalysts

Given that most pharmaceuticals and biological-active molecules are chiral, the development of chiral catalysts for stereoselective hydroamination has drawn increasing attention in the last decade.

The asymmetric hydroamination of nonactivated alkenes with *N*-unprotected amines remains elusive. In 2001, first chiral rare earth metal-catalyzed hydroamination of terminal unactivated olefins with primary alkylamines was reported by Hultzsch and colleagues. Moderate enantioselectivities up to 61% ee may be achieved in the presence of the yttrium binaphtholate catalyst (Scheme 3.16).[122] Secondary amines and internal alkenes are unreactive; moreover, sterically hindered terminal olefins, such as vinyl cyclohexane, also display low reactivity.

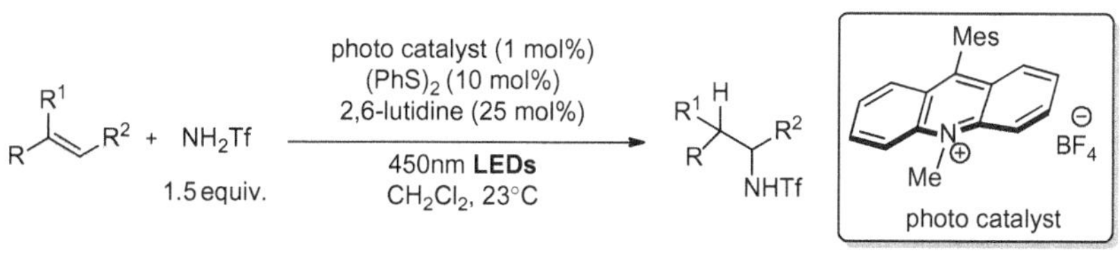

SCHEME 3.15 Anti-Markovnikov intermolecular hydroamination reaction employing organic photoredox catalyst system.

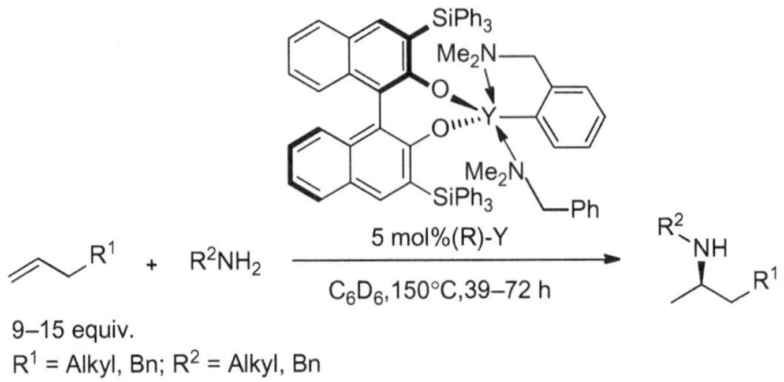

SCHEME 3.16 Asymmetric intermolecular hydroamination of unactivated alkenes.

The catalytic activity of rare earth metal complexes in intermolecular hydroaminations is influenced by the electron deficiency of the metal center. Therefore catalysts comprised of a C1-symmetric NOBIN-based aminodiolate ligand exhibit lower catalytic activity and lower enantioselectivities (up to 40% ee) compared to the binaphtholate complex.[123] The decrease in catalytic activity can be rationalized by the presence of an additional amine donor in the aminodiolate ligand framework.

3.3.2 Late Transition Metal-Based Catalysts

3.3.2.1 Ir Catalyst System

The first chiral iridium-based catalyst system was reported by Togni in 1997.[124] The activity and enantioselectivity of this catalyst system is significantly enhanced with the addition of Schwesinger's "naked" fluoride $[N\{P(NMe_2)_3\}_2]F$ (Scheme 3.17). In some cases, the addition of fluoride also results in a reversal of absolute product configuration. However, the reason for the strong fluoride effect remains unclear, but it has been suggested that hydrogen bridging of fluoride might play a role.

This chemistry has been extended to a number of bicyclic alkenes and dienes utilizing various chelating axially chiral bisphosphine iridium-catalysts.[125] Unfortunately, alkylamines and sterically encumbered aniline derivatives do not undergo hydroamination reactions under these conditions. In addition, greater than 90% enantioselectivity can be obtained if amides and sulfonamides were used as starting materials.[72]

Cationic iridium catalysts also facilitate the Markovnikov hydroamination of vinyl arenes with 2-aminopyridines, 2-aminopyrimidines, 2-aminopyrazins, and related heteroaromatic amines with moderate to good enantioselectivities up to 83% ee using C_3-TUNEPHOS as chiral ligand.[126] A variety of electron-donating and electron-withdrawing substituents are tolerated in the vinyl arene. However, *ortho*-substitution does diminish the yield and enantioselectivity.

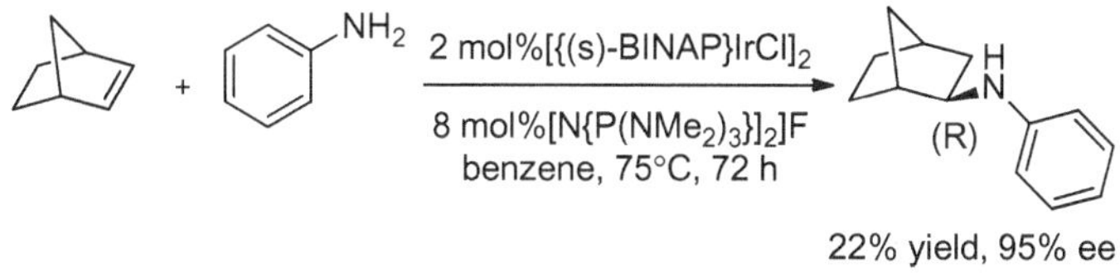

SCHEME 3.17 Iridium-catalyzed asymmetric hydroamination of norbornene.

Additionally, an iridium-catalyzed addition of amides to unreactive α-olefins has been reported to generate a mixture of hydroamination product and oxidative amination product. The *N*-alkyl amide is formed in low enantioselectivities up to 19% ee with a large excess of alkene.[72] The enamide by-product can be reduced to the *N*-alkyl amide via transfer hydrogenation using *i*-PrOH.

The tentative mechanism for this iridium-catalyzed hydroamination of alkenes with heteroaromatic amines has been proposed to proceed via oxidative addition of the amine to afford the four-member ring intermediate. Alkene insertion into the iridium-amide bond yields the more favorable six-member ring intermediate, which releases the hydroamination product via reductive elimination. The chelation of the pyridine nitrogen to iridium appears to be crucial for the control of regioselectivity. Steric hindrance in the *ortho*-position to the pyridine nitrogen results in the formation of a mixture of Markovnikov and anti-Markovnikov product (Scheme 3.18).

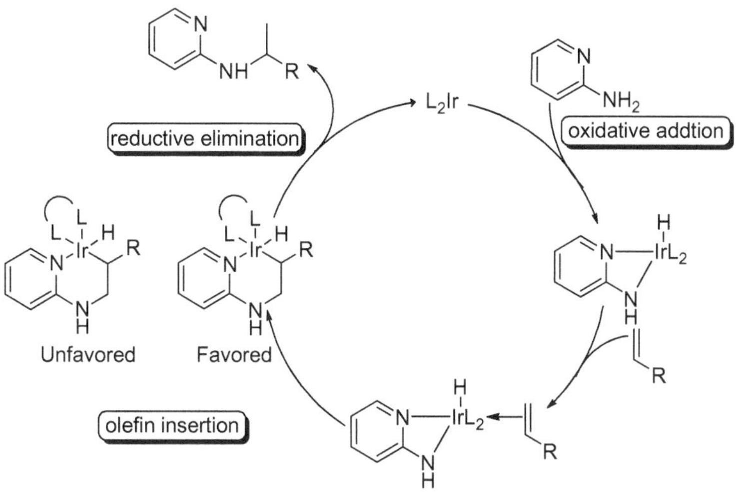

SCHEME 3.18 Proposed mechanism for the iridium-catalyzed regioselective intermolecular hydroamination of alkenes with 2-aminopyridine.

3.3.2.2 Pd Catalyst System

Since more reactive alkenes, such as vinyl arenes or sterically strained polycycles, can react more readily in the hydroamination reaction, several asymmetric hydroamination reactions utilizing these substrates have been explored. Weakly basic anilines react with vinyl arenes to give the Markovnikov addition products with good yields and moderate to high enantioselectivities in the presence of chiral palladium complexes utilizing BINAP[67] or the 4,4′-disubstituted SEGPHOS[127] as ligands (Scheme 3.19).

The scope of amine substrates can be expanded to more basic alkylamines. For example, arylethanamine is obtained in moderate yields and enantioselectivity utilizing the (*R*,*R*)-Et-FerroTANE ligand. Unfortunately, almost stoichiometric amounts of Brønsted acid are required to afford the Markovnikov hydroamination product of the vinyl arene.[128]

The palladium-catalyzed asymmetric hydroamination of cyclohexadiene with arylamines utilizing a variant of Trost's ligand proceeds with high enantioselectivities under mild conditions.[62] This catalyst system can be applied to a broader range of anilines with consistent high enantioselectivities under neutral conditions at room temperature with moderate activity. Other chiral bidentate phosphine ligands, such as (*R*)-BINAP, (*S*,*S*)-DIOP, or (*S*,*S*)-BDPP, produce only inferior enantioselectivities.

Mechanistic investigations[68] using stoichiometric amounts of the enantio- and diastereomerically pure η^3-benzyl palladium complex [{(*R*)-Tol-BINAP}-{η^3-1-(2-naphthyl)ethyl}Pd](OTf) reveal that the reaction with aniline produces predominantly (*R*)-*N*-1-(2-naphthyl)ethylaniline, consistent with external nucleophilic attack. However, the catalytic reaction of [{(*R*)-Tol-BINAP}Pd(OTf)$_2$] with vinyl arenes and amines more likely produces the opposite enantiomeric (*S*)-amine hydroamination product, indicating that the minor diastereomer of the catalytic active benzyl palladium species might be responsible for the majority of product formed in the catalytic process.

3.3.2.3 Au Catalyst System

Although several significant contributions to the field of intermolecular hydroamination of simple alkenes have been made within the last decade, asymmetric hydroamination of nonactivated alkenes with unprotected amines remains one

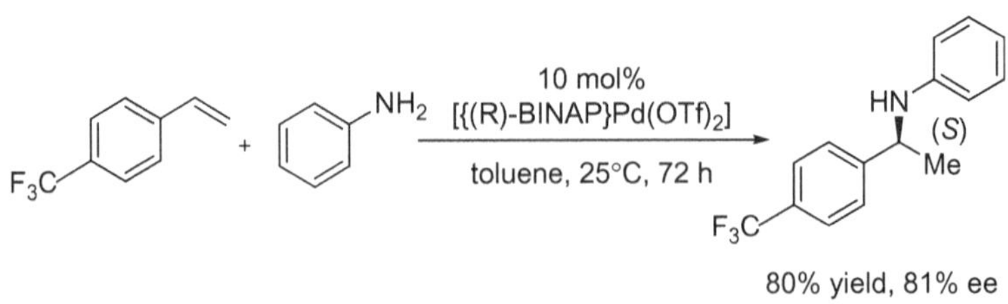

SCHEME 3.19 Palladium-catalyzed asymmetric Markovnikov addition of aniline to vinyl arenes.

of the most challenging problems in the field. Recently, the asymmetric intermolecular hydroamination of terminal, nonactivated alkenes with cyclic ureas has been reported.[129] The Markovnikov addition proceeds with up to 78% ee utilizing the axial chiral MeOBIPHEP-ligated bis(gold(I))-catalyst system (Scheme 3.20). Unfortunately, the reaction requires a large excess of alkene substrate, which leads to diminished enantioselectivities.

3.4 INTRAMOLECULAR HYDROAMINATION OF AMINOALKENES

3.4.1 Alkali and Alkaline Earth Metal-Based Catalysts

In 1992, the first example of intramolecular hydroamination of nonactivated alkene was reported by Suginome and colleagues.[130] Treatment of akenylamines with a catalytic amount of butyllithium produced pyrrolidines in 49%–99% yields. This catalyst system is further optimized through using a 1:1 mixture of THP/toluene solvent at 110°C (Scheme 3.21). Cyclization of Thorpe-Ingold-activated aminoalkenes and even unsubstituent to the chain linking the aminoalkenes produce substituted pyrrolidines and piperidines, as well as fused and bridged bicyclic amine derivatives in good yields.[131,132]

The β-diketiminato heavier alkaline earth metals complexes also show good catalytic activity in the hydroamination/cyclization of aminoalkenes.[133,134] The trends in reactivity are Ca > Sr > Ba for the studied family of these catalysts. In addition, phenoxyamine[135] and cyclopentadienyl-based Mg complexes[136] are identified as effective catalysts for this cyclization reaction. DFT study indicates that the mechanistic scenario involves a reversible olefin insertion step followed by rate-determining protonolysis.[137,138] The concerted proton-assisted cyclization pathway is energetically prohibitive.

3.4.2 Rare Earth Metal-Based Catalysts

In 1989, rare-earth-metal-based catalysts for the hydroamination/cyclization of primary aminoalkenes were first introduced by Marks and colleagues. By applying $(Cp^*{}_2LaH)_2$ as catalyst, the cyclization of aminoalkenes to generate heterocycle products was realized with TOF up to 140 (Scheme 3.22).[139] Following this report, a variety of related $Cp^*{}_2LnR$ complexes were shown to be

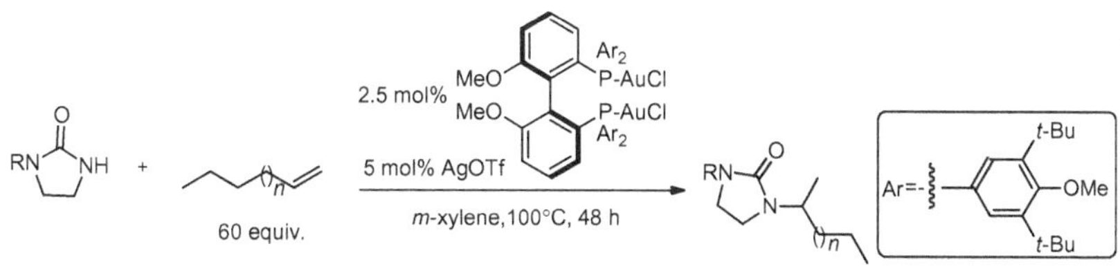

SCHEME 3.20 Gold-catalyzed asymmetric hydroamination of alkenes with cyclic ureas.

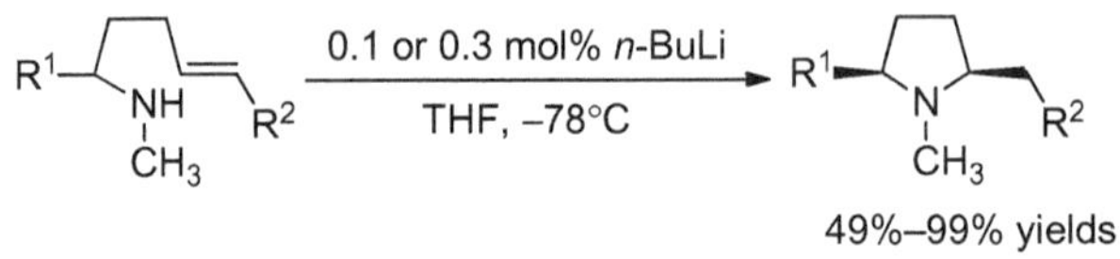

SCHEME 3.21 *n*-BuLi-catalyzed intramolecular hydroamination of a nonactivated alkene.

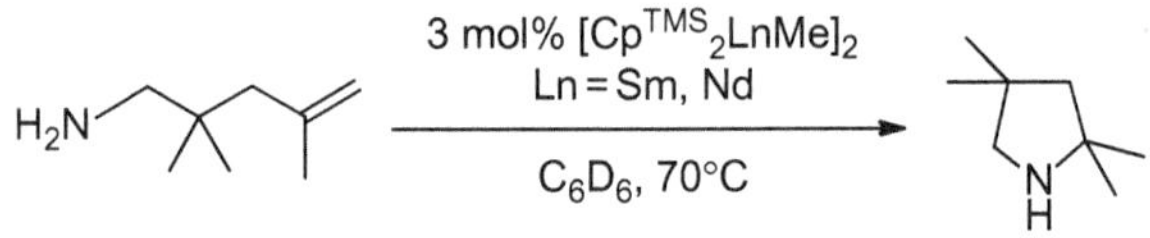

SCHEME 3.22 Catalytic intramolecular hydroamination of hindered alkenes using organolanthanide complexes.

active for the cyclohydroamination of aminoalkenes and conjugated aminodienes,[140–144] and $[Cp_2^{TMS}LnMe]_2$ (Ln) = Sm, Nd) allow facile cyclization of hindered amino olefins to provide products containing quaternary centers.[145,146]

Subsequently, a series of silyl-linked amido cyclopentadienyl ligand-based organolanthanide complexes have been developed for the hydroamination/cyclization of primary aminoalkenes.[147–151] The high activity of these complexes permit limitations to be overcome in the elusive cyclization of amines tethered to 1,2-disubstituted alkenes.[149,150]

Further investigation shows that catalytic activity can be dramatically increased and the cyclization diastereoselectivity can be improved by coordination of the organolanthanide to simple chelating diamide or homoleptic trisamido ligands.[152–156] The diamidoamine ligand provides a good steric and electronic environment for the catalytic hydroamination reaction.

The mechanism of the hydroamination/cyclization proceeds through a metal-amido species, which is formed upon protonolysis of a metal-amido, metal-alkyl, or metal-aryl bond in the precatalyst, followed by insertion of the olefin into the metal-amido bond with a seven-membered chair-like transition state. The roughly thermoneutral insertion step is usually rate-determining, giving rise to a zero-order rate dependence on substrate concentration and first-order rate dependence on catalyst concentration. The resulting metal-alkyl species undergoes fast protonolysis with a second amine molecule, regenerating the metal-amido species and releasing the heterocyclic product.[144,157,158]

3.4.3 Group 4/5 Metal-Based Catalysts

Following an initial report in 2000 by Hosomi and colleagues on $TiCl_4$-catalyzed cyclization of vinylsilanes bearing an N-Ts protected amino group

SCHEME 3.23 $TiCl_4$-catalyzed cyclization of vinylsilanes.

to 2-(silylmethyl)pyrrolidines (Scheme 3.23),[159] $Ti(NMe_2)_4$ and $ZrCl_4$ have been used effectively as precatalysts for the intramolecular hydroamination of geminally substituted primary aminoalkenes to yield pyrrolidine and piperidine heterocyclic products.[160,161]

In 2004, it was shown that the Cp-based alkyl zirconocene cations are promising catalysts for the hydroamination/cyclization of secondary aminoalkenes.[162] It should be noted that this cyclization does not require gem-disubstituents. Subsequently, a variety of related Cp-based Ti, Zr, and Hf complexes are reported with primary aminoalkenes cyclization in good yields.[163–166] Since Ti catalysts are prone to form side-products by C-H activation processes, Zr complexes show the highest catalytic activity of the group-IV metal catalysts for the formation of piperidines.

Subsequent investigations are focusing on the synthesis of bis(amido) precatalyst for the hydroamination of the aminoalkenes.[167–171] In 2009, an important breakthrough for the hydroamination/cyclization of aminoalkenes with Group 4 complexes was achieved by Schafer and coworkers.[170] A tethered ureate-based zirconium complexes was effective with and did not require gem-disubstituents for cyclization. Two years later, similar outstanding results were obtained when cationic [N,O,S]zirconium complex was employed.[172] Moreover, pincer-type complexes were also prepared and tested for their activity in intramolecular hydroamination of aminoalkenes.[173,174] Unfortunately, these catalysts failed to reach the levels comparable to above two examples.

Based on the DFT study of $[Ind_2TiMe_2]$-catalyst system, the mechanism of the hydroamination/cyclization proceeds through an imido-amido complex, followed by insertion of the alkene moiety into the Ti—N single bond of this imido-amido complex. The insertion step is usually rate-determining, giving rise to a first-order dependence on the aminoalkene concentration. The resulting metal-alkyl species undergoes fast protonolysis with a second amine molecule, regenerating the metal-amido species and releasing the heterocyclic product.[175] The ability of some complexes to promote HA/cyclization of both primary and secondary aminoalkene substrates with comparable rates and activation parameters indicates M=NR intermediates [2+2]-cycloaddition with the alkene are unlikely and that a pathway involving insertion at an M—N σ-bonded intermediate is kinetically viable.[176]

3.4.4 Late Transition Metal-Based Catalysts

3.4.4.1 Pt Catalyst System

In 1975, Zambonelli and colleagues described the first example of Pt-promoted hydroamination of 4-pentenylammonium ions to produce 2-methylpyrrolidine.[177–179] Thirty years later, the platinum-catalyzed intramolecular hydroamination of unactivated olefins with secondary alkylamines was achieved by Widenhoefer and colleagues.[180] With $[PtCl_2(H_2C{=}CH_2)]_2$ and PPh_3 as the catalyst system, γ-amino olefins underwent cyclization to generate the substituted pyrrolidines in 59%–92% yields. A 57% yield was obtained if δ-amino olefin was used as a starting material. Further investigation indicated that significantly higher catalytic activity was achieved with sterically hindered mono(phosphines) as supporting ligands (Scheme 3.24).[181]

Mechanism exploration[84] showed that the platinum-catalyzed hydroamination initiates the formation of platinum amine complex, followed by outer-sphere attack of the pendant amine on the olefin. Deprotonation/chloride displacement followed intermolecular protonolysis of the Pt-C bond, to produce Pt(IV) hydride intermediate, with reductive elimination giving Pt-amine complex. Finally, Ligand exchange from Pt-amine complex would release product and regenerate active catalyst.

3.4.4.2 Pd Catalyst System

In 1981, Venanzi and coworker reported a protocol for $PdCl_2(PhCN)_2$ promoted cyclization of the aminoalkenes to produce the corresponding 2-methylpyrrolidines and 2-methylpiperidine.[182] Two decades later, hydroamination of aminopropyl vinyl ether in the presence of PdI_2 to give tetrahydro-2-methyl-1,3-oxazine with TON of 1500 was reported by Westcott and coworkers (Scheme 3.25).[183] Subsequently, the SiO_2, Al_2O_3, and TiO_2-supported Pd catalysts also showed good catalytic activity for this reaction with up to TOF of 22.6.[184]

In 2005, the Pd-catalyzed cyclization of carbamate-protected aminoalkenes was reported to produce the heterocyclic products in 60%–96% yields.[185] It should be noted that the formation of both five- and six-membered rings is facile

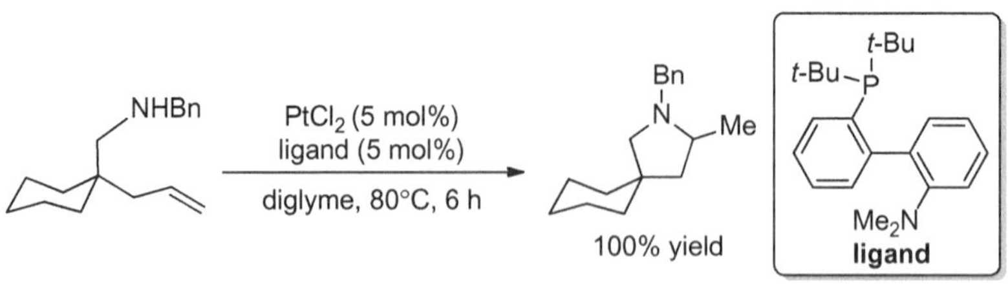

SCHEME 3.24 Pt-catalyzed hydroamination with sterically hindered mono(phosphines) as supporting ligands.

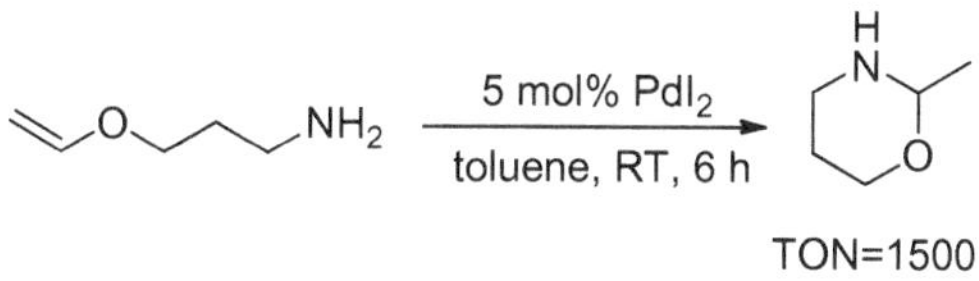

SCHEME 3.25 PdI_2-catalyzed hydroamination of aminopropyl vinyl ether.

and the presence of gem-dimethyl substituents on the alkyl chain to promote ring closure is not required.

The proposed mechanism for this process involves coordination of the aminoalkene substrate to Pd complex, followed by a reversible intramolecular nucleophilic attack to form alkylpalladium intermediate. Subsequent protonation regenerates the Pd species and releases the heterocyclic product. The palladium alkyl complex is the resting state and protonolysis of the Pd—C bond is the turnover-limiting step.[186]

3.4.4.3 Au Catalyst System

In 2006, gold-catalyzed intramolecular hydroamination of alkenyl tosylamides and carbamates was achieved by He and Widenhoefer, respectively.[91,187] With $Au(PR_3)Cl$ and AgOTf as catalyst system, the reaction proceeds smoothly to form protected nitrogen heterocycles in good to excellent yields (Scheme 3.26).

Following these reports, *N*-alkenyl ureas and alkyl ammonium salts have been cyclized using similar catalyst system.[188–190] Alkyl-gold(I) complexes are formed from the gold(I)-promoted intramolecular addition of various amine nucleophiles to alkenes, which provide the direct experimental evidence for the elementary step of gold-promoted nucleophilic addition to an alkene.[191] In addition to homogeneous catalysis, gold nanoclusters stabilized by PVP also display good activity in the cyclization of gem-disubstituted alkenyl tosylamides.[192]

3.4.4.4 Rh Catalyst System

In 2006, rhodium-catalyzed intramolecular hydroamination of unactivated olefins with secondary alkylamines was reported by Hartwig and colleagues.[193]

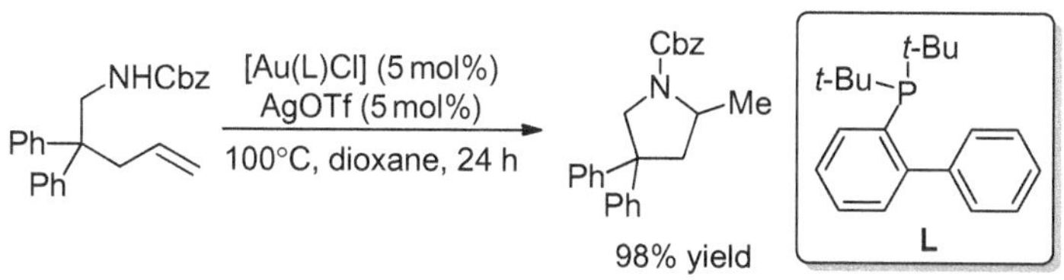

SCHEME 3.26 Gold-catalyzed intramolecular hydroamination of alkenyl tosylamides and carbamates.

Note that anti-Markovnikov products 3-arylpiperidines are exclusively formed with $[Rh(COD)(DPPB)]BF_4$ as catalyst (Scheme 3.27).

Subsequently, using biarylphosphine ligand L_2 as ligand, the intramolecular hydroamination of unactivated terminal and internal alkenes with primary and secondary alkylamines is established to form five- or six-membered ring products in good yields.[194] Further investigation of the effects of ligand showed that the applying of diaminophosphine ligand allow facile cyclization of unprotected primary aminoalkenes lacking substituents on the alkyl chain.[194] Computational studies suggest that the catalytic activity is correlated to the relative binding energies of the ligand to rhodium,[195] and the faster cyclization of gem-disubstituted aminoalkenes resulted from destabilization of an off-cycle complex, rather than an increased rate of nucleophilic attack.

Based on a mechanistic study of achiral rhodium complexes, the mechanism involves the generation of cationic rhodium species where the biaryldialkylphosphine ligand binds in a κ^1,η^6 chelating fashion.[196] Olefin coordination to this 16 valence electron species is followed by reversible nucleophilic attack of the amine to the coordinated alkene. The olefin complex is the resting state and protonation of the β-aminoalkyl-rhodium species is a turnover limiting step.[196] The mechanism of the turnover-limiting protonolysis is a stepwise proton transfer to rhodium, followed by reductive elimination to regenerate the catalyst and release the hydroamination product.

3.4.4.5 Ir Catalyst System

In 2009, a protocol for $Ir(COD)Cl]_2$ catalyzed intramolecular hydroamination of unactivated alkenes with secondary alkyl- and arylamines was disclosed by Stradiotto and coworkers (Scheme 3.28).[197,198] Notably, no ligands or additives are required in order to promote the hydroamination of these substrates. If employing $[Ir(COD)Cl]_2$ and $HNEt_3Cl$ (1:2) as a precatalyst mixture, primary aminoalkenes can be cyclized in good yields.

On the basis of DFT calculations, the mechanistic scenario involves: (1) smooth and reversible nucleophilic attack of the amine unit on the metal coordinated C=C double bond to afford a zwitterionic intermediate; (2) Ir—C bond protonolysis via stepwise proton transfer from the ammonium unit to the metal and ensuing reductive elimination; and (3) final irreversible regeneration of M1 through associative cycloamine expulsion by new substrate.[198] Notably, such an

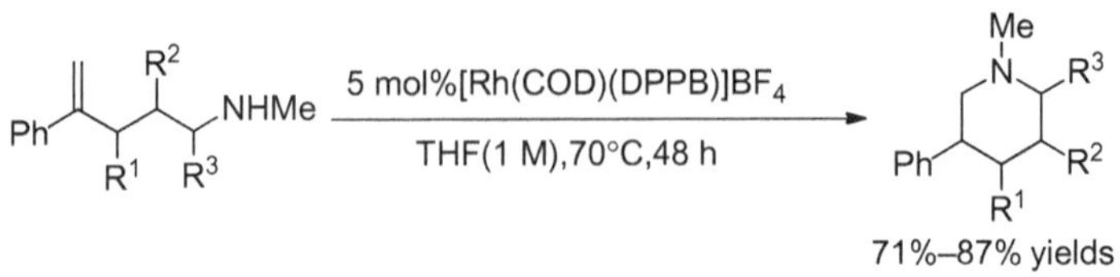

SCHEME 3.27 Rhodium-catalyzed intramolecular anti-Markovnikov hydroamination of unactivated olefins.

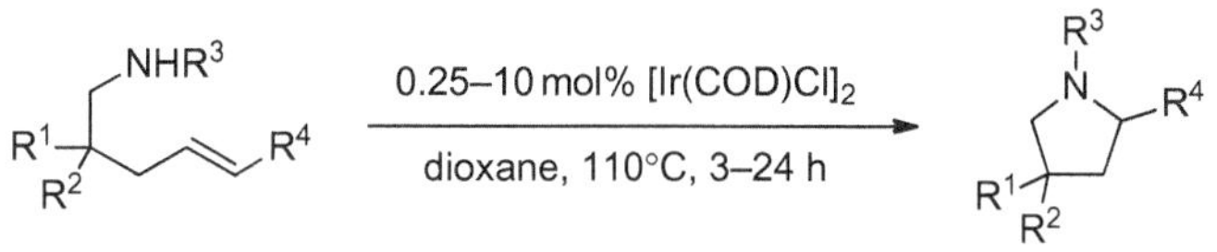

SCHEME 3.28 Ir-catalyzed intramolecular hydroamination of unactivated alkenes with secondary amines.

olefin activation mechanism had not been previously documented for Ir-catalyzed alkene hydroamination.

Moreover, pincer[199] and half-sandwich Iridium complexes[200] also facilitate the cyclization of aminoalkenes.

3.4.4.6 Cu and Fe Catalyst Systems

In 2009, the Cu-catalyzed intramolecular hydroamination of unactivated terminal alkenes bearing primary amines and secondary alkylamines was achieved by Sawamura and coworkers.[201] Utilizing Cu(O-*t*-Bu)/Xantphos as the catalyst, aminoalkenes undergo cyclization to generate pyrrolidine and piperidine derivatives in excellent yields.

A $FeCl_3$-catalyzed hydroamination/cyclization of the Ts-protected aminoalkenes was reported to produce the heterocyclic products in 81%–99% yields.[202] Subsequent investigation extended the scope of aminoalkenes to unprotected primary aminoalkenes with a four-coordinated β-diketiminatoiron(II) alkyl complex.[203]

3.4.5 Acid Catalysts

In 1999, Akiyama and coworkers described the TsOH-catalyzed cyclization of chiral *N*-tosyl-3-silyl-4-pentenamine by way of a 3-silyl carbocation intermediate to furnish 2,3,5-trisubstituted pyrrolidines in enantiomerically pure form with 74%–92% yields (Scheme 3.29).[204] The silyl group incorporated in the organic structure turn out to play a key role in the stereochemical control in this cyclization reaction.

The intramolecular hydroaminations of N-Ts protected alkenylamines are achieved for the synthesis of pyrrolidines and piperidines using TsOH,[205] BF_3Et_2O,[206] and HI[207] as catalysts. Note that aminoalkenes lacking substituents on the alkyl chain can also be cyclized to the hydroamination products with up

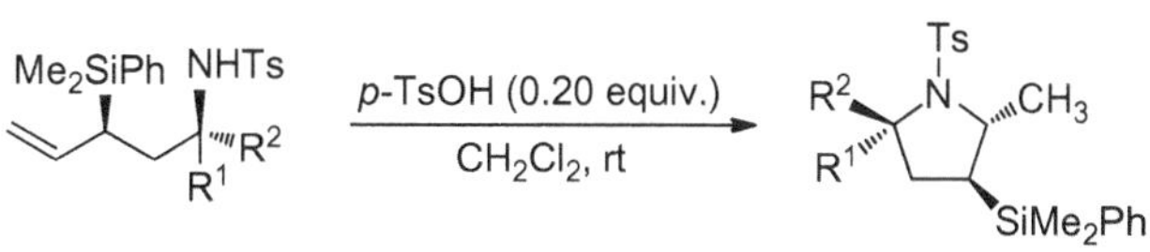

SCHEME 3.29 TsOH-catalyzed cyclization of chiral *N*-tosyl-3-silyl-4-pentenamine.

to 95% yield. Using this strategy, substituted indolines can be synthesized from olefinic sulfonamides with heptadecafluorooctanesulfonic acid ($C_8F_{17}SO_3H$)[208] and $Bi(OTf)_3$[209] as catalysts.

Following an initial report by Roesky and Blechert in the intramolecular hydroamination of unactivated terminal alkenes bearing primary amines with [(^{i}Pr)$_2$ATI]-ZnMe as catalyst,[210] a series of zinc complexes has been prepared and investigated in the intramolecular hydroamination of nonactivated alkenes.[211–217] The results indicate that complexes bearing ligands with cyclic alkyl groups and small size leaving the group on the Zn as well as donor substituents at the backbone of the ligand to show superior activity in these reactions.

Phenylene-diamine-based aluminum complexes[218] and cationic dialkylaluminum complexes[219] have been synthesized as active catalyst for the intramolecular hydroamination of unactivated terminal alkenes bearing primary amines and secondary alkyl amines.

A variety of Brønsted acid, such as TFA, 3-hydroxy-2-naphthoic acid, and 2-(trifluoromethanesulfonamido)benzoic acid, have also been reported as catalysts for the cyclization of gem-dimethyl-substituted secondary alkyl aminoalkenes with moderate to good yields.[220–222]

3.4.6 Other Catalysts

In 2014, Nicewicz and coworkers disclosed a visible light promoted intramolecular anti-Markovnikov hydroamination of aminoalkenes with thiophenol and 9-mesityl-10-methylacridinium tetrafluoroborate at catalysts. The Ts protected alkenylamines undergo cyclization to provide the corresponding pyrrolidines and piperidines in 54%–89% yields (Scheme 3.30).[223]

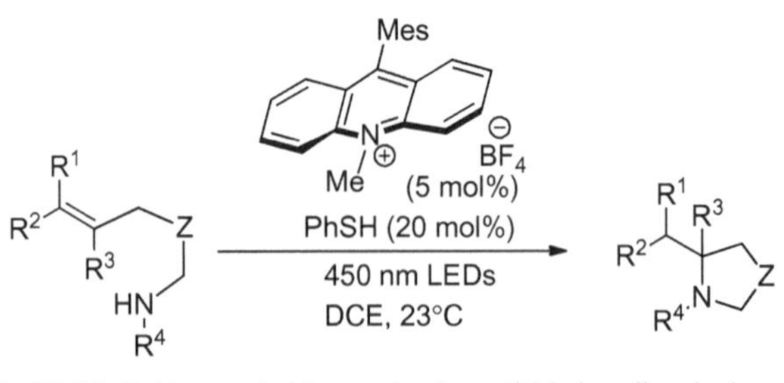

SCHEME 3.30 Visible light promoted intramolecular anti-Markovnikov hydroamination of the aminoalkenes.

3.5 ASYMMETRIC INTRAMOLECULAR HYDROAMINATION OF AMINOALKENES

Chiral pyrrolidine and piperidine derivatives are an important class of organic compounds that can serve as key intermediates for the synthesis of a variety of biologically active molecules. The catalytic asymmetric intramolecular hydroamination of aminoalkenes represents an attractive strategy, offering a waste-free, highly atom efficient pathway to produce these valuable synthetic building blocks. This section will cover the recent advances in the area of stereoselective hydroamination of aminoalkenes and aminodialkenes and is divided according to the type of catalysts used in the reactions. Since the ligand plays an important role in determining the development of alkali metal, alkaline earth metal, rare earth metal-catalyzed asymmetric intramolecular hydroamination of aminoalkenes, each section below will be divided by the type of ligand used in the reactions. The late transition metal catalyzed asymmetric intramolecular hydroamination is divided according to the type of metal used in the reactions.

3.5.1 Alkali Metal and Alkaline Earth Metal-Based Catalysts

3.5.1.1 *N,N*-Ligand

Although it has been known for more than fifty years that alkali metals are hydroamination catalysts, the application of chiral alkali and alkaline earth metal complexes in asymmetric hydroamination of nonactivated aminoalkenes remains limited. The first chiral alkali metal catalyst system for asymmetric hydroamination cyclization of aminoalkenes was based on proline derived diamidobinaphthyl dilithium salt, which is a suitable catalyst for cyclization of aminoalkenes with enantioselectivities up to 74% ee at or below ambient temperatures (Scheme 3.31).[224]

In 2007, a lithium-based catalyst system utilizing chiral bisoxazoline ligand was shown to catalyze the asymmetric hydroamination/cyclization of amino-substituted stilbenes with up to 91% ee.[225] Subsequent investigation extended

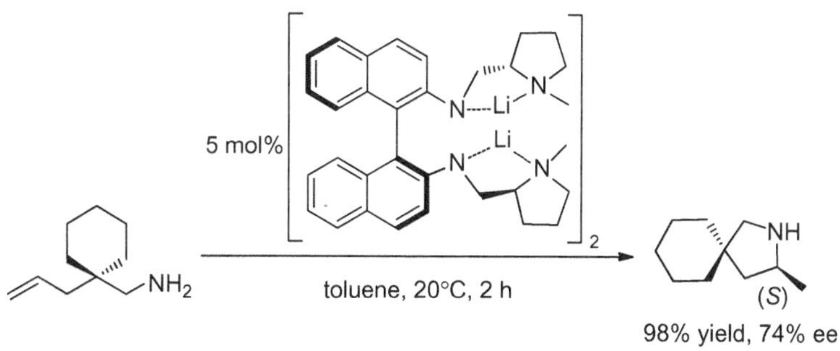

SCHEME 3.31 Lithium-catalyzed asymmetric hydroamination/cyclization of aminoalkenes.

the asymmetric hydroamination/cyclization to aminodienes utilizing chiral diamidobinaphthyl dilithium salts with up to 72% ee.[226]

The first chiral alkaline earth metal-catalyst system was reported by Hultzsch and coworkers in 2009.[227] A magnesium-based catalyst system utilizing proline-modified diamidobinaphthyl ligands has been shown to catalyze the asymmetric hydroamination/cyclization of Thorpe-Ingold-activated aminoalkenes with low selectivity and up to 14% ee. Following this report, a set of chiral 1,2-diamines,[228] tris(oxazolinyl)borato,[229] bisimidazoline,[230] amido-bisoxazoline,[231] and amino-phosphine-/amido-phosphine-based[232] magnesium or calcium complexes were applied in chiral alkaline earth metal catalysts elaborating for asymmetric hydroaminations, which result in enantioselectivities exceeding 50% ee.[231] The low selectivity apparently owing to the presence of facile ligand redistribution processes occurs under catalytic conditions.

3.5.1.2 N,O-Ligand

In 2012, a big breakthrough for asymmetric olefin hydroamination/cyclization was achieved by Hultzsch and coworkers.[233] Much higher enantioselectivities up to 93% ee was achieved for enantioselective hydroamination/cyclization of aminoalkenes with a chiral magnesium (amino)phenolate catalyst (Scheme 3.32). This promising example suggests that alkaline earth metal catalysts can exhibit adequate reactivity and selectivity comparable to rare earth metal systems.

3.5.2 Rare Earth Metal-Based Catalysts

3.5.2.1 Clopentadienyl-Based Ligand

Chiral rare-earth-metal-based catalysts for the asymmetric hydroamination/cyclization of aminoalkenes were first introduced in 1992 by Marks and coworkers (Scheme 3.33).[234,235] The C1-symmetric chiral ansa-lanthanocene complexes with (+)-neomenthyl, (−)-menthyl, or (−)-phenylmenthyl substituents attached to one of the cyclopentadienyl ligands exist in two diastereomeric forms, depending on which diastereotopic face of the cyclopentadienyl ligand

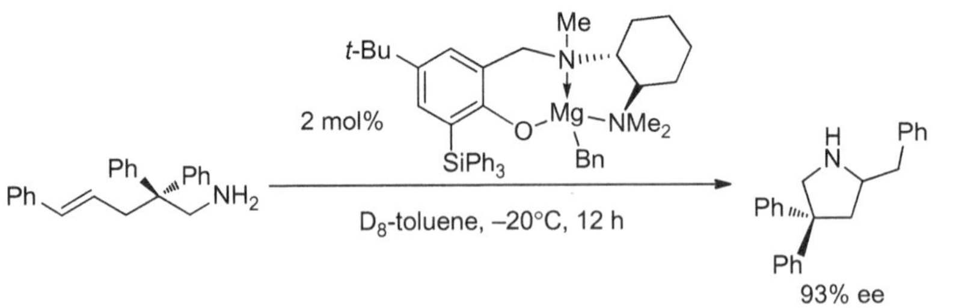

SCHEME 3.32 Asymmetric intramolecular hydroamination/cyclization of aminoalkenes catalyzed by a chiral magnesium (amino)phenolate catalyst.

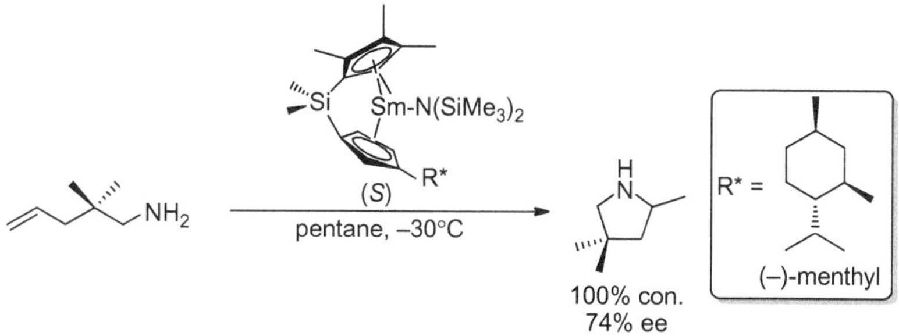

SCHEME 3.33 Chiral lanthanocene precatalysts for asymmetric hydroamination.

coordinates to the metal center. Up to 74% enantioselectivity can be obtained under the optimized reaction conditions.

Unfortunately, the complexes undergo facile epimerization during the catalytic hydroamination via reversible protolytic cleavage of the metal cyclopentadienyl bond.[235–237] Thus, the product enantioselectivity is limited by the catalyst's epimeric ratio in solution and the absolute configuration of the hydroamination product is independent of the diastereomeric purity of the precatalyst. The octahydrofluorenyl ligand in complex is supposed to provide an increased enantiofacial discrimination for prochiral substrates.[236]

Subsequent investigation by Marks and coworkers has been extended to intramolecular hydroamination/cyclization of conjugated aminodienes. Utilizing (*S*)-$Me_2Si(OHF)(CpR^*)SmN(TMS)_2$, the cyclization of conjugated aminodienes produce the (*E*)-olefins with 69% ee in a benzene solution at 0°C.[144,238]

Internal 1,1- or 1,2-disubstituted olefins are much less reactive for hydroamination, except aminoalkenes bearing a phenyl substituent attached to the unsaturated carbon-carbon linkage. The formation of pyrrolidines and piperidines often proceeds at comparable rates, contrasting with the general trend of significant faster five-membered ring formation observed with terminal aminoalkenes, and up to 58% ee at 100°C (up to 68% at 60°C) are observed.[237]

Heteroscorpionate-containing lutetium catalysts have been identified as efficient catalysts for the intramolecular hydroamination of aminoalkenes, and up to 70% enantioselectivities were achieved in the cyclization of the 1,2-disubstituted olefin.[239]

The zwitterionic oxazolinylborato yttrium complex displays good activity and excellent enantioselectivity in the cyclization of a variety of gem-disubstituted aminopentenes (Scheme 3.34).[240] The observed kinetic isotope effect and the isotopic perturbation of enantioselectivity was initially interpreted as an indication of a concerted insertion/protonolysis mechanism; however, a computational study suggests that a stepwise mechanism involving a reversible olefin insertion and subsequent irreversible and rate determining protonolysis prevails energetically.

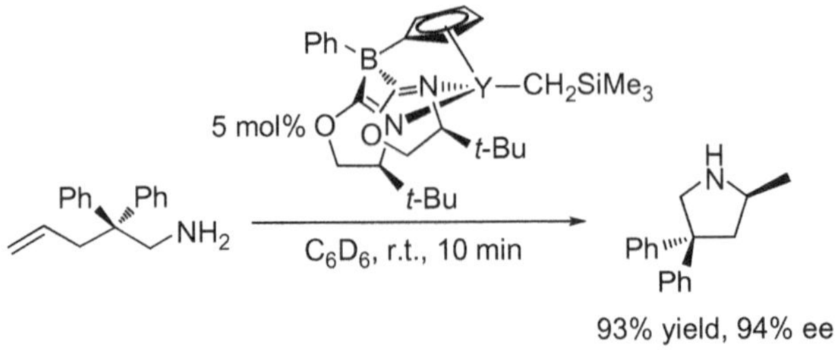

SCHEME 3.34 Hydroamination/cyclization of gem-disubstituted aminopentenes using a chiral zwitterionic yttrium cyclopentadienyl-bis(oxazolidinyl)borate complex.

3.5.2.2 N,N-Ligand

Although significant progress in this area has been made since 2003 utilizing noncyclopentadienyl-based ligand sets, only a few catalyst systems display high catalytic activity as lanthanocene catalysts and show significantly improved enantioselectivities. A variety of bis(oxazolinato)lanthanide complexes served as precatalysts for efficient enantioselective intramolecular hydroamination/cyclization of aminoalkenes and aminodienes.[241] The precatalysts are conveniently generated in situ from $[Ln\{N(SiMe_3)_2\}_3]$ (Ln = La, Nd, Sm, Y, Lu) and the corresponding bisoxazoline, resulting in minimal reactivity loss, but with similar enantioselectivity, in comparison to the isolated precatalysts (Scheme 3.35). A new family of structurally defined heterobimetallic rare earth lithium ate complexes based on *N*-substituted binaphthylamido ligands were discovered that promoted the hydroamination/cyclization of aminoolefins with up to 87% ee.[242–247]

A number of chiral bisaryl or bialkyl ate-complexes have been synthesized and are shown to be efficient catalysts for the enantioselective intramolecular hydroamination of primary amines tethered to sterically demanding alkenes at high reaction temperatures.[248–254] The high activity of the bisalkyl ate complexes (Scheme 3.36a) allow facile cyclization of 1,2-disubstituted and trisubstituted aminoalkenes with enantioselectivities of up to 77% ee at temperatures of 40–110°C.[253]

Moderate to high (69%–89%) enantioselectivities for a wide range of aminoalkene substrates, including internal olefins or secondary amines, are achieved using the aminothiophenolate catalyst system (Scheme 3.36B), which is generated in situ. Enantiomeric excess of up to 89% has been achieved at 30°C, though reactions at this temperature require a long period, i.e., 23 days, to reach completion.[255]

The tetrazane precatalysts may be activated with alkyl lithium reagents (Scheme 3.37), such as *n*-BuLi or $LiCH_2SiMe_3$, to generate a moderately active hydroamination catalyst. Cyclization of Thorpe-Ingold-activated substrates proceeds with up to 90% ee.[256]

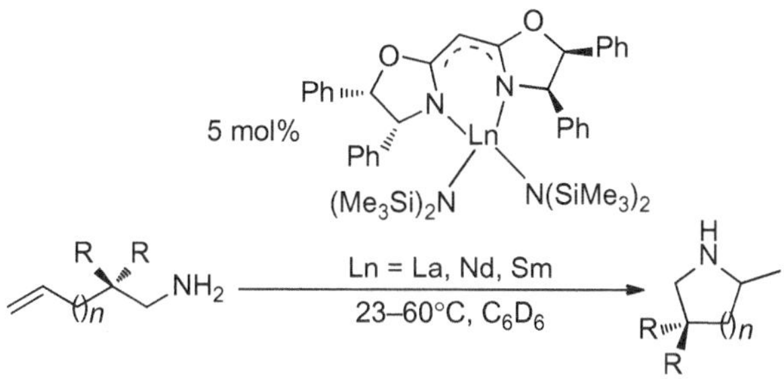

SCHEME 3.35 Bisoxazolinato rare earth metal catalysts in the hydroamination/cyclization of aminoalkenes.

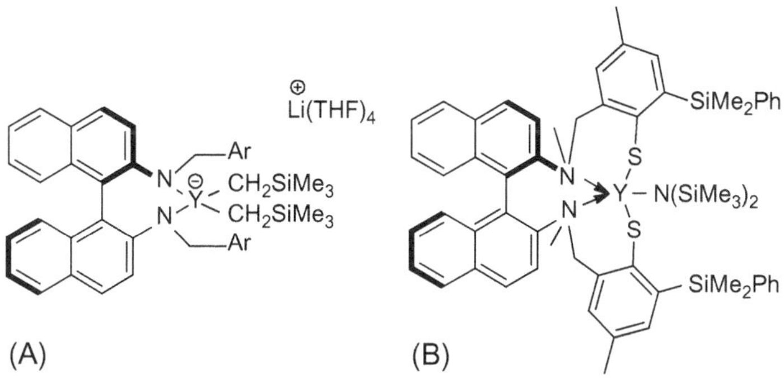

SCHEME 3.36 Diamidobinaphthyl yttrium ate complexes and aminothiophenolate catalyst.

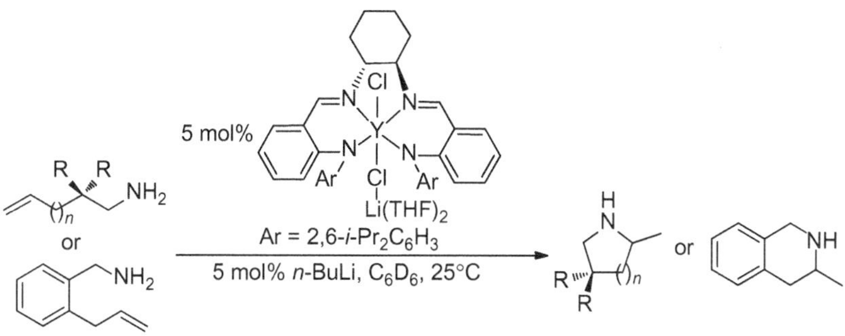

SCHEME 3.37 Asymmetric hydroamination/cyclization of aminoalkenes catalyzed by a tetrazane rare earth metal complex.

3.5.2.3 O,O-Ligand

Significantly higher catalytic activities are achieved when more electron deficient ligand sets are employed. A variety of binaphtholate rare earth metal complexes have been studied with regard to their hydroamination/cyclization catalytic activity.[257–261] Among them, binaphtholate aryl complexes, (Ln = Sc, Y, Lu) with sterically demanding tris(aryl)silyl substituents in the 3 and 3′ position show excellent catalytic activity at room temperature, comparable in magnitude to lanthanocene catalysts, and achieve up to 95% ee in hydroamination/cyclization reactions of aminoalkenes (Scheme 3.38).[258,260] Sterically demanding tris(aryl)silyl substituents in the diolate complexes play a pivotal role not only to achieve high enantioselectivities but also to prevent undesired complex aggregation and reduce detrimental amine binding of the substrate and product to the catalytic active metal centers.

3.5.2.4 *N,O-Ligand*

The NOBIN-derived aminodiolate complexes have been reported to cyclize (2,2-diphenylpent-4-enyl)amine with enantioselectivities of up to 92% ee at 25°C (Scheme 3.39).[123] A variety of other chiral rare-earth complexes[262–265] have also been studied with regard to their hydroamination/cyclization catalytic activity; however, only moderate enantioselectivities of up to 76% ee are achieved.

3.5.3 Group 4/5 Metal-Based Catalysts

3.5.3.1 *O,O-Ligand*

With only a few exceptions, reactivity of Group 4/5 metal-based catalysts is significantly lower than that of rare earth, alkali, and alkaline earth metal-based catalysts. In most instances gem-dialkyl activation of the aminoalkene substrate is required for catalytic turnover. The first chiral Group 4 metal catalyst system for asymmetric hydroamination/cyclization of aminoalkenes was based on the cationic aminophenolate complex.[266] Secondary aminoalkenes react readily to yield hydroamination products with enantioselectivities up to 82% ee. For catalyst solubility reasons, reactions are commonly performed at 100°C in bromobenzene using 10 mol% catalyst loading (Scheme 3.40). The mechanism of this

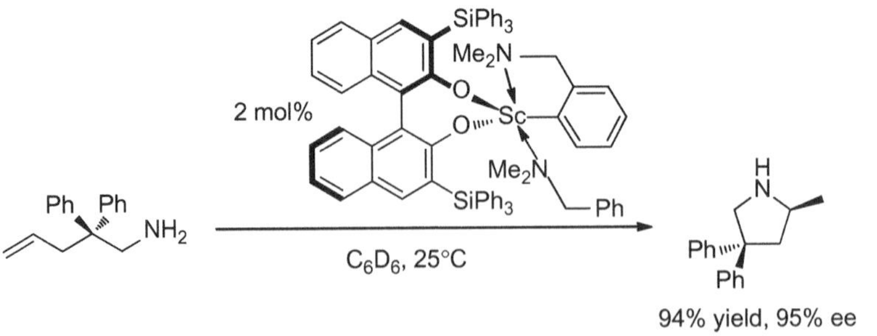

SCHEME 3.38 Asymmetric hydroamination/cyclization of aminoalkenes using 3,3-bis (trisaryl-silyl)-substituted binaphtholate rare earth metal complexes.

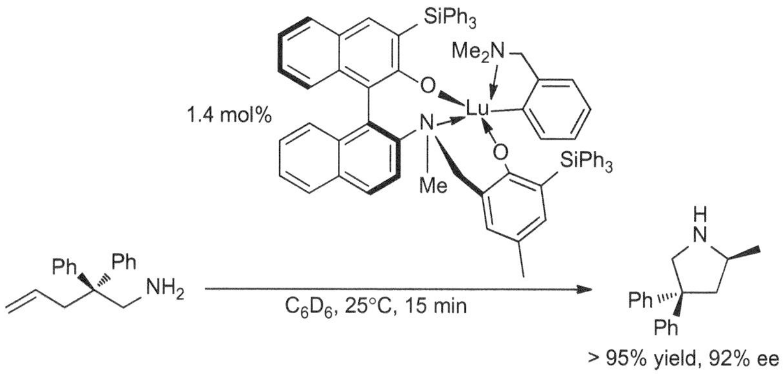

SCHEME 3.39 Asymmetric hydroamination/cyclization of aminoalkenes using NOBIN-derived aminodiolate complexes.

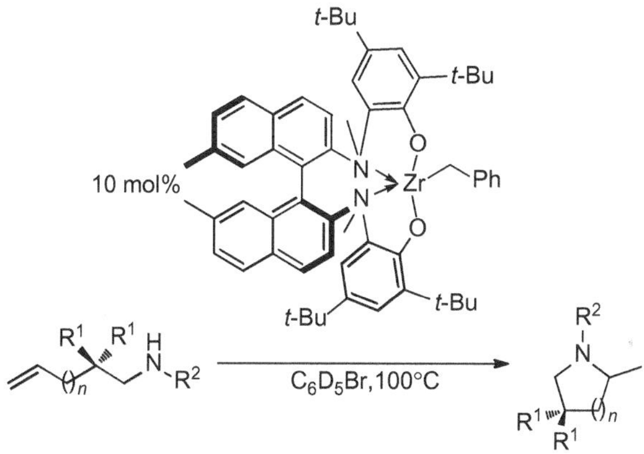

SCHEME 3.40 Hydroamination/cyclization of secondary aminoalkenes using a cationic chiral zirconium catalyst system.

cationic system is thought to proceed similar to the σ-bond metathesis mechanism of rare earth metal-based catalyst systems.

In addition, the half-sandwich zirconium salicyloxazoline complexes have also been studied for cyclization of gem-dimethyl-substituted primary aminoalkenes with enantioselectivities up to 70% ee at 110°C. The significantly higher reactivity of binaphthalenedicarboxamide zirconium complexes allows cyclization of aminoheptene to give the azepane in 60% ee (Scheme 3.41),[267] which may be attributed to the k^1 binding mode of the amidate moieties leading to more electron-deficient metal and the more active catalyst system.[268] The increased reactivity may be utilized to reduce catalyst loadings down to 0.5 mol% at 70°C in the formation of pyrrolidines.

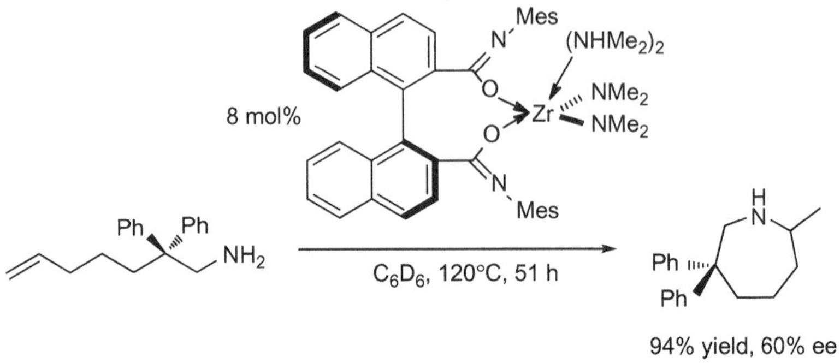

SCHEME 3.41 Enantioselective azepane synthesis via zirconium-catalyzed asymmetric hydroamination/cyclization of an aminoheptene.

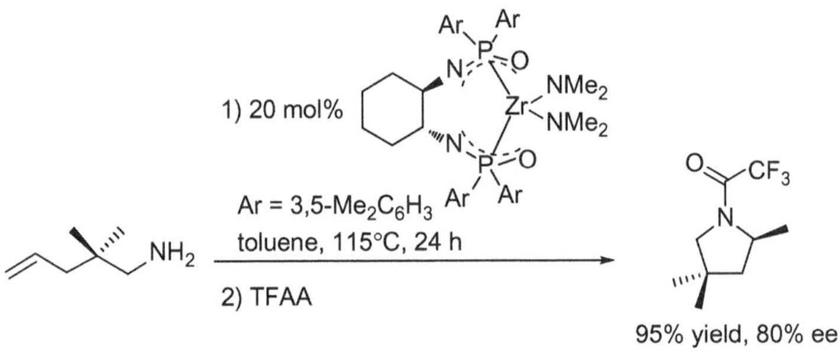

SCHEME 3.42 Hydroamination/cyclization of primary aminoalkenes using a neutral bis(phosphinic amido) zirconium catalyst system.

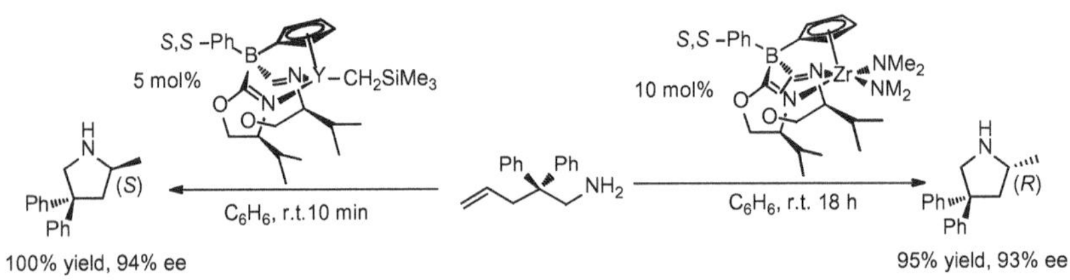

SCHEME 3.43 Hydroamination/cyclization of gem-disubstituted aminoalkenes using a chiral zwitterionic zirconium cyclopentadienyl-bis(oxazolidinyl)borate complex.

The 3,3′-silyl-disubstituted binaphtholate tantalum complex is shown to be an efficient catalyst for the exclusive asymmetric hydroamination of the Thorpe-Ingold-activated aminoalkenes in enantioselectivities up to 81% ee.[269]

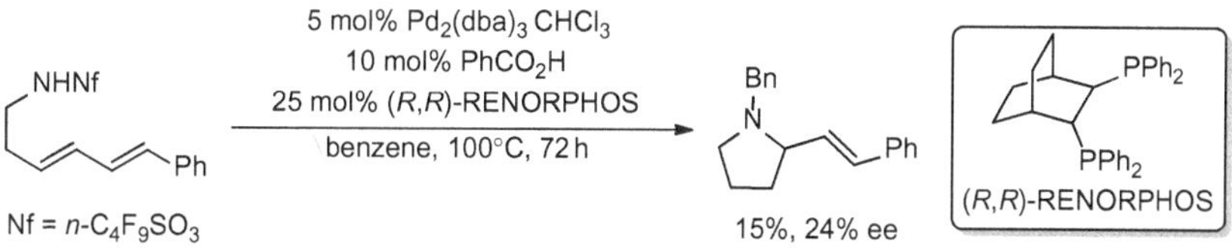

SCHEME 3.44 Palladium-catalyzed asymmetric hydroamination/cyclization of a dienylsulfonamide.

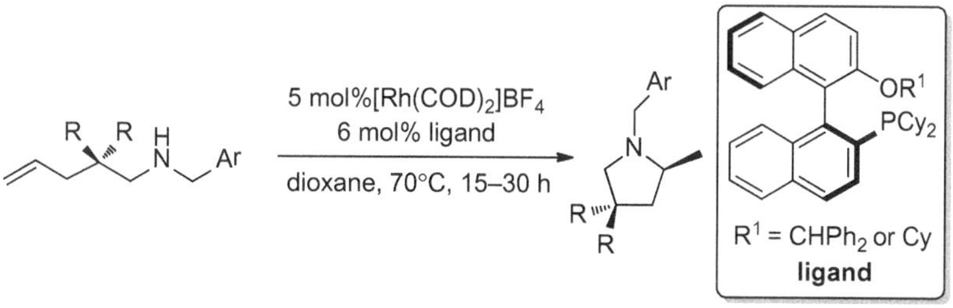

SCHEME 3.45 Rhodium-catalyzed enantioselective hydroamination/cyclization of aminopentenes.

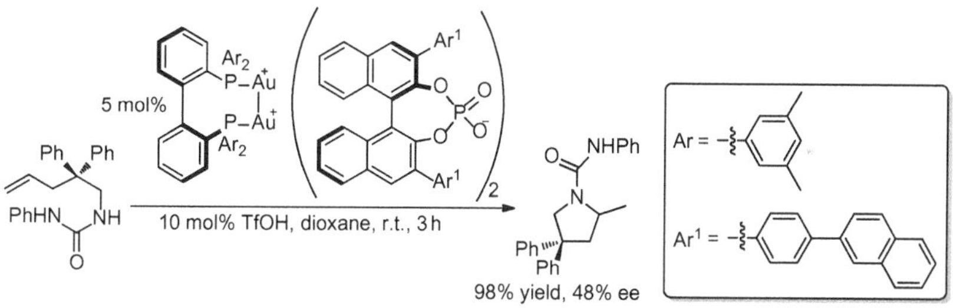

SCHEME 3.46 Asymmetric hydroamination/cyclization of an alkenyl urea with a diastereomerically pure BIPHEP-gold complex.

3.5.3.2 N,N-Ligand

In situ combination of diphosphinic amides and $Zr(NMe_2)_4$ affords active chiral bis(phosphinic amido) zirconium complex, which is a competent catalyst for intramolecular asymmetric alkene hydroamination to provide piperidines and pyrrolidines in up to 80% ee (Scheme 3.42).[270] In contrast to the cationic Group 4 metal hydroamination catalysts, their neutral counterparts will generally react only with primary aminoalkenes. Subsequently, much higher enantioselectivities up to 93% ee were achieved for aminopentene substrates using the chiral bis(amidate) zirconium complex.[271–275]

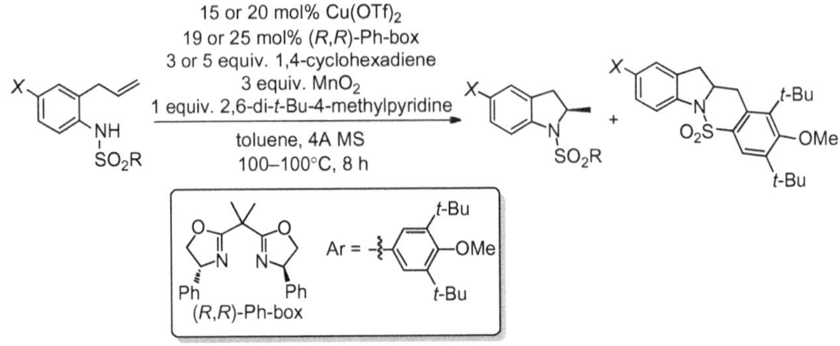

SCHEME 3.47 Copper-catalyzed enantioselective hydroamination/cyclization of *N*-sulfonyl-2-allylanilines.

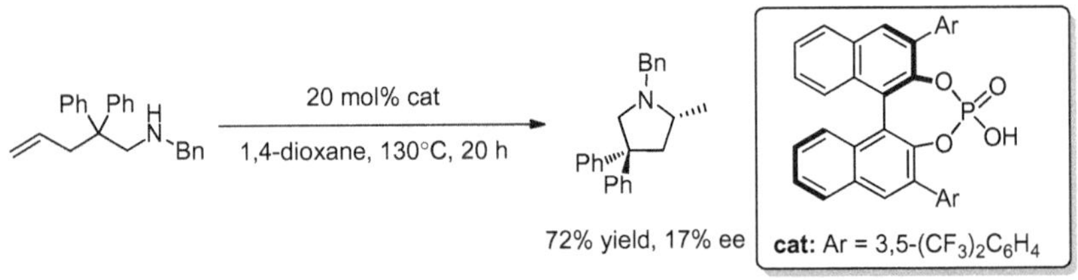

SCHEME 3.48 Organocatalytic asymmetric hydroamination/cyclization of a secondary aminoalkene.

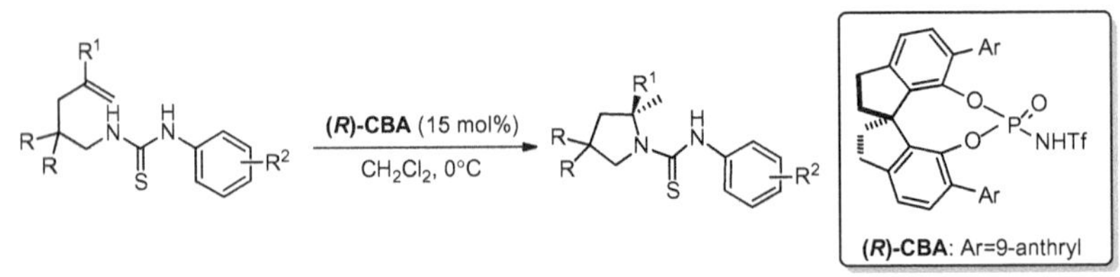

SCHEME 3.49 Chiral phosphoric acid-catalyzed hydroamination of *N*-thioureas-protected Aminoalkenes.

In 2011, an important breakthrough for the asymmetric olefin hydroamination/cyclization was achieved by Sadow and colleagues.[276,277] The high activity of the zwitterionic zirconium cyclopentadienyl-bis(oxazolinyl) borate complex allows facile cyclization of Thorpe-Ingold-activated substrates with enantioselectivities up to 99% ee at temperatures as low as −30°C (Scheme 3.43). It should be noted that 1,2-disubstituted alkenes can also be cyclized at room temperature with 89% enantioselectivity.

3.5.4 Late Transition Metal-Based Catalysts

The limitations of late transition metal catalysts in the intramolecular hydroamination of aminoalkenes pose a significant challenge with respect to enantioselective transformations. Few examples exist.

3.5.4.1 Pd Catalyst System

In 2004, Yamamoto and colleagues introduced the first examples of the cyclization of dienylsulfonamide catalyzed by a chiral palladium phosphine complex (Scheme 3.44).[278] Unfortunately, the resulting enantioselectivities remained low, and a significant excess of the bidentate RENORPHOS ligand was required.

3.5.4.2 Rh Catalyst System

In 2010, Buchwald and coworkers disclosed the first example of rhodium-catalyzed asymmetric intramolecular addition of amines to olefins.[279] Using a catalyst system derived from $[Rh(cod)_2]BF_4$ and binaphthyl-based electron-rich phosphine ligand, a variety of enantioenriched 2-methylpyrrolidines were synthesized with enantioselectivities up to 91% ee (Scheme 3.45). Unfortunately, unprotected primary aminoalkenes and secondary aminohexenes displayed less reactivity.

3.5.4.3 Au Catalyst System

In 2012, enantioselective intramolecular hydroamination of *N*-alkenyl ureas was reported with moderate enantioselectivities utilizing the diasteromerically pure tropos BIPHEP-gold(I) complex at room temperature (Scheme 3.46).[280] It should be noted that the Au-Au interaction can have a pronounced influence on product enantioselectivities.

Following this report, the gold-catalyzed asymmetric intramolecular hydroamination of aminoalkenes has been studied by screening a series of phosphoramidite-[281] and diphosphine-[282,283]based gold complexes in combination with silver salts. Among them, a mixture of diphosphine $(AuCl)_2$ and $AgBF_4$ displays good enantioselectivity in the cyclization of a variety of gem-disubstituted alkenes with carbamates and ureas to form pyrrolidine derivatives with up to 85% ee.[283]

In addition, gold catalysts are also quite efficient in the enantioselective cyclization of aminodienes.[284] The efficiency and regioselectivity in favor of the (*E*)-alkenylpyrrolidines and (*E*)-alkenylpiperidines are significantly improved with the addition of stoichiometric amounts of a (−)-menthol.

3.5.4.4 Cu Catalyst System

Moreover, copper-catalyzed enantioselective alkene hydroamination/cyclization of *N*-sulfonyl-2-allylanilines has also been achieved.

The 2-methylindolines are obtained with up to 90% ee using bisoxazoline copper(II) triflate as catalyst and stoichiometric amounts of cyclohexadiene as hydrogen donor, MnO_2 as oxidant (Scheme 3.47).[285] However, formation of the carbamination by-product for arylsulfonamide substrates and the necessity to use stoichiometric amounts reagents, significantly reduce the appeal of this protocol.

The reaction is believed to proceed via addition of the sulfonamide nitrogen to the alkene in a *cis*-aminocupration transition state.[286] The resulting unstable copper(II)-alkyl undergoes homolysis. Subsequently the primary organic radical is trapped by hydrogen atom abstraction from 1,4-cyclohexadiene and the copper(I) species is reoxidized by MnO_2.

3.5.5 Acid Catalyst Systems

In 2008, Althammer and coworkers described the chiral phosphoric acid diester catalyzed asymmetric hydroamination of an *N*-benzyl-protected aminopentene (Scheme 3.48).[287] It represents the first example of a chiral acid catalyst for asymmetric intramolecular hydroamination reactions of aminoalkenes. Although both selectivity and catalytic activity for the hydroamination were rather low, the method itself remains very promising.

Seven years later, a chiral phosphoric acid-catalyzed hydroamination of *N*-thioureas protected aminoalkenes was reported to provide a variety of pyrrolidine derivatives with α-tetrasubstituted carbon stereocenter with enantioselectivities up to 96% ee (Scheme 3.49).[288] The crucial factor to success in high levels of stereoinduction is the use of an athiourea group acting as both the activating and the directing group through cooperative multiple hydrogen bonding with a Brønsted acid and the double bond.

The chiral dithiophosphoric acids-catalyzed hydroamination of *N*-sulfonyl-protected aminodienes has been achieved to afford five-membered rings with enantioselectivities as high as 99% ee.[289] The dithiophosphoric acid-catalyzed reaction proceeds via an initial *syn*-addition of the dithiophosphoric acid to the distal alkene. The resulting allylic dithiophosphate ester intermediate undergoes intramolecular *syn*-S_N2'-type substitution by the sulfonylamide nucleophile, displacing and releasing the dithiophosphoric acid catalyst. The high degree of organization of this cyclization transition state is responsible for the high level of enantioselectivity.

3.6 CONCLUSIONS AND OUTLOOK

This chapter summarizes developments in the field of intermolecular and intramolecular hydroamination of alkenes over the past decades. A great number of catalyst systems have been reported and they show excellent catalytic properties in hydroamination with yields exceeding 90%. Moreover, many asymmetric hydroaminations can now be performed with enantioselectivities exceeding

90% ee. These elegant procedures provide a facile and highly atom-economical strategy for the synthesis of amines.

Although significant progress has been made in this field, many challenges still remain to be addressed. First of all, the majority of these catalyst systems are confined to a limited set of substrates and fewer catalyst systems have been reported to combine high catalytic activity and consistent high level of enantioselectivity for a wide range of substrates. Moreover, the development of efficient protocols for intermolecular and intramolecular hydroamination reaction of more hindered trisubstituted or tetrasubstituted internal olefin with amines remains a challenge. Furthermore, selective anti-Markovnikov hydroamination reactions of terminal nonactivated alkenes have not yet been achieved. Finally, further mechanistic studies and deep understandings of the underlying mechanistic steps of these transformations would be welcome and will continue to play an important role in the rational design of more effective catalysts for both the inter- and intramolecular alkene hydroamination reactions.

REFERENCES

1. Gasc, M. B.; Lattes, A.; Perie, J. J. *Tetrahedron* **1983**, *39*, 703.
2. Brunet, J. J.; Neibecker, D.; Niedercorn, F. *J. Mol. Catal.* **1989**, *49*, 235.
3. Muller, T. E.; Beller, M. *Chem. Rev.* **1998**, *98*, 675.
4. Nobis, M.; Driessen-Holscher, B. *Angew. Chem. Int. Ed.* **2001**, *40*, 3983.
5. Seayad, J.; Tillack, A.; Hartung, C. G.; Beller, M. *Adv. Synth. Catal.* **2002**, *344*, 795.
6. Pohlki, F.; Doye, S. *Chem. Soc. Rev.* **2003**, *32*, 104.
7. Roesky, P. W.; Muller, T. E. *Angew. Chem. Int. Ed.* **2003**, *42*, 2708.
8. Beller, M.; Seayad, J.; Tillack, A.; Jiao, H. *Angew. Chem. Int. Ed.* **2004**, *43*, 3368.
9. Hultzsch, K. C. *Adv. Synth. Catal.* **2005**, *347*, 367.
10. Hultzsch, K. C. *Org. Biomol. Chem.* **2005**, *3*, 1819.
11. Aillaud, I.; Collin, J.; Hannedouche, J.; Schulz, E. *Dalton Trans.* **2007**, 5105.
12. Brunet, J.-J.; Chu, N.-C.; Rodriguez-Zubiri, M. *Eur. J. Inorg. Chem.* **2007**, *2007*, 4711.
13. Severin, R.; Doye, S. *Chem. Soc. Rev.* **2007**, *36*, 1407.
14. Muller, T. E.; Hultzsch, K. C.; Yus, M.; Foubelo, F.; Tada, M. *Chem. Rev.* **2008**, *108*, 3795.
15. Hesp, K. D.; Stradiotto, M. *ChemCatChem* **2010**, *2*, 1192.
16. Taylor, J. G.; Adrio, L. A.; Hii, K. K. *Dalton Trans.* **2010**, *39*, 1171.
17. Nishina, N.; Yamamoto, Y. *Top. Organomet. Chem.* **2012**, *43*, 115.
18. Hannedouche, J.; Schulz, E. *Chem. Eur. J.* **2013**, *19*, 4972.
19. Reznichenko, A. L.; Nawara-Hultzsch, A. J.; Hultzsch, K. C. *Top. Curr. Chem.* **2014**, *343*, 191.
20. Bernoud, E.; Lepori, C.; Mellah, M.; Schulz, E.; Hannedouche, J. *Catal. Sci. Technol* **2015**, *5*, 2017.
21. Huang, L.; Arndt, M.; Goossen, K.; Heydt, H.; Goossen, L. J. *Chem. Rev.* **2015**, *115*, 2596.
22. Pirnot, M. T.; Wang, Y. M.; Buchwald, S. L. *Angew. Chem. Int. Ed.* **2016**, *55*, 48.
23. Wang, J.; Cui, D. *Chin. J. Org. Chem.* **2016**, *36*, 1163.
24. Steinborn, D.; Taube, R. *Z. Chem.* **1986**, *26*, 349.
25. Hii, K. K. *Pure Appl. Chem.* **2006**, *78*, 341.
26. Reznichenko, A. L.; Nawara-Hultzsch, A. J.; Hultzsch, K. C. *Topics in current chemistry*. Springer: Berlin Heidelberg, 2013; p. 191.

27. Cooper, N. J.; Knight, D. W. *Tetrahedron* **2004**, *60*, 243.
28. Beauchemin, A. M. *Org. Biomol. Chem.* **2013**, *11*, 7039.
29. Howk, B. W.; Little, E. L.; Scott, S. L.; Whitman, G. M. *J. Am. Chem. Soc.* **1954**, *76*, 1899.
30. Pez, G. P.; Galle, J. E. *Pure Appl. Chem.* **1985**, *57*, 1917.
31. Khedkar, V.; Tillack, A.; Benisch, C.; Melder, J.-P.; Beller, M. *J. Mol. Catal. A Chem.* **2005**, *241*, 175.
32. Horrillo-Martínez, P.; Hultzsch, K. C.; Gil, A.; Branchadell, V. *Eur. J. Org. Chem.* **2007**, *2007*, 3311.
33. Closson, R. D.; Napolitano, J. P.; Ecke, G. G.; Kolka, A. J. *J. Org. Chem.* **1957**, *22*, 646.
34. Schlott, R. J.; Falk, J. C.; Narducy, K. W. *J. Org. Chem.* **1972**, *37*, 4243.
35. Beller, M.; Breindl, C. *Tetrahedron* **1998**, *54*, 6359.
36. Beller, M.; Breindl, C.; Riermeier, T. H.; Eichberger, M.; Trauthwein, H. *Angew. Chem. Int. Ed.* **1998**, *37*, 3389.
37. Hartung, C. G.; Breindl, C.; Tillack, A.; Beller, M. *Tetrahedron* **2000**, *56*, 5157.
38. Kumar, K.; Michalik, D.; Garcia Castro, I.; et al. *Chem. Eur. J.* **2004**, *10*, 746.
39. Tzalis, D.; Koradin, C.; Knochel, P. *Tetrahedron Lett.* **1999**, *40*, 6193.
40. Barrett, A. G.; Brinkmann, C.; Crimmin, M. R.; Hill, M. S.; Hunt, P.; Procopiou, P. A. *J. Am. Chem. Soc.* **2009**, *131*, 12906.
41. Brinkmann, C.; Barrett, A. G.; Hill, M. S.; Procopiou, P. A. *J. Am. Chem. Soc.* **2012**, *134*, 2193.
42. Liu, B.; Roisnel, T.; Carpentier, J. F.; Sarazin, Y. *Angew. Chem. Int. Ed.* **2012**, *51*, 4943.
43. Li, Y. W.; Marks, T. J. *Organometallics* **1996**, *15*, 3770.
44. Ryu, J. S.; Li, G. Y.; Marks, T. J. *J. Am. Chem. Soc.* **2003**, *125*, 12584.
45. Kissel, A. A.; Mahrova, T. V.; Lyubov, D. M.; et al. *Dalton Trans.* **2015**, *44*, 12137.
46. Coulson, D. R. *Tetrahedron Lett.* **1971**, 429.
47. Diamond, S. E.; Szalkiewicz, A.; Mares, F. *J. Am. Chem. Soc.* **1979**, *101*, 490.
48. Baudequin, C.; Brunet, J. J.; Rodriguez-Zubiri, M. *Organometallics* **2007**, *26*, 5264.
49. Ickes, A. R.; Ensign, S. C.; Gupta, A. K.; Hull, K. L. *J. Am. Chem. Soc.* **2014**, *136*, 11256.
50. Ensign, S. C.; Vanable, E. P.; Kortman, G. D.; Weir, L. J.; Hull, K. L. *J. Am. Chem. Soc.* **2015**, *137*, 13748.
51. Brunet, J.-J.; Neibecker, D.; Philippot, K. *J. Chem. Soc. Chem. Commun.* **1992**, 1215.
52. Brunet, J.-J.; Commenges, G.; Neibecker, D.; Philippot, K. *J. Organomet. Chem.* **1994**, *469*, 221.
53. Brunet, J.-J.; Chu, N. C.; Diallo, O.; Mothes, E. *J. Mol. Catal. A Chem.* **2003**, *198*, 107.
54. Brunet, J.-J.; Neibecker, D.; Philippot, K. *Tetrahedron Lett.* **1993**, *34*, 3877.
55. Beller, M.; Trauthwein, H.; Eichberge, M.; Breindl, C.; Herwig, J.; Müller, T. E.; Thiel, O. R. *Chem. Eur. J* **1999**, *5*, 1306.
56. Beller, M.; Trauthwein, H.; Eichberger, M.; Breindl, C.; Muller, T. E. *Eur. J. Inorg. Chem* **1999**, *1999*, 1121.
57. Utsunomiya, M.; Kuwano, R.; Kawatsura, M.; Hartwig, J. F. *J. Am. Chem. Soc.* **2003**, *125*, 5608.
58. Akermark, B.; Backvall, J. E.; Hegedus, L. S.; Zetterbe, K.; Siiralah, K.; Sjoberg, K. *J. Organomet. Chem.* **1974**, *72*, 127.
59. Hegedus, L. S.; Akermark, B.; Zetterberg, K.; Olsson, L. F. *J. Am. Chem. Soc.* **1984**, *106*, 7122.
60. Gurak, J. A., Jr.; Yang, K. S.; Liu, Z.; Engle, K. M. *J. Am. Chem. Soc.* **2016**, *138*, 5805.
61. Armbruster, R. W.; Morgan, M. M.; Schmidt, J. L.; et al. *Organometallics* **1986**, *5*, 234.
62. Lober, O.; Kawatsura, M.; Hartwig, J. F. *J. Am. Chem. Soc.* **2001**, *123*, 4366.
63. Minami, T.; Okamoto, H.; Ikeda, S.; Tanaka, R.; Ozawa, F.; Yoshifuji, M. *Angew. Chem. Int. Ed.* **2001**, *40*, 4501.

64. Behr, A.; Johnen, L.; Rentmeister, N. *Adv. Synth. Catal.* **2010**, *352*, 2062.
65. Baig, T.; Jenck, J.; Kalck, P. *J. Chem. Soc. Chem. Commun.* **1992**, 1552.
66. Cheng, X. H.; Hii, K. K. *Tetrahedron* **2001**, *57*, 5445.
67. Kawatsura, M.; Hartwig, J. F. *J. Am. Chem. Soc.* **2000**, *122*, 9546.
68. Nettekoven, U.; Hartwig, J. *J. Am. Chem. Soc.* **2002**, *124*, 1166.
69. Johns, A. M.; Utsunomiya, M.; Incarvito, C. D.; Hartwig, J. F. *J. Am. Chem. Soc.* **2006**, *128*, 1828.
70. Li, K. L.; Horton, P. N.; Hursthouse, M. B.; Hii, K. K. *J. Organomet. Chem.* **2003**, *665*, 250.
71. Casalnuovo, A. L.; Calabrese, J. C.; Milstein, D. *J. Am. Chem. Soc.* **1988**, *110*, 6738.
72. Sevov, C. S.; Zhou, J.; Hartwig, J. F. *J. Am. Chem. Soc.* **2012**, *134*, 11960.
73. Sevov, C. S.; Zhou, J. S.; Hartwig, J. F. *J. Am. Chem. Soc.* **2014**, *136*, 3200.
74. Schaffrath, H.; Keim, W. *J. Mol. Catal. A Chem.* **2001**, *168*, 9.
75. Ovchinnikov, M. V.; LeBlanc, E.; Guzei, I. A.; Angelici, R. J. *J. Am. Chem. Soc.* **2001**, *123*, 11494.
76. Utsunomiya, M.; Hartwig, J. F. *J. Am. Chem. Soc.* **2004**, *126*, 2702.
77. Takaya, J.; Hartwig, J. F. *J. Am. Chem. Soc.* **2005**, *127*, 5756.
78. Yi, C. S.; Yun, S. Y. *Org. Lett.* **2005**, *7*, 2181.
79. Pawlas, J.; Nakao, Y.; Kawatsura, M.; Hartwig, J. F. *J. Am. Chem. Soc.* **2002**, *124*, 3669.
80. Brunet, J. J.; Cadena, M.; Chu, N. C.; Diallo, O.; Jacob, K.; Mothes, E. *Organometallics* **2004**, *23*, 1264.
81. Brunet, J.-J.; Chu, N. C.; Diallo, O. *Organometallics* **2005**, *24*, 3104.
82. Rodriguez-Zubiri, M.; Anguille, S.; Brunet, J.-J. *J. Mol. Catal. A Chem.* **2007**, *271*, 145.
83. Dub, P. A.; Béthegnies, A.; Poli, R. *Organometallics* **2012**, *31*, 294.
84. McBee, J. L.; Bell, A. T.; Tilley, T. D. *J. Am. Chem. Soc.* **2008**, *130*, 16562.
85. Dub, P. A.; Rodriguez-Zubiri, M.; Daran, J.-C.; Brunet, J.-J.; Poli, R. *Organometallics* **2009**, *28*, 4764.
86. Dub, P. A.; Poli, R. *J. Am. Chem. Soc.* **2010**, *132*, 13799.
87. Wang, X.; Widenhoefer, R. A. *Organometallics* **2004**, *23*, 1649.
88. Karshtedt, D.; Bell, A. T.; Tilley, T. D. *J. Am. Chem. Soc.* **2005**, *127*, 12640.
89. Cao, P.; Cabrera, J.; Padilla, R.; Serra, D.; Rominger, F.; Limbach, M. *Organometallics* **2012**, *31*, 921.
90. Qian, H.; Widenhoefer, R. A. *Org. Lett.* **2005**, *7*, 2635.
91. Zhang, J.; Yang, C.-G.; He, C. *J. Am. Chem. Soc.* **2006**, *128*, 1798.
92. Brouwer, C.; He, C. *Angew. Chem. Int. Ed.* **2006**, *45*, 1744.
93. Giner, X.; Najera, C. *Org. Lett.* **2008**, *10*, 2919.
94. Kovacs, G.; Ujaque, G.; Lledos, A. *J. Am. Chem. Soc.* **2008**, *130*, 853.
95. Giner, X.; Najera, C.; Kovacs, G.; Lledos, A.; Ujaque, G. *Adv. Synth. Catal.* **2011**, *353*, 3451.
96. Deeba, M.; Ford, M. E.; Johnson, T. A. *J. Chem. Soc. Chem. Commun.* **1987**, 562.
97. Deeba, M.; Ford, M. E. *J. Org. Chem.* **1988**, *53*, 4594.
98. Mizuno, N.; Tabata, M.; Uematsu, T.; Iwamoto, M. *J. Chem. Soc. Faraday Trans.* **1993**, *89*, 3513.
99. Mizuno, N.; Tabata, M.; Uematsu, T.; Iwamoto, M. *J. Catal.* **1994**, *146*, 249.
100. Motokura, K.; Nakagiri, N.; Mori, K.; et al. *Org. Lett.* **2006**, *8*, 4617.
101. Huang, J.-M.; Wong, C.-M.; F-X, X.; Loh, T.-P. *Tetrahedron Lett.* **2007**, *48*, 3375.
102. Michaux, J.; Terrasson, V.; Marque, S.; Wehbe, J.; Prim, D.; Campagne, J.-M. *Eur. J. Org. Chem.* **2007**, *2007*, 2601.
103. Wei, H.; Qian, G.; Xia, Y.; Li, K.; Li, Y.; Li, W. *Eur. J. Org. Chem.* **2007**, *2007*, 4471.
104. Cheng, X.; Xia, Y.; Wei, H.; et al. *Eur. J. Org. Chem.* **2008**, *2008*, 1929.

105. Ackermann, L.; Kaspar, L. T.; Gschrei, C. J. *Org. Lett.* **2004**, *6*, 2515.
106. Kaspar, L. T.; Fingerhut, B.; Ackermann, L. *Angew. Chem. Int. Ed.* **2005**, *44*, 5972.
107. Taylor, J. G.; Whittall, N.; Hii, K. K. *Org. Lett.* **2006**, *8*, 3561.
108. Munro-Leighton, C.; Delp, S. A.; Alsop, N. M.; Blue, E. D.; Gunnoe, T. B. *Chem. Commun.* **2008**, 111.
109. Fei, J.; Wang, Z.; Cai, Z.; Sun, H.; Cheng, X. *Adv. Synth. Catal.* **2015**, *357*, 4063.
110. Moran, J.; Cebrowski, P. H.; Beauchemin, A. M. *J. Am. Chem. Soc.* **2008**, *73*, 1004.
111. Liu, P. N.; Xia, F.; Zhao, Z. L.; Wang, Q. W.; Ren, Y. J. *Tetrahedron Lett.* **2011**, *52*, 6113.
112. Ng, E.-P.; Law, S.-P.; Mukti, R. R.; Juan, J.-C.; Adam, F. *Chem. Eng. J.* **2014**, *243*, 99.
113. Anderson, L. L.; Arnold, J.; Bergman, R. G. *J. Am. Chem. Soc.* **2005**, *127*, 14542.
114. Lapis, A. A. M.; DaSilveira Neto, B. A.; Scholten, J. D.; Nachtigall, F. M.; Eberlin, M. N.; Dupont, J. *Tetrahedron Lett.* **2006**, *47*, 6775.
115. Qin, H.; Yamagiwa, N.; Matsunaga, S.; Shibasaki, M. *J. Am. Chem. Soc.* **2006**, *128*, 1611.
116. Yang, L.; L-W, X.; Xia, C.-G. *Tetrahedron Lett.* **2008**, *49*, 2882.
117. Qureshi, Z. S.; Deshmukh, K. M.; Tambade, P. J.; Dhake, K. P.; Bhanage, B. M. *Eur. J. Org. Chem.* **2010**, *2010*, 6233.
118. Gb, K.; Lledós, A.; Ujaque, G. *Organometallics* **2010**, *29*, 5919.
119. Talluri, S. K.; Sudalai, A. *Org. Lett.* **2005**, *7*, 855.
120. Yadav, J. S.; Reddy, B. V. S.; Rao, T. S.; Krishna, B. B. M. *Tetrahedron Lett.* **2009**, *50*, 5351.
121. Nguyen, T. M.; Manohar, N.; Nicewicz, D. A. *Angew. Chem. Int. Ed.* **2014**, *53*, 6198.
122. Reznichenko, A. L.; Nguyen, H. N.; Hultzsch, K. C. *Angew. Chem. Int. Ed.* **2010**, *49*, 8984.
123. Reznichenko, A. L.; Hultzsch, K. C. *Organometallics* **2013**, *32*, 1394.
124. Dorta, R.; Egli, P.; Zurcher, F.; Togni, A. *J. Am. Chem. Soc.* **1997**, *119*, 10857.
125. Zhou, J. S.; Hartwig, J. F. *J. Am. Chem. Soc.* **2008**, *130*, 12220.
126. Pan, S.; Endo, K.; Shibata, T. *Org. Lett.* **2012**, *14*, 780.
127. Hu, A.; Ogasawara, M.; Sakamoto, T.; et al. *Adv. Synth. Catal.* **2006**, *348*, 2051.
128. Utsunomiya, M.; Hartwig, J. F. *J. Am. Chem. Soc.* **2003**, *125*, 14286.
129. Zhang, Z.; Lee, S. D.; Widenhoefer, R. A. *J. Am. Chem. Soc.* **2009**, *131*, 5372.
130. Fujita, H.; Tokuda, M.; Nitta, M.; Suginome, H. *Tetrahedron Lett.* **1992**, *33*, 6359.
131. Ates, A.; Quinet, C. *Eur. J. Org. Chem.* **2003**, *2003*, 1623.
132. Quinet, C.; Jourdain, P.; Hermans, C.; Ates, A.; Lucas, I.; Markó, I. E. *Tetrahedron* **2008**, *64*, 1077.
133. Crimmin, M. R.; Casely, I. J.; Hill, M. S. *J. Am. Chem. Soc.* **2005**, *127*, 2042.
134. Arrowsmith, M.; Crimmin, M. R.; Barrett, A. G. M.; Hill, M. S.; Kociok-Köhn, G.; Procopiou, P. A. *Organometallics* **2011**, *30*, 1493.
135. Zhang, X.; Emge, T. J.; Hultzsch, K. C. *Organometallics* **2010**, *29*, 5871.
136. Gallegos, C.; Camacho, R.; Valiente, M.; Cuenca, T.; Cano, J. *Catal. Sci. Technol.* **2016**, *6*, 5134.
137. Tobisch, S. *Chem. Eur. J.* **2015**, *21*, 6765.
138. Zhang, X.; Tobisch, S.; Hultzsch, K. C. *Chem. Eur. J.* **2015**, *21*, 7841.
139. Gagne, M. R.; Marks, T. J. *J. Am. Chem. Soc.* **1989**, *111*, 4108.
140. Gagne, M. R.; Nolan, S. P.; Marks, T. J. *Organometallics* **1990**, *9*, 1716.
141. Li, Y. W.; Marks, T. J. *J. Am. Chem. Soc.* **1998**, *120*, 1757.
142. Gilbert, A. T.; Davis, B. L.; Emge, T. J.; Broene, R. D. *Organometallics* **1999**, *18*, 2125.
143. Molander, G. A.; Dowdy, E. D.; Pack, S. K. *J. Am. Chem. Soc.* **2001**, *66*, 4344.
144. Hong, S.; Kawaoka, A. M.; Marks, T. J. *J. Am. Chem. Soc.* **2003**, *125*, 15878.
145. Molander, G. A.; Dowdy, E. D. *J. Org. Chem.* **1998**, *63*, 8983.
146. Molander, G. A.; Dowdy, E. D. *J. Org. Chem.* **1999**, *64*, 6515.

147. Tian, S.; Arredondo, V. M.; Stern, C. L.; Marks, T. J. *Organometallics* **1999**, *18*, 2568.
148. Kim, Y. K.; Livinghouse, T.; Bercaw, J. E. *Tetrahedron Lett.* **2001**, *42*, 2933.
149. Ryu, J.-S.; Marks, T. J.; McDonald, F. E. *Org. Lett.* **2001**, *3*, 3091.
150. Stubbert, B. D.; Marks, T. J. *J. Am. Chem. Soc.* **2007**, *129*, 4253.
151. Chai, Z.; Hua, D.; Li, K.; Zhou, S.; Chu, J.; Yang, G. *J. Organomet. Chem.* **2014**, *768*, 136.
152. Kim, Y. K.; Livinghouse, T. *Angew. Chem. Int. Ed.* **2002**, *41*, 3645.
153. Kim, Y. K.; Livinghouse, T.; Horino, Y. *J. Am. Chem. Soc.* **2003**, *125*, 9560.
154. Hultzsch, K. C.; Hampel, F.; Wagner, T. *Organometallics* **2004**, *23*, 2601.
155. Lauterwasser, F.; Hayes, P. G.; Brase, S.; Piers, W. E.; Schafer, L. L. *Organometallics* **2004**, *23*, 2234.
156. Kim, J. Y.; Livinghouse, T. *Org. Lett.* **2005**, *7*, 4391.
157. Gagne, M. R.; Stern, C. L.; Marks, T. J. *J. Am. Chem. Soc.* **1992**, *114*, 275.
158. Alessandro Motta, G. L.; Fragala, I. L.; Marks, T. J. *Organometallics* **2004**, *37*, 673.
159. Miura, K.; Hondo, T.; Nakagawa, T.; Takahashi, T.; Hosomi, A. *Org. Lett.* **2000**, *2*, 385.
160. Bexrud, J. A.; Beard, J. D.; Leitch, D. C.; Schafer, L. L. *Org. Lett.* **2005**, *7*, 1959.
161. Ackermann, L.; Kaspar, L. T.; Althammer, A. *Org. Biomol. Chem* **2007**, *5*, 1975.
162. Gribkov, D. V.; Hultzsch, K. C. *Angew. Chem. Int. Ed.* **2004**, *43*, 5542.
163. Müller, C.; Loos, C.; Schulenberg, N.; Doye, S. *Eur. J. Org. Chem.* **2006**, *2006*, 2499.
164. Gräbe, K.; Pohlki, F.; Doye, S. *Eur. J. Org. Chem.* **2008**, *2008*, 4815.
165. Janssen, T.; Severin, R.; Diekmann, M.; et al. *Organometallics* **2010**, *29*, 1806.
166. Müller, C.; Saak, W.; Doye, S. *Eur. J. Org. Chem.* **2008**, *2008*, 2731.
167. Kim, H.; Lee, P. H.; Livinghouse, T. *Chem. Commun.* **2005**, 5205.
168. Thomson, R. K.; Bexrud, J. A.; Schafer, L. L. *Organometallics* **2006**, *25*, 4069.
169. Majumder, S.; Odom, A. L. *Organometallics* **2008**, *27*, 1174.
170. Leitch, D. C.; Payne, P. R.; Dunbar, C. R.; Schafer, L. L. *J. Am. Chem. Soc.* **2009**, *131*, 18246.
171. Antunes, M. A.; Munhá, R. F.; Alves, L. G.; Schafer, L. L.; Martins, A. M. *J. Organomet. Chem.* **2011**, *696*, 2.
172. Wang, X.; Chen, Z.; Sun, X. L.; Tang, Y.; Xie, Z. *Org. Lett.* **2011**, *13*, 4758.
173. Helgert, T. R.; Hollis, T. K.; Valente, E. J. *Organometallics* **2012**, *31*, 3002.
174. Luconi, L.; Klosin, J.; Smith, A. J.; et al. *Dalton Trans.* **2013**, *42*, 16056.
175. Muller, C.; Koch, R.; Doye, S. *Chem. Eur. J.* **2008**, *14*, 10430.
176. Stubbert, B. D.; Marks, T. J. *J. Am. Chem. Soc.* **2007**, *129*, 6149.
177. Ambühl, J.; Pregosin, P. S.; Venanzi, L. M.; Ughetto, G.; Zambonelli, L. *Angew. Chem. Int. Ed.* **1975**, *14*, 369.
178. Ambuehl, J.; Pregosin, P. S.; Venanzi, L. M.; Ughetto, G.; Zambonelli, L. *J. Organomet. Chem.* **1978**, *160*, 329.
179. Ambuehl, J.; Pregosin, P. S.; Venanzi, L. M.; Consiglio, G.; Bachechi, F.; Zambonelli, L. *J. Organomet. Chem.* **1979**, *181*, 255.
180. Bender, C. F.; Widenhoefer, R. A. *J. Am. Chem. Soc.* **2005**, *127*, 1070.
181. Bender, C. F.; Hudson, W. B.; Widenhoefer, R. A. *Organometallics* **2008**, *27*, 2356.
182. Pugin, B.; Venanzi, L. M. *J. Organomet. Chem.* **1981**, *214*, 125.
183. Vogels, C. M.; Hayes, P. G.; Shaver, M. P.; Westcott, S. A. *Chem. Commun.* **2000**, 51.
184. Tada, M.; Shimamoto, M.; Sasaki, T.; Iwasawa, Y. *Chem. Commun.* **2004**, 2562.
185. Michael, F. E.; Cochran, B. M. *J. Am. Chem. Soc.* **2006**, *128*, 4246.
186. Cochran, B. M.; Michael, F. E. *J. Am. Chem. Soc.* **2008**, *130*, 2786.
187. Han, X.; Widenhoefer, R. A. *Angew. Chem. Int. Ed.* **2006**, *45*, 1747.
188. Liu, X. Y.; Li, C. H.; Che, C. M. *Org. Lett.* **2006**, *8*, 2707.
189. Bender, C. F.; Widenhoefer, R. A. *Org. Lett.* **2006**, *8*, 5303.

190. Bender, C. F.; Widenhoefer, R. A. *Chem. Commun.* **2008**, 2741.
191. Lalonde, R. L.; Brenzovich, W. E., Jr.; Benitez, D.; et al. *Chem. Sci.* **2010**, *1*, 226.
192. Kitahara, H.; Sakurai, H. *J. Organomet. Chem.* **2011**, *696*, 442.
193. Takemiya, A.; Hartwig, J. F. *J. Am. Chem. Soc.* **2006**, *128*, 6042.
194. Liu, Z.; Hartwig, J. F. *J. Am. Chem. Soc.* **2008**, *130*, 1570.
195. Strom, A. E.; Balcells, D.; Hartwig, J. F. *ACS Catal.* **2016**, *6*, 5651.
196. Liu, Z.; Yamamichi, H.; Madrahimov, S. T.; Hartwig, J. F. *J. Am. Chem. Soc.* **2011**, *133*, 2772.
197. Hesp, K. D.; Stradiotto, M. *Org. Lett.* **2009**, *11*, 1449.
198. Hesp, K. D.; Tobisch, S.; Stradiotto, M. *J. Am. Chem. Soc.* **2010**, *132*, 413.
199. Bauer, E. B.; Andavan, G. T.; Hollis, T. K.; et al. *Org. Lett.* **2008**, *10*, 1175.
200. Kashiwame, Y.; Kuwata, S.; Ikariya, T. *Chem. Eur. J.* **2010**, *16*, 766.
201. Ohmiya, H.; Moriya, T.; Sawamura, M. *Org. Lett.* **2009**, *11*, 2145.
202. Komeyama, K.; Morimoto, T.; Takaki, K. *Angew. Chem. Int. Ed.* **2006**, *45*, 2938.
203. Bernoud, E.; Oulie, P.; Guillot, R.; Mellah, M.; Hannedouche, J. *Angew. Chem. Int. Ed.* **2014**, *53*, 4930.
204. Akiyama, T.; Ishida, Y.; Kagoshima, H. *Tetrahedron Lett.* **1999**, *40*, 4219.
205. Schlummer, B.; Hartwig, J. F. *Org. Lett.* **2002**, *4*, 1471.
206. Yang, C.-H.; Fan, W.-W.; Liu, G.-Q.; Duan, L.; Li, L.; Li, Y.-M. *RSC Adv.* **2015**, *5*, 61081.
207. Leger, P. R.; Murphy, R. A.; Pushkarskaya, E.; Sarpong, R. *Chem. Eur. J.* **2015**, *21*, 4377.
208. Yin, Y.; Zhao, G. *J. Fluor. Chem.* **2007**, *128*, 40.
209. Youn, S. W.; Jang, S. S.; Lee, S. R. *Tetrahedron* **2016**, *72*, 4902.
210. Zulys, A.; Dochnahl, M.; Hollmann, D.; Löhnwitz Dipl, K.; Herrmann Dipl, J.-S.; Roesky, P. W.; Blechert, S. *Angew. Chem. Int. Ed* **2005**, *44*, 7794.
211. Dochnahl, M.; Pissarek, J. W.; Blechert, S.; Lohnwitz, K.; Roesky, P. W. *Chem. Commun.* **2006**, 3405.
212. Dochnahl, M.; Lohnwitz, K.; Pissarek, J. W.; Biyikal Dipl, M.; Schulz Dipl, S. R.; Schön, S.; Meyer Dipl, N.; Roesky, P. W.; Blechert, S. *Chem. Eur. J* **2007**, *13*, 6654.
213. Löhnwitz, K.; Molski, M. J.; Lühl, A.; Roesky, P. W.; Dochnahl, M.; Blechert, S. *Eur. J. Inorg. Chem.* **2009**, *2009*, 1369.
214. Dochnahl, M.; Löhnwitz, K.; Lühl, A.; et al. *Organometallics* **2010**, *29*, 2637.
215. Jenter, J.; Lühl, A.; Roesky, P. W.; Blechert, S. *J. Organomet. Chem.* **2011**, *696*, 406.
216. Duncan, C.; Biradar, A. V.; Asefa, T. *ACS Catal.* **2011**, *1*, 736.
217. Liu, G.-Q.; Li, W.; Wang, Y.-M.; Ding, Z.-Y.; Li, Y.-M. *Tetrahedron Lett.* **2012**, *53*, 4393.
218. Koller, J.; Bergman, R. G. *Chem. Commun.* **2010**, *46*, 4577.
219. Khandelwal, M.; Wehmschulte, R. J. *J. Organomet. Chem.* **2012**, *696*, 4179.
220. Liu, G.-Q.; Cui, B.; Sun, H.; Li, Y.-M. *Tetrahedron* **2014**, *70*, 5696.
221. Wang, Y.-M.; Li, T.-T.; Liu, G.-Q.; et al. *RSC Adv.* **2014**, *4*, 9517.
222. Li, T.-T.; Liu, G.-Q.; Wang, Y.-M.; Cui, B.; Sun, H.; Li, Y.-M. *Tetrahedron* **2015**, *71*, 7003.
223. Nguyen, T. M.; Nicewicz, D. A. *J. Am. Chem. Soc.* **2013**, *135*, 9588.
224. Martinez, P. H.; Hultzsch, K. C.; Hampel, F. *Chem. Commun.* **2006**, 2221.
225. Ogata, T.; Ujihara, A.; Tsuchida, S.; Shimizu, T.; Kaneshige, A.; Tomioka, K. *Tetrahedron Lett.* **2007**, *48*, 6648.
226. Deschamp, J.; Collin, J.; Hannedouche, J.; Schulz, E. *Eur. J. Org. Chem.* **2011**, *2011*, 3329.
227. Horrillo-Martínez, P.; Hultzsch, K. C. *Tetrahedron Lett.* **2009**, *50*, 2054.
228. Wixey, J. S.; Ward, B. D. *Chem. Commun.* **2011**, *47*, 5449.
229. Neal, S. R.; Ellern, A.; Sadow, A. D. *J. Organomet. Chem.* **2011**, *696*, 228.
230. Wixey, J. S.; Ward, B. D. *Dalton Trans.* **2011**, *40*, 7693.
231. Nixon, T. D.; Ward, B. D. *Chem. Commun.* **2012**, *48*, 11790.

232. Schmid, B.; Friess, S.; Herrera, A.; et al. *Dalton Trans.* **2016**, *45*, 12028.
233. Zhang, X.; Emge, T. J.; Hultzsch, K. C. *Angew. Chem. Int. Ed.* **2012**, *51*, 394.
234. Gagne, M. R.; Brard, L.; Conticello, V. P.; Giardello, M. A.; Stern, C. L.; Marks, T. J. *Organometallics* **1992**, *11*, 2003.
235. Giardello, M. A.; Conticello, V. P.; Brard, L.; Gagne, M. R.; Marks, T. J. *J. Am. Chem. Soc.* **1994**, *116*, 10241.
236. Douglass, M. R.; Ogasawara, M.; Hong, S.; Metz, M. V.; Marks, T. J. *Organometallics* **2002**, *21*, 283.
237. Ryu, J. S.; Marks, T. J.; McDonald, F. E. *J. Am. Chem. Soc.* **2004**, *69*, 1038.
238. Hong, S.; Marks, T. J. *J. Am. Chem. Soc.* **2002**, *124*, 7886.
239. Otero, A.; Lara-Sánchez, A.; Nájera, C.; Fernández-Baeza, J.; Márquez-Segovia, I.; Castro-Osma, J. A.; Martínez, J.; Sánchez-Barba, L. F.; Rodríguez, A. M. *Organometallics* **2012**, *31*, 2244.
240. Manna, K.; Kruse, M. L.; Sadow, A. D. *ACS Catal.* **2011**, *1*, 1637.
241. Hong, S.; Tian, S.; Metz, M. V.; Marks, T. J. *J. Am. Chem. Soc.* **2003**, *125*, 14768.
242. Collin, J.; Daran, J.-C.; Schulz, E.; Trifonov, A. *Chem. Commun.* **2003**, 3048.
243. Collin, J.; Daran, J. C.; Jacquet, O.; Schulz, E.; Trifonov, A. *Chem. Eur. J.* **2005**, *11*, 3455.
244. Riegert, D.; Collin, J.; Meddour, A.; Schulz, E.; Trifonov, A. *J. Am. Chem. Soc.* **2006**, *71*, 2514.
245. Aillaud, I.; Collin, J.; Duhayon, C.; et al. *Chem Eur J* **2008**, *14*, 2189.
246. Aillaud, I.; Wright, K.; Collin, J.; Schulz, E.; Mazaleyrat, J.-P. *Tetrahedron Asymmetry* **2008**, *19*, 82.
247. Hannedouche, J.; Collin, J.; Trifonov, A.; Schulz, E. *J. Organomet. Chem.* **2011**, *696*, 255.
248. O'Shaughnessy, P. N.; Scott, P. *Tetrahedron Asymmetry* **2003**, *14*, 1979.
249. O'Shaughnessy, P. N.; Knight, P. D.; Morton, C.; Gillespie, K. M.; Scott, P. *Chem. Commun.* **2003**, 1770.
250. Riegert, D.; Collin, J.; Daran, J. C.; et al. *Eur. J. Inorg. Chem.* **2007**, *2007*, 1159.
251. Hannedouche, J.; Aillaud, I.; Collin, J.; Schulz, E.; Trifonov, A. *Chem. Commun.* **2008**, 3552.
252. Chapurina, Y.; Hannedouche, J.; Collin, J.; Guillot, R.; Schulz, E.; Trifonov, A. *Chem. Commun.* **2010**, *46*, 6918.
253. Chapurina, Y.; Ibrahim, H.; Guillot, R.; et al. *J. Am. Chem. Soc.* **2011**, *76*, 10163.
254. Lovick, H. M.; An, D. K.; Livinghouse, T. S. *Dalton Trans.* **2011**, *40*, 7697.
255. Kim, J. Y.; Livinghouse, T. *Org. Lett.* **2005**, *7*, 1737.
256. Zhang, Y.; Yao, W.; Li, H.; Mu, Y. *Organometallics* **2012**, *31*, 4670.
257. Gribkov, D. V.; Hultzsch, K. C.; Hampel, F. *Chem. Eur. J.* **2003**, *9*, 4796.
258. Gribkov, D. V.; Hultzsch, K. C. *Chem. Commun.* **2004**, 730.
259. Hultzsch, K. C.; Gribkov, D. V.; Hampel, F. *J. Organomet. Chem.* **2005**, *690*, 4441.
260. Gribkov, D. V.; Hultzsch, K. C.; Hampel, F. *J. Am. Chem. Soc.* **2006**, *128*, 3748.
261. Yu, X.; Marks, T. J. *Organometallics* **2007**, *26*, 365.
262. Meyer, N.; Zulys, A.; Roesky, P. W. *Organometallics* **2006**, *25*, 4179.
263. Heck, R.; Schulz, E.; Collin, J.; Carpentier, J.-F. *J. Mol. Catal. A Chem.* **2007**, *268*, 163.
264. Benndorf, P.; Jenter, J.; Zielke, L.; Roesky, P. W. *Chem. Commun.* **2011**, *47*, 2574.
265. Vitanova, D. V.; Hampel, F.; Hultzsch, K. C. *J. Organomet. Chem.* **2011**, *696*, 321.
266. Knight, P. D.; Munslow, I.; O'Shaughnessy, P. N.; Scott, P. *Chem. Commun.* **2004**, 894.
267. Bexrud, J. A.; Schafer, L. L. *Dalton Trans.* **2010**, *39*, 361.
268. Ayinla, R. O.; Gibson, T.; Schafer, L. L. *J. Organomet. Chem.* **2011**, *696*, 50.
269. Reznichenko, A. L.; Emge, T. J.; Audörsch, S.; Klauber, E. G.; Hultzsch, K. C.; Schmidt, B. *Organometallics* **2011**, *30*, 921.

270. Watson, D. A.; Chiu, M.; Bergman, R. G. *Organometallics* **2006**, *25*, 4731.
271. Wood, M. C.; Leitch, D. C.; Yeung, C. S.; Kozak, J. A.; Schafer, L. L. *Angew. Chem. Int. Ed.* **2007**, *46*, 354.
272. Gott, A. L.; Clarke, A. J.; Clarkson, G. J.; Scott, P. *Organometallics* **2007**, *26*, 1729.
273. Zi, G.; Zhang, F.; Xiang, L.; Chen, Y.; Fang, W.; Song, H. *Dalton Trans.* **2010**, *39*, 4048.
274. Zhang, F.; Song, H.; Zi, G. *Dalton Trans.* **2011**, *40*, 1547.
275. Yonson, N.; Yim, J. C. H.; Schafer, L. L. *Inorg. Chim. Acta* **2014**, *422*, 14.
276. Manna, K.; Xu, S.; Sadow, A. D. *Angew. Chem. Int. Ed.* **2011**, *50*, 1865.
277. Manna, K.; Everett, W. C.; Schoendorff, G.; Ellern, A.; Windus, T. L.; Sadow, A. D. *J. Am. Chem. Soc.* **2013**, *135*, 7235.
278. Lutete, L. M.; Kadota, I.; Yamamoto, Y. *J. Am. Chem. Soc.* **2004**, *126*, 1622.
279. Shen, X.; Buchwald, S. L. *Angew. Chem. Int. Ed.* **2010**, *49*, 564.
280. Kojima, M.; Mikami, K. *Synlett* **2012**, *2012*, 57.
281. Michon, C.; Abadie, M.-A.; Medina, F.; Agbossou-Niedercorn, F. *Catal. Today* **2014**, *235*, 2.
282. Abadie, M.-A.; Trivelli, X.; Medina, F.; Capet, F.; Roussel, P.; Agbossou-Niedercorn, F.; Michon, C. *ChemCatChem* **2014**, *6*, 2235.
283. Lee, S. D.; Timmerman, J. C.; Widenhoefer, R. A. *Adv. Synth. Catal.* **2014**, *356*, 3187.
284. Kanno, O.; Kuriyama, W.; Wang, Z. J.; Toste, F. D. *Angew. Chem. Int. Ed.* **2011**, *50*, 9919.
285. Turnpenny, B. W.; Hyman, K. L.; Chemler, S. R. *Organometallics* **2012**, *31*, 7819.
286. Paderes, M. C.; Belding, L.; Fanovic, B.; Dudding, T.; Keister, J. B.; Chemler, S. R. *Chem. Eur. J.* **2012**, *18*, 1711.
287. Ackermann, L.; Althammer, A. *Synlett* **2008**, *2008*, 995.
288. Lin, J. S.; Yu, P.; Huang, L.; Zhang, P.; Tan, B.; Liu, X. Y. *Angew. Chem. Int. Ed.* **2015**, *54*, 7847.
289. Shapiro, N. D.; Rauniyar, V.; Hamilton, G. L.; Wu, J.; Toste, F. D. *Nature* **2011**, *470*, 245.

Chapter 4

N-Alkyl Amine Synthesis by Hydroaminomethylation

4.1 INTRODUCTION

Amines and their derivatives are an important class of molecules in the chemical industry and are produced on a multimillion-ton scale per year. In particular, aliphatic amines are not only recognized as fundamental building blocks in myriad of natural products, pharmaceuticals, functional materials, and agrochemicals but also serve as attractive precursors in a variety of organic transformations. Given the importance of aliphatic amines, many methodologies have been developed for the preparation of these amines, including hydrocyanation of alkenes/reduction reactions, nucleophilic substitution of alkyl halides, and reductive amination, to name a few. However, these methodologies often suffer from expensive starting materials, low generality, and selectivity, and are environmentally unfriendly. Therefore, development of general and selective methods for the preparation of aliphatic amines is of vital importance. Hydroaminomethylation reaction[1–12] represents one of the most important catalytic protocols for the preparation of amines with respect to atom economy and an environmentally benign aspect, which gives amines from the readily available and economical olefins in the presence of syngas. It can be formally divided into three reactions: the first is hydroformylation, followed by condensation to give the imine or enamine, and subsequent reduction to form the amine (Scheme 4.1).

Although hydroaminomethylation is an elegant protocol for the preparation of amines, two challenges persist. The first challenge is achieving high chemo- and regioselectivity to give the amines products. Because of the hydroaminomethylation reaction could afford linear amines and branched amines, it is difficult to produce the expected amines with high *l*/*b* ratio. The second major challenge is avoiding the side reaction. Some side reactions usually occur in hydroaminomethylation reaction. For example, hydrogenation of starting alkene and the aldehydes intermediates could take place under reduction conditions to provide alkane and alcohols, respectively.

This chapter will describe developments and recent advances in the area of hydroaminomethylation reaction. The sections are divided according to the type of catalyst system used in the reactions.

Catalytic Amination for *N*-Alkyl Amine Synthesis. https://doi.org/10.1016/B978-0-12-812284-6.00004-4

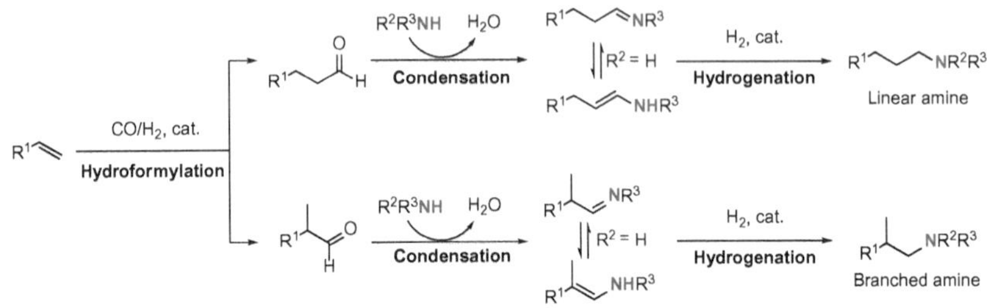

SCHEME 4.1 Three successive steps of hydroaminomethylation.

4.2 INTERMOLECULAR HYDROAMINOMETHYLATION

4.2.1 Fe Catalyst System

In 1949, Reppe et al. described the first example of hydroaminomethylation at BASF, whereby $Fe(CO)_5$ was shown to be an effective catalyst for the reaction of CO with acetylenic compounds in the presence of ammonia and water to afford amine products.[13–14] However, this reaction suffers from high pressure (up to 150 bar) and temperature ($T > 300°C$) as well as requires a stoichiometric amount of $Fe(CO)_5$. In 1994 it was observed that hydroaminomethylation could be proceeded in the absence of CO upon further optimization of this reaction.[15] With $Fe(CO)_5$ as catalyst and source of carbon monoxide, hydroaminomethylation of piperidine with ethylene (20 bar) at 170°C for 14 h has been achieved to afford *N-n*-propylpiperidine in 75% yield. It should be noted that >3 of the 5 carbon monoxide ligands of $Fe(CO)_5$ can be efficiently used to introduce CH_2 moieties in the aminomethylation product.

4.2.2 Rh Catalyst System

4.2.2.1 Unmodified Rhodium Catalyst System

In 1971, Iqbal described the first example of Rh-catalyzed hydroaminomethylation of olefins with CO, H_2O, and secondary amines. With Rh_2O_3 as catalyst, the reaction proceeds at 170°C for 3 h to give the corresponding products in 50%–80% yields. Importantly, the addition of $Fe(CO)_5$ as cocatalyst could improve the yields of products over 90% (Scheme 4.2).[16]

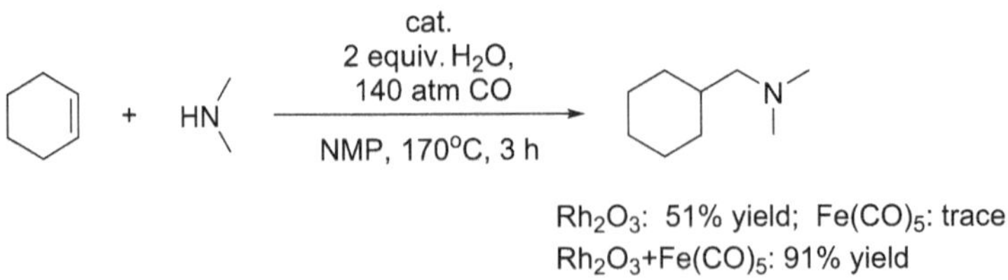

SCHEME 4.2 Rhodium-catalyzed aminomethylation of cyclohexene.

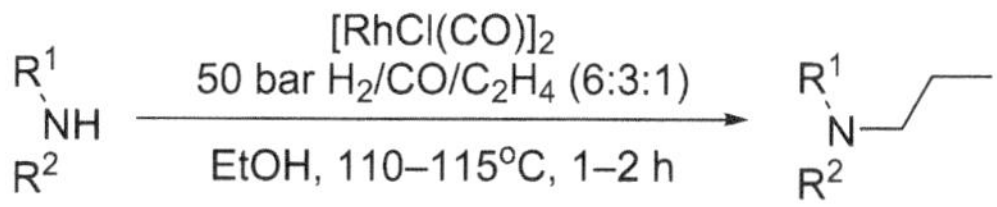

SCHEME 4.3 Rhodium-catalyzed reaction of ethylene and syngas with alkylamines.

Nearly three decades later, $[RhCl(CO)]_2$-catalyzed hydroaminomethylation of ethylene with primary amine and CO/H_2 for synthesis of *n*-propylamines has been achieved by Jones and colleagues (Scheme 4.3).[17] Subsequently, many efforts in $[RhCl(COD)]_2$-catalyzed hydroaminomethylation have been contributed by Eilbracht and colleagues. A variety of alkenes, including vinylarenes,[18–20] inactive alkenes,[18] enamines,[21] vinylsilanes,[21] allyl ethers,[22–23] allyl silanes,[22–23] allyl amines,[22–27] dienes,[28] and 1,1-diaryl-allyl-alcohols,[29] were applied to the synthesis of biological activity compounds, 12- to 36-membered polyheterocyeles,[28] and polyamines.[26] In addition to $[RhCl(COD)]_2$, Zwitterionic rhodium complex[22,30] and $\{Rh[(CH_3(CH_2)_6CO_2)]_2\}_2$[31] have been shown to catalyze the hydroaminomethylation of vinylarenes and cyclopentadiene, respectively.

Moreover, in order to achieve simple separation and recycling of the catalysts, a thermomorphic multicomponent solvent system for hydroaminomethylation reactions was developed by Behr and colleagues.[32–35]

4.2.2.2 P-Ligands

As early as 1992, Kalck et al. showed that the hydroaminomethylation of alkenes was efficiently catalyzed by dinuclear rhodium complexes $[Rh_2(\mu\text{-S-Bu}^t)_2(CO)_2(PPh_3)_2]$ with a slight excess of PPh_3 under low pressures of carbon monoxide (0.4 MPa) and hydrogen (0.4 MPa) (Scheme 4.4).[36] By optimizing the reaction conditions, it is found that the nature of the solvent has a dramatic influence on the course of the reaction and tetrahydrofuran is shown to be the best choice. About 80% yield of the desired amine was formed by the reaction of oct-1-ene and diethylamine. However, a much lower yield (55%) was obtained with morpholine. For the amines with steric hindrance such as diisopropylamine and *tert*-butylamine, the amine products were formed only in <10% yields. Further, the effect of solvents, the pressure and composition of synthesis gas, reaction time, and the concentration of ethanol as a cosolvent were also investigated in detail.[37]

Following the work of Kalck, some other catalyst systems were also investigated for the hydroaminomethylation of long-chain alkenes such as 1-octene,

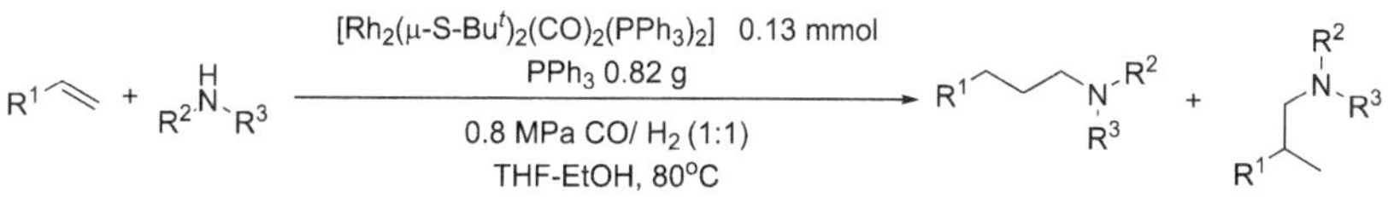

SCHEME 4.4 The Rh-catalyzed hydroaminomethylation of alkenes with PPh_3 as ligand.

1-dodecene, and 1-nonene with primary and secondary amines on the basis of the combination of Rh-complexes and monophosphorus ligands such as $[(H_2C(3,5\text{-}Me_2pz)_2)Rh(CO)L]^+$ (L = PPh_3, $PMePh_2$, or $P(OMe)_3$),[38] $RhCl(CO)(TPPTS)_2$/TPPTS-TPPDS-CTAB,[39] $RhCl(CO)(TPPTS)_2$/TPPTS-CTAB,[40] Rh(I) complex with amphiphilic triphenylphosphane functionalized poly (2-oxaz oline)s as ligands,[41] $Rh(cod)Cl]_2$/TPPTS,[42] and $Rh(acac)(CO)_2/PPh_3$.[43]

In addition to the simple long-chain alkenes, the hydroaminomethylation of other alkenes was also explored with the catalyst system composed of Rh-complexes and monophosphorus ligands. For example, the hydroaminomethylation of styrene was reported by Bühl and Clarke with the combination of $[RhCl(COD)]_2$ and different monophosphines.[44] Trös et al. described the hydroaminomethylation of steroids with aminoalcohols in the presence of $[Rh(nbd)Cl]_2$ and PBu_3.[45] A similar protocol was shown by Eilbracht et al.[46] and pharmacologically active secondary and tertiary 1-(3,3-diarylpropyl) amines were synthesized in 68%–76% yields by the reaction of 1,1-diarylethenes and primary or secondary amines in the presence of $[Rh(cod)Cl]_2/PBu_3$. Rosa and Behr independently reported the hydroaminomethylation of limonene with $HRh(CO)(PPh_3)_3$ and $[Rh(cod)Cl]_2$/TPPTS-CTAB, respectively.[47–48] Santos et al. showed an efficient transformation of naturally occurring monoterpenes into amines through hydroaminomethylation catalyzed by $[Rh(cod)(\mu\text{-}OMe)]_2$ with PPh_3 or PBn_3.[49] Santos, Gouygou, and Urrutigoïty et al. demonstrated that the hydroaminomethylation of estragole could be efficiently catalyzed by $[Rh(cod)(\mu\text{-}OMe)]_2$ with phosphorus(III) compounds.[50] Three classes of phosphorus (III) compounds, phosphine, phosphites, and phospholes, were tested and compared. The best result was obtained with phosphole-promoted system. It is worth noting that the diastereoselective hydroaminomethylation of methallylic alcohols was also achieved by Breit with $[Rh(CO)_2acac/4\ P(OPh)_3]$ and the aid of the substrate bound catalyst directing *o*-DPPB-group (o-DPPB = *ortho*-diphenylphosphanyl benzoate).[51] Moderate to good yields and excellent diastereomer ratios (*syn*: *anti* > 94: 6) were obtained.

Although the combination of Rh-complexes and monophosphorus ligands has been applied in the hydroaminomethylation of various alkenes with amines, these catalyst systems often suffer from unsatisfactory chemo- and regioselectivities. In comparison with monophosphorus ligands, bisphosphorus ligands generally give higher chemo- and regioselectivities owing to the much better chelating ability and steric environment.

In 2003, Beller group reported a highly chemo- and regioselective hydroaminomethylation of α-olefins using a cationic rhodium precatalyst together with Xantphos as ligand (Scheme 4.5).[52] Secondary and tertiary aliphatic amines were prepared with excellent yields (>99%) and regioselectivity (*n*/*iso* = 99:1) in the presence of 0.1 mol% $[Rh(cod)_2]BF_4$ and 0.4 mol% Xantphos. It is of interest that an example of aromatic amines (i.e., aniline) was also contained and similar regioselectivity (*n*/*iso* = 98:2) and a good yield (76%) was afforded. Further study revealed that the branched amine products were also selectively

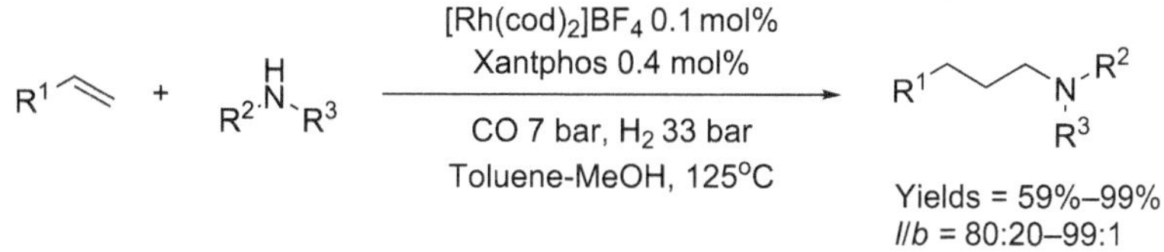

SCHEME 4.5 The Rh-catalyzed hydroaminomethylation of alkenes with Xantphos as ligand.

synthesized by changing the ligand to dppf.[53] Under the optimized reaction conditions, the branched amphetamine derivatives were synthesized with good selectivities and yields by the reaction of arylethylenes and anilines, but 10 mol% HBF_4 as additives was required.

Subsequently, some similar catalyst systems based on the combination of $[Rh(cod)_2]BF_4$ and different functionalized Xantphos as ligands such as xantphenoxaphos,[54] sulfoxantphos,[55] and xanthene-based, amino-functionalized ligands[56] were also used in the hydroaminomethylation of alkenes with amines.

Additionally, some other catalyst systems consisted of Rh-complexes and bisphosphorus ligands have also been used in the hydroaminomethylation of alkenes with amines. For example, in 2005, Whiteker et al. demonstrated combination of $Rh(CO)_2(acac)$ and bis-phosphite ligand, which was able to efficiently catalyze synthesis of biologically active tertiary amines such as Ibutilide and Aripiprazole, though a hydroaminomethylation process (Scheme 4.6).[57] Although only moderate to good yields (55%–67%) were obtained, good regioselectivity (l:b=37:1 or 48:1) was afforded. Further, they demonstrated this catalyst is active for synthesis of antihistamines such as Terfenadine and Fexofenadine via a similar process.[58] Interestingly, the sensitive hydroxyl substituent in the starting olefins was well tolerated.

An efficient biphasic hydroaminomethylation of long chain alkenes was developed by Luo et al. employing the $RhCl(CO)(TPPTS)_2$ as catalyst precursor together with water-soluble BISBIS (sulfonated 2,2′-bis(diphenylphosphinomethyl)-1.1′-biphenyl) as ligands.[59] High regioselectivity for linear amine was observed and the ratio of linear to branched amine was up to 83 in the hydroaminomethylation of 1-dodecene with dimethylamine. Further investigation showed the use of ionic liquids 1-n-alkyl-3-methylimidazolium tosylates $[Rmim][p\text{-}CH_3C_6H_4SO_3]$ (R=n-butyl, octyl, dedecyl, cetyl) as reaction

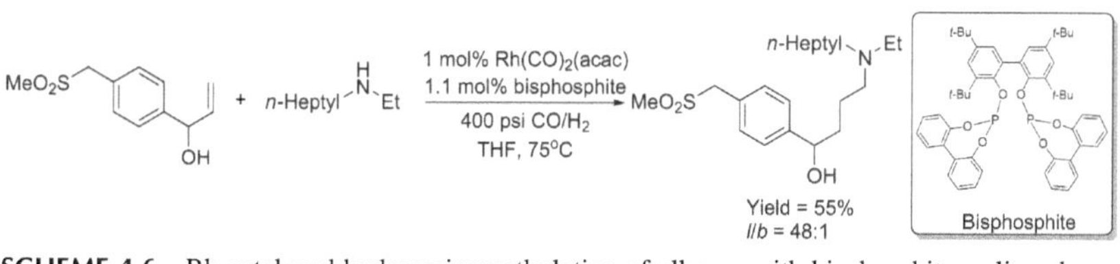

SCHEME 4.6 Rh-catalyzed hydroaminomethylation of alkenes with bisphosphite as ligand.

medium made it possible to separate and reuse the catalyst.[59] The best overall performance was obtained with [Bmim][p-$CH_3C_6H_4SO_3$] as a reaction medium and six recycles were achieved with only a slight decrease in activity. But the selectivity for linear amines significantly decreased, which was mainly due to the partial oxidation of ligand BISBIS during the separation of catalyst.

A microwave-assisted hydroaminomethylation of alkenes with secondary amines was reported by Taddei et al.[60] Under the optimized reaction conditions (EtOH, 110°C, 100 psi syngas, MW irradiation), the hydroaminomethylation of terminal alkenes can be accomplished in very short time (10–30 min) in the presence of $(PPh_3)_3RhCO(H)$ and Xantphos or Biphephos to afford linear amine products in moderate to good yields.

A solvent-free protocol was presented by Sémeril and Matt in 2010.[61] Under solvent-free conditions, neutral [Rh(acac)(1,3-calix-diphosphite)] complexes catalyzed efficiently the selective synthesis of linear amines. In comparison to the result obtained in toluene, it is found that solvent-free conditions do not change the selectivity of linear product.

Bhanage et al.[62] developed recyclable homogeneous rhodium catalyst [RhPEGD] which was prepared by anchoring rhodium polyether diphosphinite complex in polyethylene glycol. The catalyst was proved to be highly active for the hydroaminomethylation of various alkenes with primary and secondary amines and good to excellent yields were obtained. It is noted that during the course of reaction, the catalyst was soluble with reactants/products, but it could be quantitatively separated from reaction media and subsequently reused up to five times without significant decrease in catalytic performance.

The hydroaminomethylation of eugenol with di-n-butylamine was reported by Santos et al. by employing a bis[(1,5-ciclooctadiene)-(μ-methoxy)-rhodium(I)] as precatalyst.[63] It is found that the addition of phosphine ligands greatly improved the regioselectivity for linear amines but inhibited the hydrogenation of enamine intermediates, which led to moderate yield (64%) of linear amine product. By introducing triflic acid (10–20 mol%) as a promoter, the linear product was obtained up to 93% yields with 99% regioselectivity.

Although various Rh-catalyzed hydroaminomethylation have been reported, most of them suffer from rather high pressure (>100 bar) and/or temperature (>100°C). A rather mild protocol was reported by Bhanage et al. in 2013.[64] Under mild conditions (80°C, 30 bar syngas (CO/H_2=7:23)), rhodium-phosphinite complex catalyst $(Rh(cod)Cl(POPh_2))_2(C_6H_4)$ could efficiently catalyze the hydroaminomethylation of various olefins with primary and secondary amines (Scheme 4.7). By optimizing the reaction parameters such as temperature, pressure, catalyst loading, time, and solvents, aliphatic amines and arylethylamines were synthesized with excellent yields and turnover numbers up to 2500.

In 2013, Wu and Zhang et al. showed that the combination of $Rh(nbd)_2BF_4$ and NAPHOS was able to efficiently catalyze the hydroaminomethylation of 1,1-diphenylethene to synthesize the fenpiprane and related pharmaceuticals.[65]

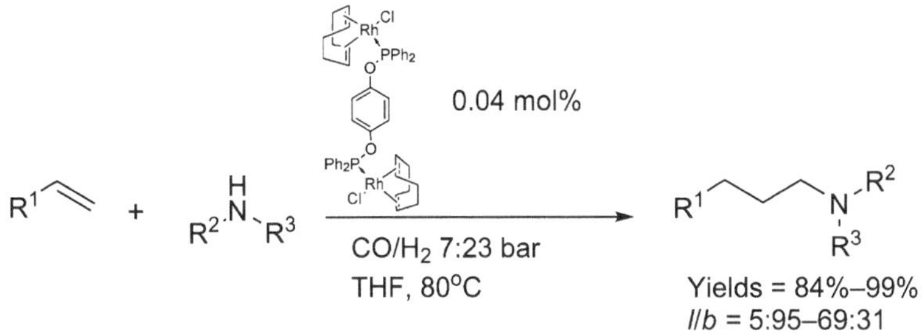

SCHEME 4.7 The hydroaminomethylation of alkenes catalyzed by rhodium-phosphinite complex.

Under relatively mild conditions ($CO/H_2 = 20$ bar/10 bar, 125°C), good to excellent yields were obtained. Noteworthy, only 0.1 mol% catalyst loadings were required.

In addition to the hydroaminomethylation of alkenes with primary and secondary amines, the hydroamidomethylation of alkenes and amides has also been investigated. For example, in 2013, Bouwman et al. reported the hydroamidomethylation of terminal alkenes and amides by employing the catalyst system of $[Rh(acac)(CO)_2]$ and Xantphos ligand in combination with a strong acid promoter such as *p*-toluenesulfonic acid (TsOH) and a weakly acidic protic promoter, hexafluoroisopropyl alcohol (HOR^F) (Scheme 4.8).[66] Under the optimized reaction conditions, the linear mono-*N*-alkylation amines products were obtained with excellent selectivity (>99%). TsOH was necessary to establish the equilibrium addition/condensation reaction between the aldehyde formed in situ and the amide. The presence of HOR^F was necessary for the generation and solvation of cationic Rh^{III}-hydride species, which were active as hydrogenation catalysts in the presence of carbon monoxide.

In 2015, Beller et al. revealed the Rh-catalyzed synthesis of sulphonamides via a similar protocol.[67] In the presence of $[Rh(acac)(CO)_2]$ together with Naphos as ligands, various olefins, including terminal, internal, and 1,1′-disubstituted alkenes, could be converted with a range of sulphonamides to generate the corresponding linear sulphonamides in good yields with excellent selectivities (Scheme 4.9).

In 2016, a bis-hydroaminomethylation protocol was presented by Vorholt et al.[68] A series of bifunctional molecules were synthesized by the reaction

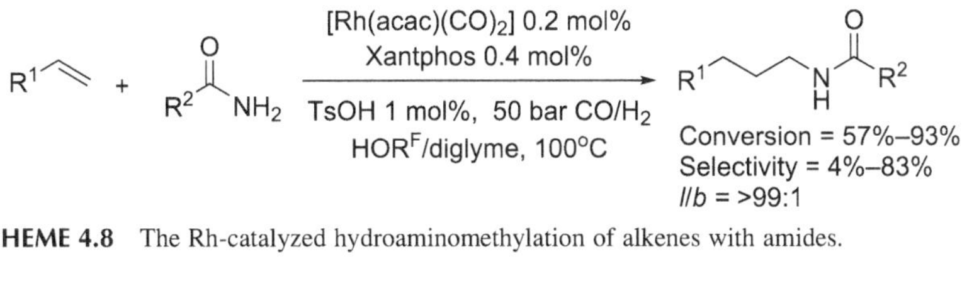

SCHEME 4.8 The Rh-catalyzed hydroaminomethylation of alkenes with amides.

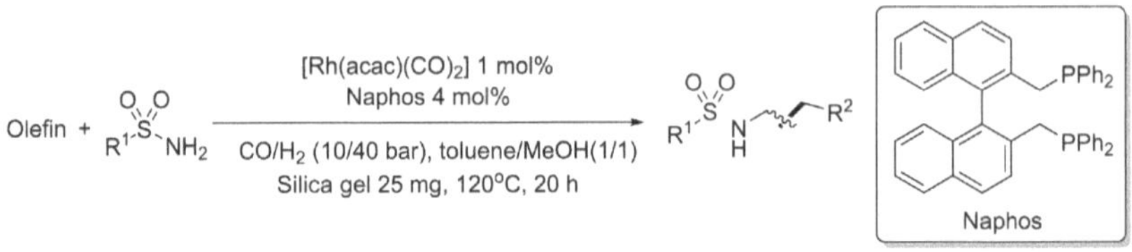

SCHEME 4.9 The Rh-catalyzed hydroaminomethylation of alkenes with sulphonamides.

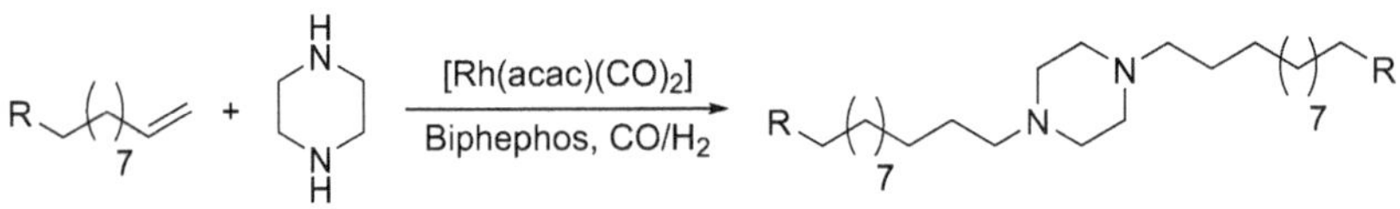

SCHEME 4.10 The Rh-catalyzed bis-hydroaminomethylation of alkenes.

of linear aliphatic alkenes and piperazine catalyzed by the combination of [Rh (acac)$(CO)_2$] and Biphephos (Scheme 4.10). Moderate to good yields and linear product selectivities were obtained under the optimized reaction conditions. Further study demonstrated that this catalyst system was also effective for the castor oil-derived renewable alkenes such as methyl 10-undecenoate and 10-undecene-1-ol.[69] Quantitative conversions and approximately 70% selectivity toward the linear product were obtained.

The hydroaminomethylation reaction of alkenes with amines under CO/H_2 generally includes three successive steps: the hydroformylation of alkene, the formation of imine or enamine, and the hydrogenation of imine or enamine. It is well known that the neutral rhodium precursors are often used in hydroformylation, while cationic precursors are known to generate active hydrogenation catalysts. It means that the interplay between cationic and neutral species may be possible. In 2012, Kalck et al. investigated the hydroaminomethylation catalyzed by $[Rh(CO)_2$-((*R*,*R*)-Ph-BPE)]BF_4 via a DFT study and confirmed the production of neutral monohydride complex [RhH$(CO)_2$((*R*,*R*)-Ph-BPE)].[70] Additionally, the mechanistic pathway from the cationic complex to the neutral monohydride species was also elucidated.

In addition to the widely used mono- and di-phosphorus ligands, some tetra-biphosphorus ligands have also been combined with Rh-complexes to catalyze the hydroaminomethylation reaction of alkenes with amines under CO/H_2. An example was shown by Wu and Zhang et al. in 2011.[71] In the presence of [Rh (acac)$(CO)_2$] and 2, 2′, 6,6′-tetrakis ((diphenylphosphino)-methyl)-1, 1′-biphenyl ligand, the hydroaminomethylation of terminal olefins with aliphatic secondary amines was efficiently promoted and the corresponding amine products were obtained in excellent yields (>98%) and selectivities (99%) (Scheme 4.11). It is worth noting that under optimized reaction conditions, the turnover number could reach 6930, and the linear/branched amine products ratio can reach up to >525. Further investigation revealed the combination of

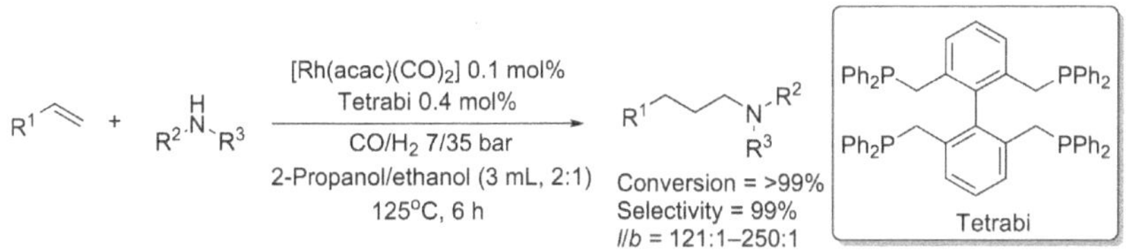

SCHEME 4.11 The Rh-catalyzed hydroaminomethylation of alkenes with tetrabi as ligands.

$[Rh(acac)(CO)_2]$ and pyrrole-based tetraphosphorus ligands to also be effective for the previous reaction although lower regioselectivity and turnover number were obtained.[72] By changing the Rh-complex to $Rh(nbd)_2SbF_6$ along with pyrrole-based 3,3′,5,5′-substituted tetraphosphorus ligands, the hydroaminomethylation of styrenes with aliphatic secondary amines was also achieved under the optimized reaction conditions. The linear amines were still the major product and the ratio of linear and branched products could reach 99.[73]

4.2.2.3 P,N-Ligands and Carbene Ligands

Excluding the phosphorus ligands widely used in the hydroaminomethylation of alkenes with amines, some P,N ligands and carbene ligands without phosphorus have been reported. For example, as early as 1999, Kostas et al. demonstrated that hydroaminomethylation of styrene with morpholine could be efficiently promoted by the combination of a cationic rhodium complex and a P,N ligand possessing a hydroxy group (Scheme 4.12).[74] Under the optimized reaction conditions, the branched amine products were formed with about 85% selectivity.

A rhodium carbene catalyst [Rh(cod)(Imes)Cl] which proved to be highly active for hydroaminomethylation of alkenes and amines was developed by Beller et al. in 2003 (Scheme 4.13).[75] Various alkenes including terminal as well as internal aliphatic and aromatic olefins with aliphatic amines could be well tolerated and good to excellent yields were generally obtained with only 0.1 mol% of catalyst. It is found that up to 99:1 regioselectivity (*n*/*iso* selectivity) is obtained in the case of substituted styrenes while terminal aliphatic olefins give mixtures of linear and branched products. Furthermore, it is found that this catalyst system is also effective for the synthesis of various biologically active pheniramines by the hydroaminomethylation of 1,1-diphenylethylene

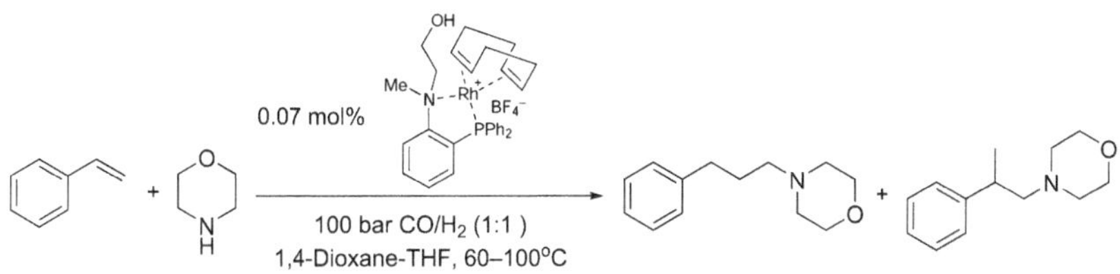

SCHEME 4.12 The Rh-catalyzed hydroaminomethylation of alkenes with a P,N ligand.

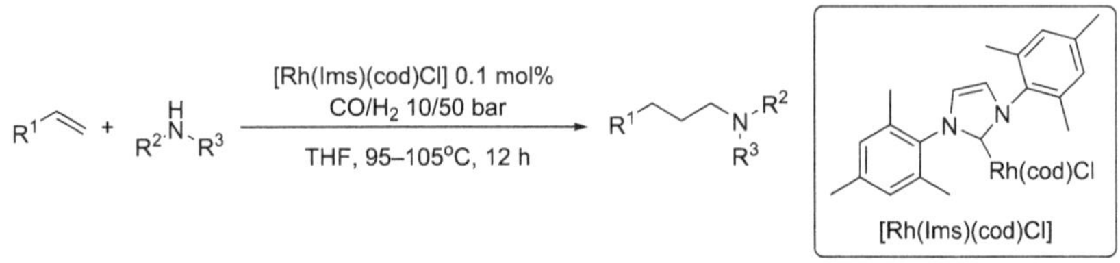

SCHEME 4.13 The hydroaminomethylation of alkenes catalyzed by a rhodium carbene catalyst.

with piperidine and good to excellent yields were obtained.[76] In general, the linear product was formed with high selectivity (*n*/*iso* ≥99:1) while the hydroaminomethylation of styrene gave preferentially the branched product in good yield.

The new bis-*N*-heterocyclic carbene (bis-NHC) complexes of rhodium was developed by Abu-Reziq et al. and applied in the hydroaminomethylation of styrene with aliphatic amines (Scheme 4.14).[77] Similar to previous reports, the branched products were selectively generated. Under the optimized reaction conditions, excellent yields (up to 98%) were obtained along with high branched selectivity (B/L = 24:1). Furthermore, a semiheterogeneous protocol was developed by immobilizing a bis-NHC ligand functionalized with silane groups on magnetic nanoparticles and coordinating with rhodium. In this case, it is easy to separate and reuse the catalyst by applying an external magnetic field, but a decrease in the branched selectivity was observed. Of note, the separated catalyst could be recycled four times without an appreciable loss in catalytic activity.

4.2.2.4 Heterogeneous and Semiheterogeneous Catalyst System

Although great progress has been made in Rh-catalyzed homogeneous hydroaminomethylation of alkenes with amines and good results have been obtained, the drawback in the separation and recycle is obvious. Thus, some heterogeneous and semiheterogeneous protocols were explored. For example, Wang et al. reported a recyclable Rh nanoparticle catalyst stabilized by thermoregulated ligand $Ph_2P(CH_2CH_2O)_{16}CH_3$, which was demonstrated as being active for the hydroaminomethylation of 1-octene with di-*n*-propyl amine in the aqueous/1-butanol biphasic system (Scheme 4.15).[78] Under the optimized conditions, 99% conversion of 1-octene and 97% of amine selectivity were obtained. Moreover, the Rh nanoparticle catalyst can be separated from products by simple phase separation and recycled directly for the next run. However, a decrease in amine selectivity was observed during the recyclable experiment, which was attributed to the leaching of Rh. ICP-AES indicated that the average leaching of Rh was 2.7 wt%, but by prolonging the reaction time, ≥80% amine selectivity could still be kept after six recycles. Further study revealed that the Rh nanoparticle catalyst was also active for the hydroaminomethylation of 1-octene with di-*n*-propyl amine in the ionic liquid

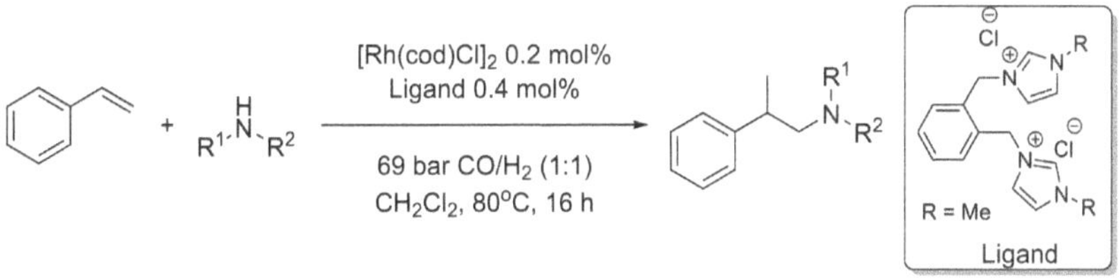

SCHEME 4.14 The hydroaminomethylation of alkenes catalyzed by bis-*N*-heterocyclic carbene complexes.

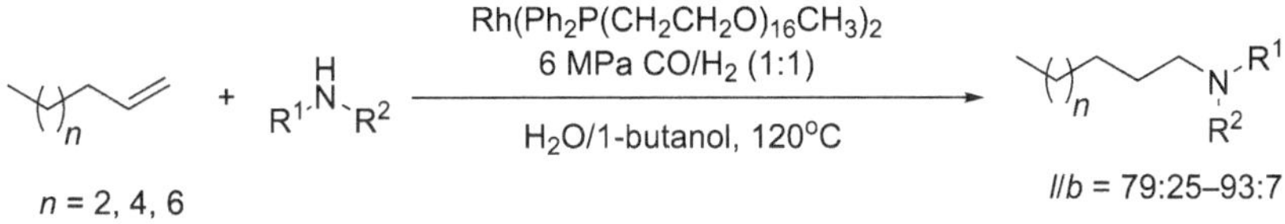

SCHEME 4.15 The hydroaminomethylation of alkenes catalyzed by Rh nanoparticle.

$[CH_3(OCH_2CH_2)_{22}N^+Et_3][CH_3SO_3^-]$ ($IL_{PEG1000}$) and organic biphasic system and 100% conversion of 1-octene and 90% amine selectivity were obtained under the optimized conditions,[78] which is slightly lower than the result obtained in the aqueous/1-butanol biphasic system. Similarly, the $IL_{PEG1000}$-stabilized Rh nanoparticle catalyst could be easily separated and four recycles were performed without evident loss in activity.

In 2013, Wasserscheid et al. described the hydroaminomethylation in a continuous gas-phase reaction using supported ionic liquid phase (SILP) materials as microscopically homogeneous but macroscopically heterogeneous catalysts (Scheme 4.16).[79] The hydroaminomethylation of ethylene with diethylamine was investigated by employing SILP catalyst containing $Rh(acac)(CO)_2$ and Xantphos (L/Rh = 5:1). Different ionic liquid coatings and supports were explored and the best result was obtained with the highly porous carbon support PBSAC (PBSAC = polymer-based spherical activated carbon) and [MMMIM] $[NTf_2]$ ionic liquid coatings. Under the optimized conditions, the selectivity of the amine product could reach >99% along with 72% conversion of ethylene. It is worth noting that this catalyst system showed a >18 days' stability and, except for a slight deactivation (by approximately 20%) at the early stage,

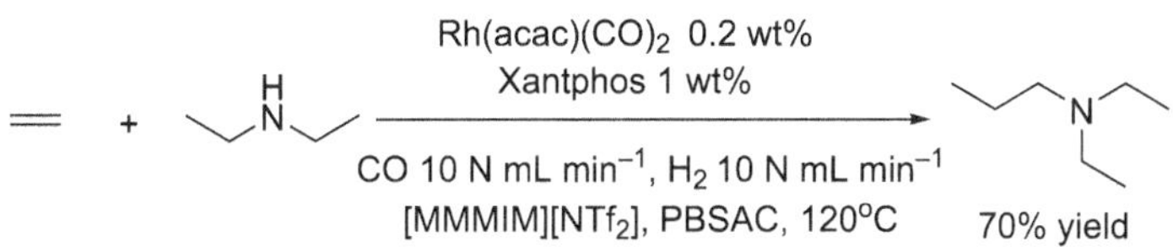

SCHEME 4.16 The hydroaminomethylation of alkenes catalyzed by supported ionic liquid phase (SILP) catalyst.

the catalyst activity remained almost stable during this extended period at a level of TOF = 450 h^{-1} with an average amine product selectivity of >99%.

A heterogeneous hydroaminomethylation was presented by Shukla et al. in 2013 by employing a rhodium complex encapsulated functionalized hexagonal mesoporous silica (HMS) as catalyst (Scheme 4.17).[80] Under optimized conditions, the heterogeneous catalyst prepared by encapsulating $HRh(CO)(PPh_3)_3$ complex into functionalized hexagonal mesoporous silica efficiently promoted the hydroaminomethylation of aliphatic alkenes with aliphatic primary and secondary amines and afforded the corresponding amine products in excellent yields. The result is comparable with the homogeneous protocol reported although the regioselectivity obtained is generally much lower. Moreover, in contrast to the homogeneous protocol, the catalyst in this protocol could be reused five times without obvious loss in its activity and selectivity. Furthermore, it is found that the rhodium exchanged ETS-10 and ETS-4 exhibited better catalytic performances for the above reaction than the rhodium complex encapsulated functionalized hexagonal mesoporous silica since similar results were obtained with lower reaction temperature and catalyst loadings, and shorter reaction time.[81] Besides, the recycle of the rhodium exchanged ETS-10 is also comparable to the rhodium complex encapsulated functionalized hexagonal mesoporous silica, and it can be reused three times without obvious loss in activity and selectivity.

Although some heterogeneous hydroaminomethylation protocols have been reported, most focus on the reaction of alkenes and aliphatic amines. An example of heterogeneous hydroaminomethylation of vinylarenes with anilines was shown by Blum et al. in 2014 (Scheme 4.18).[82] By immobilizating 6 mol% [Rh(cod)Cl]$_2$ in a hydrophobic silica sol-gel matrix, 2-arylpropylamines was synthesized in good to excellent yields with good regioselectivity under mild

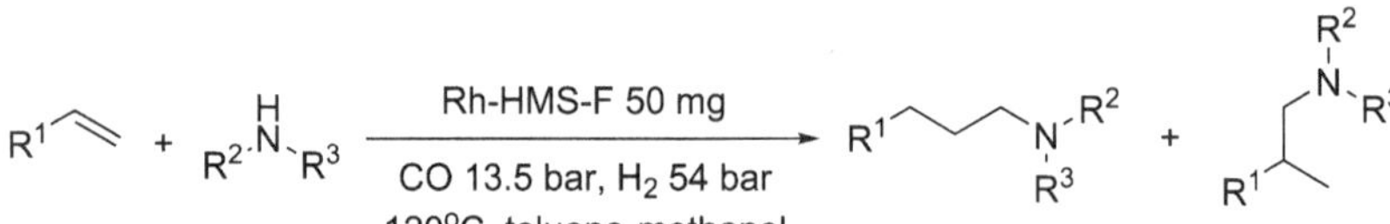

SCHEME 4.17 The hydroaminomethylation of alkenes catalyzed by Rh-HMS.

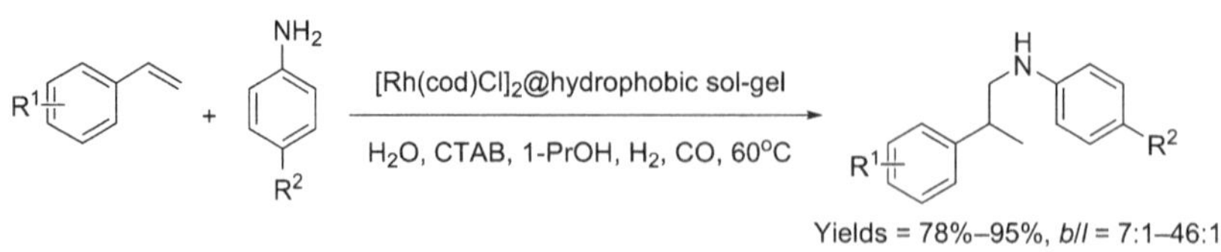

SCHEME 4.18 The Rh-catalyzed hydroaminomethylation of alkenes with anilines.

conditions (60°C, H_2 200 psi, CO 100 psi). It is found that the regioselectivity was mainly dependent on the hydrophobicity of the catalyst support.

4.2.3 Co Catalyst System

The first cobalt-catalyzed hydroaminomethylation was reported in 1984 by Murata and colleagues. With $Co_2(CO)_8$ and dppe as catalyst system, reaction of propylene with morpholine in the presence of CO (10 Mpa) and H_2O at 165°C produces *N*-butylmorpholines in 89% yield with 66.4% selectivity of *N*-*n*-butylmorpholine.[83] Subsequently, the Knifton group have extended the $Co_2(CO)_8$/phosphine ligand catalyst system to the synthesis of amines with syngas. The reaction of olefin, syngas ($CO/H_2 = 1:2$, 140 bar), and ammonia at 200° C generates corresponding aliphatic primary amine in 85% yield with 32% selectivity.[84]

4.2.4 Ru Catalyst System

In 1982, one example of a Ru-catalyzed hydroaminomethylation under water gas shift conditions was reported by Jachimowicz and colleagues.[85] Using $Ru_3(CO)_{12}$ as catalyst, 1-decene reacted with dimethylamine in the presence of CO(11 MPa) and H_2O at 160°C to yield tertiary amines with only 38% conversion. More than 30 years later, a major breakthrough was realized by Beller and colleagues (Scheme 4.19).[86] Using $Ru_3(CO)_{12}/K_2CO_3$ as catalyst system, olefins reacted with various amines in the presence of CO (4 MPa) and H_2O to give corresponding tertiary amines in high yields of up to 92% with excellent regioselectivities of *n*/*iso* up to 99:1. Note that full chemoselectivity favors the terminal olefins. However, primary amines such as aniline, 1-hexaneamine, or cyclohexylamine gave somewhat lower yields (22%–45%), and styrene afforded a 30:70 *n*/*iso* ratio of regioselectivity.

In 1999, Keim and Schaffrath reported the $Ru_3(CO)_{12}$-catalyzed hydroaminomethylation reaction of propene using CO and H_2 (1:2, 55 bar).[87] A much higher yield of 95% with selectivities of up to 99% was obtained for piperidine. However, *N*-*n*-butylformamide was observed as a major side product with selectivities of 30% for *N*-butylamine.

In 2013, an important breakthrough for the Ru-catalyzed hydroaminomethylation reaction of alkenes was achieved by Beller and colleagues (Scheme 4.20).[88] Using a combination of $Ru_3(CO)_{12}$ and 2-phosphino-substituted

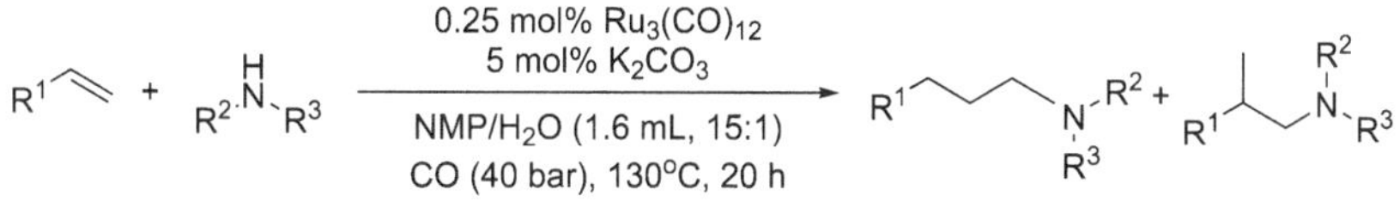

SCHEME 4.19 Ruthenium-catalyzed hydroaminomethylation of olefins without phosphine and hydrogen.

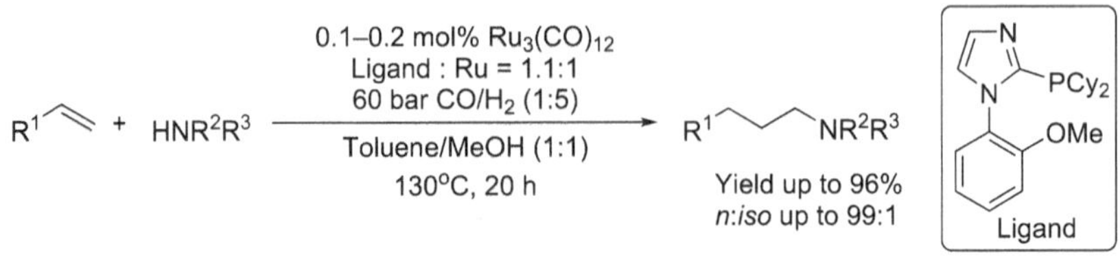

SCHEME 4.20 Ruthenium-catalyzed hydroaminomethylation of olefins.

imidazole ligand as catalyst system, aliphatic olefins react with primary and secondary amines to give the secondary and tertiary amines generally in high yields of up to 96% with excellent regioselectivities of *n*/iso up to 99:1. Unfortunately, an almost 1:1 mixture of *n*- and *iso*-amine was obtained when the styrene was subjected to this hydroaminomethylation protocol. Subsequently, this chemistry has been extended to hydroaminomethylation of olefins with ammonia for the direct synthesis of tertiary amines.[89]

In addition, extension of hydroaminomethylation from carbon monoxide- to carbon dioxide-based versions was reported by Eilbracht and colleagues in 2009 (Scheme 4.21). $Ru_3(CO)_{12}$ also shows appreciable catalytic activity in hydroaminomethylation of olefins with CO_2/H_2 (1:3, 80 bar), providing the amine products in 35%–98% yields.[90] Recently, Dupont et al. discovered that $Ru_3(CO)_{12}$/imidazolium salt ionic liquid (IL) was capable of the hydroaminomethylation of alkenes with primary or secondary amines at 120°C under CO_2/H_2 (1:1, 60 bar).[91] Much higher conversions of up to 99% and selectivities of up to 96% may be achieved using this catalyst system.

In addition to previously mentioned methods, methylformate could also be used as formylating agent instead of syngas for hydroaminomethylation of olefins in the presence of $Ru_3(CO)_{12}/Rh(acac)(CO)_2$[43] catalyst system.[92]

4.2.5 Ir Catalyst System

In 1999, Beller and coworkers described the hydroaminomethylation of olefins with ammonia to form linear primary and secondary aliphatic amines with a Rh/Ir system (Scheme 4.22). With BINAS as ligand, C_3-C_5 alkenes react with ammonia and syngas at 130°C to give the primary amines with up to 87% selectivity in 75%–90% yields and excellent *n:iso* selectivities (99:1).[93]

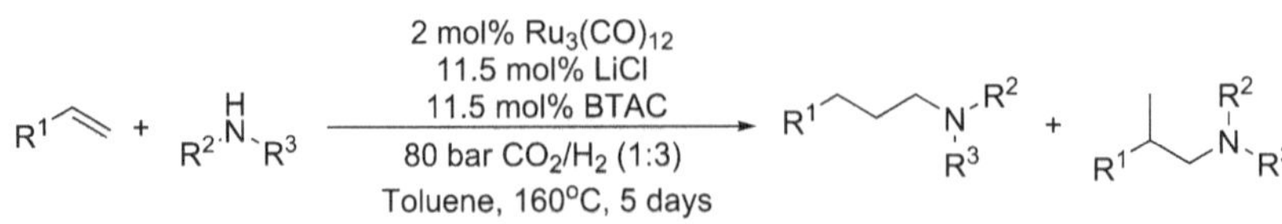

SCHEME 4.21 Ruthenium-catalyzed hydroaminomethylation of olefins with CO_2.

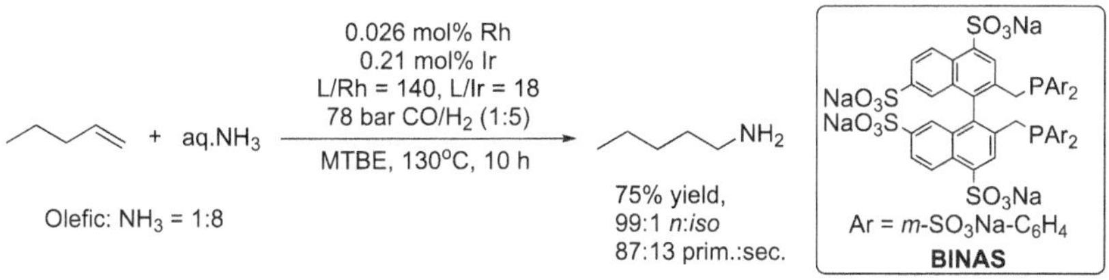

SCHEME 4.22 Rh-/Ir-catalyzed hydroaminomethylation with ammonia to primary amines.

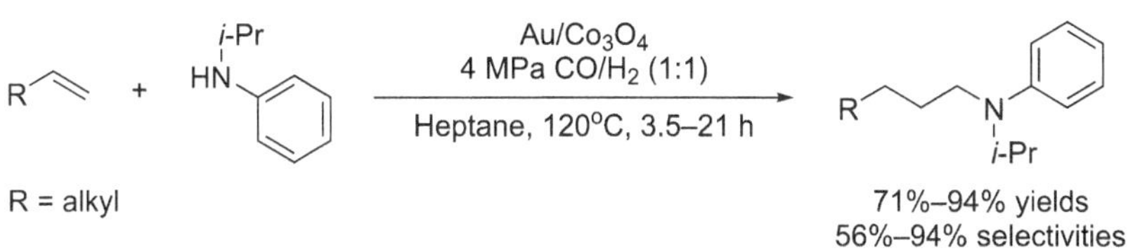

SCHEME 4.23 Au/Co_3O_4-catalyzed hydroaminomethylation of various olefins with *N*-isopropylaniline.

4.2.6 Au Catalyst System

In 2013, heterogeneous Au/Co_3O_4-catalyzed hydroaminomethylation of olefins with anilines was reported by Tokunaga (Scheme 4.23). The Au/Co_3O_4 was prepared by the coprecipitation method and shown to be an effective catalyst for the reaction of various olefins with *N*-isopropylaniline and syngas, affording hydroaminomethylated products in 71%–94% yields with up to 94% selectivity.[94]

4.3 INTRAMOLECULAR HYDROAMINOMETHYLATION

In 1992, Alper and colleagues described the first Rh-catalyzed intramolecular hydroaminomethylation.[95] *N*-allylic arylamines react with carbon monoxide (34.5 atm) and sodium borohydride in the presence of Zwitterionic complex η^6-$C_6H_6BPh_3$-Rh(COD)$^+$ at 100 °C to afford pyrrolidines in 27%–87% yields. In addition, using η^6-$C_6H_6BPh_3$-Rh(COD)$^+$ and dppb as catalyst system, CO/$NaBH_4$ can be replaced by cheap synthesis gas CO/H_2 (1:1, 48 atm) in hydroaminomethylation of *N*-allylic amines giving pyrrolidines in 55%–68% yields. Following this report, Saylik and colleagues developed a protocol for the Rh-catalyzed intramolecular hydroaminomethylation of aminoalkenes with CO/H_2 (Scheme 4.24).[96] Using $[Rh(OAc)_2]_2$ and biphephos as catalyst system, *N*-substituted alkenylamines were converted to cyclic amines with a range of medium and large ring sizes in yields up to 85% with high regioselectivity.

In further investigating the Rh-catalyzed hydroaminomethylation of allylic amine reaction, it was discovered that mixtures of [(cod)Rh(hfacac)]with

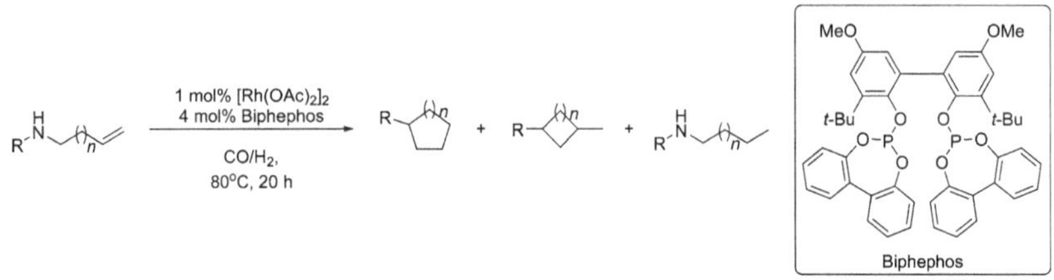

SCHEME 4.24 Intramolecular hydroaminomethylation with biphephos as the ligand.

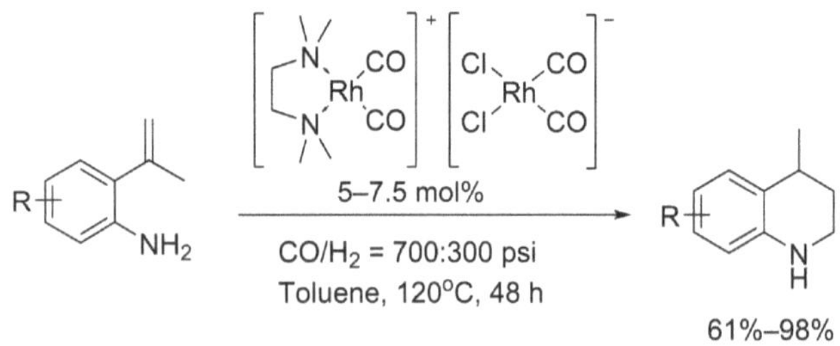

SCHEME 4.25 Rhodium-catalyzed hydroaminomethylation of 2-isopropenylanilines.

4-H^2F^6-TPP under supercritical carbon dioxide ($scCO_2$) exhibited catalytic activity for hydroaminomethylation of ethyl methallylic amine.[97] The observed preference for cyclic amines vs cyclic amide is probably due to reversible formation of the carbamic acid in the solvent CO_2. In addition, Zhang and colleagues identified that mixtures of $Rh(acac)CO_2$ and PPh_3 exhibited catalytic activity for the hydroaminomethylation of cinnamylamine with CO/H_2 (1:1, 20 bar), providing the 4-aryl-2,3-dihydropyrroles in up to 99% yield.[98]

In 2007, Alper et al. employed an ionic diamino rhodium catalyst in intramolecular hydroaminomethylation of 2-isopropenylanilines (Scheme 4.25).[99] A range of 2-isopropenylanilines was converted to give 1,2,3,4-tetrahydroquinolines in 61%–98% yields. Later, this methodology was further extended to the synthesis of seven-membered ring heterocycles benzazepines and 1,2,3,4-tetrahydroquinolines under similar reaction conditions.[100–101] The power of intramolecular hydroaminomethylation was demonstrated with the synthesis of Lycopodium alkaloid lycopladine H by using the intramolecular hydroaminomethylation reaction as key step.[102–103]

4.4 ASYMMETRIC INTERMOLECULAR HYDROAMINOMETHYLATION

The first attempt of an asymmetric one-pot hydroaminomethylation reaction with rhodium/chiral diphosphine systems was introduced in 2014 by Kalcket and colleagues.[104] A variety of chiral diphosphine ligands combined to neutral or cationic rhodium complexes have been studied, with a focus on their performance in asymmetric hydroaminomethylation reaction of styrene with piperidine. Branched amines are obtained in good yields with good chemo- and regioselectivity. However, the complete lack of enantioselectivity, whatever the rhodium-chiral ligand system involved, was observed. The results were explained by DFT calculations which suggested that the hydrogenation of the enamine intermediates has a similar energetic pathway regardless of the chiral phosphine ligands used.

In 2015, Xiao and colleagues developed a protocol for asymmetric hydroaminomethylation of styrenes by using dual metal and organocatalysts (Scheme 4.26).[105] The combination of $Rh(acac)(CO)_2/P(4\text{-}MeOC_6H_5)_3$ and a chiral phosphoric acid as catalyst system, styrenes are converted into β-chiral amines in 45%–87% yields with good enantioselectivities up to 91% utilizing Hantzsch ester as the hydrogen source. This is the first example of an asymmetric hydroaminomethylation where significant enantioselectivities have been achieved. In this dual catalyst system, Rh-phosphine species is responsible for the nonenantioselective hydroformylation while the chiral organocatalyst is responsible for the enantioselective reductive amination via DKR. Unfortunately, an equivalent amount of Hantzsch ester as the hydrogen source is required and only styrenes are applicable. After further optimization of chiral ligand and phosphoric acid,[106] a variety of alkenes including styrenes acrylamides, aliphatic alkenes, and vinyl esters have been shown to be compatible with this reaction. Much higher yields up to 99% with high enantioselectivities up to 99% ee were obtained under 1 bar of syngas.

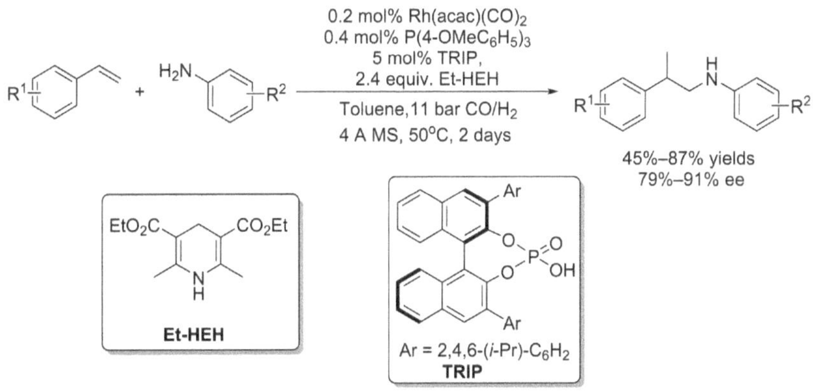

SCHEME 4.26 Metal- and organo-catalyzed asymmetric hydroaminomethylation of styrenes.

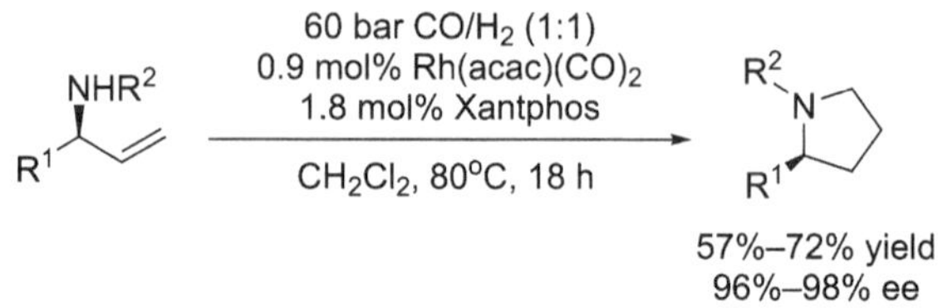

SCHEME 4.27 Enantioselective syntheses of 2-substituted pyrrolidines from allylamines.

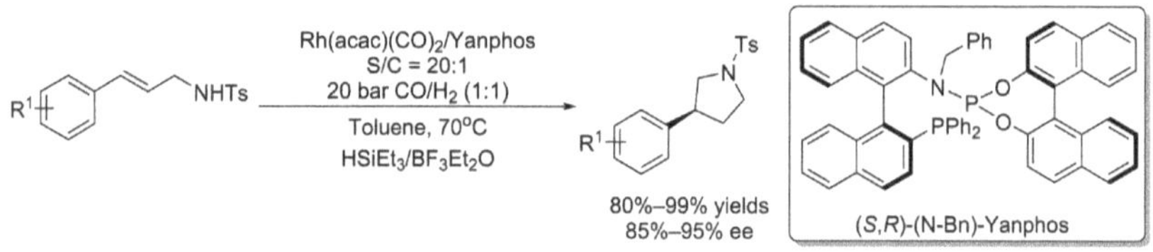

SCHEME 4.28 Rhodium-/Yanphos-catalyzed asymmetric interrupted intramolecular hydroaminomethylation.

4.5 ASYMMETRIC INTRAMOLECULAR HYDROAMINOMETHYLATION

In 2009, the use of substrate control in rhodium-catalyzed intramolecular hydroaminomethylation was reported by Helmchen and coworker (Scheme 4.27).[107] With $Rh(acac)(CO)_2$ and Xantphos as catalyst system, reaction of enantiomerically pure of allylamines was accomplished with complete retention of configuration providing the chiral 2-substituted pyrrolidines in >96% ee. To demonstrate the synthetic utility of this protocol, (*S*)-nicotine and the alkaloid 225°C are synthesized.

In 2016, the first intramolecular interrupted asymmetric hydroaminomethylation reaction was developed by Zhang and colleagues (Scheme 4.28).[108] This reaction proceeds by formation of stabilization of the chiral hemiacetal intermediate, followed by oxidant or reductant to generate the valuable chiral amides or amines. The challenging *trans*-1,2-disubstituted olefins were converted into chiral pyrrolidines in 80%–99% yields with up to 95% ee. It should be noted that the ee remained the same regardless of whether the substrate was *cis* or *trans*.

4.6 CONCLUSIONS AND OUTLOOK

As reviewed in this chapter, transition metal-catalyzed hydroaminomethylation of alkenes has received increasing attention over the past decades. A number of transition metal catalysts (Rh, Ru, Co, Ir, and Au) have been explored for hydroaminomethylation reaction, and Rh has been proven to be the more effective catalyst for this reaction. In addition to unmodified catalyst, catalyst bearing phosphine, nitrogen ligands, and carbenes ligands has also been developed for achieving better activity and selectivity. In order to recycle catalyst, biphasic conditions, thermomorphic solvent systems, and heterogeneous catalyst systems are developed for hydroaminomethylations.

Although tremendous progress has been made in this filed, many challenges still remain to be addressed. First, there is no general catalyst system for the selective hydroaminomethylations. It is expected that future research in this area will focus on the development of more generality and selectivity catalyst systems for the hydroaminomethylations. Moreover, only racemic products are obtained by direct asymmetric hydroaminomethylation with chiral phosphine ligands due to the racemization of the chiral iminium or imine intermediates. Development of new strategies for the highly enantioselective hydroaminomethylation reaction is still in high demand. Furthermore, the direct and selective syntheses of primary amines from ammonia and secondary amines from primary amines still represent a considerable challenge in certain cases. Finally, detailed mechanistic studies will form part of future studies, which are likely to be beneficial for discovery of more efficient protocol for hydroaminomethylation reaction.

REFERENCES

1. Beller, M.; Cornils, B.; Frohning, C. D.; Kohlpaintner, C. W. *J. Mol. Catal. A Chem.* **1995**, *104*, 17.
2. Eilbracht, P.; Barfacker, L.; Buss, C.; et al. *Chem. Rev.* **1999**, *99*, 3329.
3. Fogg, D. E.; dos Santos, E. N. *Coord. Chem. Rev.* **2004**, *248*, 2365.
4. Eilbracht, P.; Schmidt, A. M. *Catalytic carbonylation reactions.* Springer: Berlin Heidelberg, 2006; p.65.
5. Veige, A. S. *Polyhedron* **2008**, *27*, 3177.
6. Crozet, D.; Urrutigoïty, M.; Kalck, P. *ChemCatChem* **2011**, *3*, 1102.
7. Behr, A.; Vorholt, A. J. *Organometallics and renewables.* Springer: Berlin Heidelberg, 2012; p.103.
8. Diebolt, O.; van Leeuwen, P. W. N. M.; Kamer, P. C. J. *ACS Catal.* **2012**, *2*, 2357.
9. Petricci, E.; Cini, E. *Top. Curr. Chem.* **2013**, *342*, 117.
10. Raoufmoghaddam, S. *Org. Biomol. Chem.* **2014**, *12*, 7179.
11. Wu, X. F.; Fang, X.; Wu, L.; Jackstell, R.; Neumann, H.; Beller, M. *Acc. Chem. Res.* **2014**, *47*, 1041.
12. Chen, C.; Dong, X.-Q.; Zhang, X. *Org. Chem. Front.* **2016**, *3*, 1359.
13. Reppe, W. *Experientia* **1949**, *5*, 93.
14. Reppe, W.; Vetter, H. *Justus Liebig Ann. Chem.* **1953**, *582*, 133.
15. Brunet, J. J.; Neibecker, D.; Agbossou, F.; Srivastava, R. S. *J. Mol. Catal.* **1994**, *87*, 223.
16. Iqbal, A. F. M. *Helv. Chim. Acta* **1971**, *54*, 1440.
17. Jones, M. D. *J. Organomet. Chem.* **1989**, *366*, 403.
18. Rische, T.; Eilbracht, P. *Synthesis* **1997**, *1997*, 1331.
19. Rische, T.; Kitsos-Rzychon, B.; Eilbracht, P. *Tetrahedron* **1998**, *54*, 2723.
20. Rische, T.; Eilbracht, P. *Tetrahedron* **1999**, *55*, 7841.

21. Barfacker, L.; Rische, T.; Eilbracht, P. *Tetrahedron* **1999**, *55*, 7177.
22. Rische, T.; Barfacker, L.; Eilbracht, P. *Eur. J. Org. Chem.* **1999**, *1999*, 653.
23. Eilbracht, P.; Kranemann, C. L.; Bärfacker, L. *Eur. J. Org. Chem.* **1999**, *1999*, 1907.
24. Rische, T.; Muller, K. S.; Eilbracht, P. *Tetrahedron* **1999**, *55*, 9801.
25. Koç, F.; Eilbracht, P. *Tetrahedron* **2004**, *60*, 8465.
26. Muller, K. S.; Koc, F.; Ricken, S.; Eilbracht, P. *Org. Biomol. Chem.* **2006**, *4*, 826.
27. Subhani, M. A.; Müller, K.-S.; Eilbracht, P. *Adv. Synth. Catal.* **2009**, *351*, 2113.
28. Kranemann, C. L.; Costisella, B.; Eilbracht, P. *Tetrahedron Lett.* **1999**, *40*, 7773.
29. Schmidt, A.; Marchetti, M.; Eilbracht, P. *Tetrahedron* **2004**, *60*.
30. Lin, Y. S.; El Ali, B.; Alper, H. *Tetrahedron Lett.* **2001**, *42*, 2423.
31. Behr, A.; Levikov, D.; Nürenberg, E. *RSC Adv.* **2015**, *5*, 60667.
32. Behr, A.; Roll, R. *J. Mol. Catal. A Chem.* **2005**, *239*, 180.
33. Martin, A.; Kant, M.; Klein, H.; Jackstell, R.; Beller, M. *J. Supercrit. Fluids* **2007**, *42*, 325.
34. Behr, A.; Seidensticker, T.; Vorholt, A. J. *Eur. J. Lipid Sci. Technol.* **2014**, *116*, 477.
35. Behr, A.; Kleyensteiber, A.; Becker, M. *Chem. Eng. Process. Process Intensif.* **2013**, *69*, 15.
36. Baig, T.; Kalck, P. *J. Chem. Soc. Chem. Commun.* **1992**, 1373.
37. Baig, T.; Molinier, J.; Kalck, P. *J. Organomet. Chem.* **1993**, *455*, 219.
38. Teuma, E.; Loy, M.; Le Berre, C.; Etienne, M.; Daran, J.-C.; Kalck, P. *Organometallics* **2003**, *22*, 5261.
39. Chen, H.; Li, Y.; Chen, J.; Cheng, P.; Li, X. *Catal. Today* **2002**, *74*, 131.
40. Wang, Y. Y.; Luo, M. M.; Li, Y. Z.; Chen, H.; Li, X. J. *Appl. Catal. A Gen.* **2004**, *272*, 151.
41. Gall, B.; Bortenschlager, M.; Nuyken, O.; Weberskirch, R. *Macromol. Chem. Phys.* **2008**, *209*, 1152.
42. Behr, A.; Becker, M.; Reyer, S. *Tetrahedron Lett.* **2010**, *51*, 2438.
43. Karakhanov, E. A.; Runova, E. A.; Kardasheva, Y. S.; Losev, D. V.; Maksimov, A. L.; Terenina, M. V. *Pet. Chem.* **2012**, *52*, 179.
44. Fuentes, J. A.; Wawrzyniak, P.; Roff, G. J.; Bühl, M.; Clarke, M. L. *Catal. Sci. Technol.* **2011**, *1*, 431.
45. Nagy, E.; Heil, B.; Toros, S. *J. Organomet. Chem.* **1999**, *586*, 101.
46. Rische, T.; Eilbracht, P. *Tetrahedron* **1999**, *55*, 1915.
47. Graebin, C. S.; Eifler-Lima, V. L.; da Rosa, R. G. *Catal. Commun.* **2008**, *9*, 1066.
48. Behr, A.; Wintzer, A. *Chem. Eng. Technol.* **2015**, *38*, 2299.
49. Melo, D. S.; Pereira-Júnior, S. S.; dos Santos, E. N. *Appl. Catal. A Gen.* **2012**, *411–412*, 70.
50. Oliveira, K. C. B.; Carvalho, S. N.; Duarte, M. F.; et al. *Appl. Catal. A Gen.* **2015**, *497*, 10.
51. Breit, B. *Tetrahedron Lett.* **1998**, *39*, 5163.
52. Ahmed, M.; Seayad, A. M.; Jackstell, R.; Beller, M. *J. Am. Chem. Soc.* **2003**, *125*.
53. Routaboul, L.; Buch, C.; Klein, H.; Jackstell, R.; Beller, M. *Tetrahedron Lett.* **2005**, *46*, 7401.
54. Ahmed, M.; Bronger, R. P.; Jackstell, R.; Kamer, P. C.; van Leeuwen, P. W.; Beller, M. *Chem. Eur. J.* **2006**, *12*, 8979.
55. Hamers, B.; Bäuerlein, P. S.; Müller, C.; Vogt, D. *Adv. Synth. Catal.* **2008**, *350*, 332.
56. Hamers, B.; Kosciusko-Morizet, E.; Müller, C.; Vogt, D. *ChemCatChem* **2009**, *1*, 103.
57. Briggs, J. R.; Klosin, J.; Whiteker, G. T. *Org. Lett.* **2005**, *7*, 4795.
58. Whiteker, G. T. *Top. Catal.* **2010**, *53*, 1025.
59. Wang, Y.; Chen, J.; Luo, M.; Chen, H.; Li, X. *Catal. Commun.* **2006**, *7*, 979.
60. Petricci, E.; Mann, A.; Salvadori, J.; Taddei, M. *Tetrahedron Lett.* **2007**, *48*, 8501.
61. Monnereau, L.; Sémeril, D.; Matt, D. *Green Chem.* **2010**, *12*, 1670.
62. Khan, S. R.; Khedkar, M. V.; Qureshi, Z. S.; Bagal, D. B.; Bhanage, B. M. *Catal. Commun.* **2011**, *15*, 141.

63. Oliveira, K. C. B.; Santos, A. G.; dos Santos, E. N. *Appl. Catal. A Gen.* **2012**, *445–446*, 204.
64. Khan, S. R.; Bhanage, B. M. *Appl. Organomet. Chem.* **2013**, *27*, 711.
65. Li, S.; Huang, K.; Zhang, J.; Wu, W.; Zhang, X. *Org. Lett.* **2013**, *15*, 1036.
66. Raoufmoghaddam, S.; Drent, E.; Bouwman, E. *ChemSusChem* **2013**, *6*, 1759.
67. Dong, K.; Fang, X.; Jackstell, R.; Beller, M. *Chem. Commun.* **2015**, *51*, 5059.
68. Seidensticker, T.; Vosberg, J. M.; Ostrowski, K. A.; Vorholt, A. J. *Adv. Synth. Catal.* **2016**, *358*, 610.
69. Seidensticker, T.; Busch, H.; Diederichs, C.; von Dincklage, J. J.; Vorholt, A. J. *ChemCatChem* **2016**, *8*, 2890.
70. Crozet, D.; Gual, A.; McKay, D.; Dinoi, C.; Godard, C.; Urrutigoïty, M.; Daran, J.-C.; Maron, L.; Claver, C.; Kalck, P. *Chem. Eur. J.* **2012**, *18*, 7128.
71. Liu, G.; Huang, K.; Ca, C.; Cao, B.; Chang, M.; Wu, W.; Zhang, X. *Chem. Eur. J.* **2011**, *17*.
72. Liu, G.; Li, Z.; Geng, H.; Zhang, X. *Catal. Sci. Technol.* **2014**, *4*, 917.
73. Li, S.; Huang, K.; Zhang, J.; Wu, W.; Zhang, X. *Org. Lett.* **2013**, *15*, 3078.
74. Kostas, I. D. *J. Chem. Res.* **1999**, 630.
75. Seayad, A. M.; Selvakumar, K.; Ahmed, M.; Beller, M. *Tetrahedron Lett.* **2003**, *44*, 1679.
76. Ahmed, M.; Buch, C.; Routaboul, L.; Jackstell, R.; Klein, H.; Spannenberg, A.; Beller, M. *Chem. Eur. J.* **2007**, *13*, 1594.
77. Dutta B., Schwarz R., Omar S., Natour S., Abu-Reziq R. Eur. J. Org. Chem. **2015**, 2015, 1961.
78. Li, K.; Wang, Y.; Xu, Y.; Li, W. J.; Niu, M.; Jiang, J.; Jin, Z. *Catal. Commun.* **2013**, *34*, 73.
79. Schneider, M. J.; Lijewski, M.; Woelfel, R.; Haumann, M.; Wasserscheid, P. *Angew. Chem. Int. Ed.* **2013**, *52*, 6996.
80. Sudheesh, N.; Shukla, R. S. *Appl. Catal. A Gen.* **2013**, *453*, 159.
81. Sudheesh, N.; Shukla, R. S. *Appl. Catal. A Gen.* **2014**, *473*, 116.
82. Nairoukh, Z.; Blum, J. *J Org Chem* **2014**, *79*, 2397.
83. Murata, K.; Matsuda, A.; Masuda, T. *J. Mol. Catal.* **1984**, *23*, 121.
84. Knifton, J. F. *Catal. Today* **1997**, *36*, 305.
85. Jachimowicz, F.; Raksis, J. W. *J Org Chem* **1982**, *47*, 445.
86. Gulak, S.; Wu, L.; Liu, Q.; Franke, R.; Jackstell, R.; Beller, M. *Angew. Chem. Int. Ed.* **2014**, *53*, 7320.
87. Schaffrath, H.; Keim, W. *J. Mol. Catal. A Chem.* **1999**, *140*, 107.
88. Wu, L.; Fleischer, I.; Jackstell, R.; Beller, M. *J. Am. Chem. Soc.* **2013**, *135*, 3989.
89. Wu, L.; Fleischer, I.; Zhang, M.; Liu, Q.; Franke, R.; Jackstell, R.; Beller, M. *ChemSusChem* **2014**, *7*, 3260.
90. Srivastava, V. K.; Eilbracht, P. *Catal. Commun.* **2009**, *10*, 1791.
91. Ali, M.; Gual, A.; Ebeling, G.; Dupont, J. *ChemSusChem* **2016**, *9*, 2129.
92. Karakhanov, E.; Maksimov, A.; Kardasheva, Y.; Runova, E.; Zakharov, R.; Terenina, M.; Kenneally, C.; Arredondo, V. *Catal. Sci. Technol.* **2014**, *4*, 540.
93. Zimmermann, B.; Herwig, J.; Beller, M. *Angew. Chem. Int. Ed.* **1999**, *38*, 2372.
94. Liu, X.; Hamasaki, A.; Yamane, Y.; Aikawa, S.; Ishida, T.; Haruta, M.; Tokunaga, M. *Catal. Sci. Technol.* **2013**, *3*, 3000.
95. Zhou, J. Q.; Alper, H. *J. Org. Chem.* **1992**, *57*, 3328.
96. Bergmann, D. J.; Campi, E. M.; Jackson, W. R.; Patti, A. F.; Saylik, D. *Tetrahedron Lett.* **1999**, *40*, 5597.
97. Wittmann, K.; Wisniewski, W.; Mynott, R.; Leitner, W.; Kranemann, C. L.; Rische, T.; Eilbracht, P.; Kluwer, S.; Ernsting, J. M.; Elsevier, C. J. *Chem. Eur. J.* **2001**, *7*, 4584.
98. Zheng, X.; Cao, B.; Zhang, X. *Tetrahedron Lett.* **2014**, *55*, 4489.
99. Vieira, T. O.; Alper, H. *Chem. Commun.* **2007**, 2710.

100. Vieira, T. O.; Alper, H. *Org. Lett.* **2008**, *10*, 485.
101. Okuro, K.; Alper, H. *Tetrahedron Lett.* **2010**, *51*, 4959.
102. Sacher, J. R.; Weinreb, S. M. *Org. Lett.* **2012**, *14*, 2172.
103. Chauhan, P. S.; Sacher, J. R.; Weinreb, S. M. *Org. Lett.* **2015**, *17*, 806.
104. Crozet, D.; Kefalidis, C. E.; Urrutigoïty, M.; Maron, L.; Kalck, P. *ACS Catal.* **2014**, *4*, 435.
105. Villa-Marcos, B.; Xiao, J. *Chin. J. Catal.* **2015**, *36*, 106.
106. Meng, J.; Li, X. H.; Han, Z. Y. *Org. Lett.* **2017**, *19*, 1076.
107. Helmchen, G.; Dübon, P.; Farwick, A. *Synlett* **2009**, *2009*, 1413.
108. Chen, C.; Jin, S.; Zhang, Z.; Wei, B.; Wang, H.; Zhang, K.; Lv, H.; Dong, X.-Q.; Zhang, X. *J. Am. Chem. Soc.* **2016**, *138*, 9017.

Chapter 5

N-Methyl Amine Synthesis by Reductive Amination of CO_2

5.1 INTRODUCTION

Carbon dioxide (CO_2) is a greenhouse gas and the accumulation of a huge amount of CO_2 has caused negative effects on the global environment. On the other hand, CO_2 is an abundant, nontoxic, easily available, and renewable carbon resource.[1–3] Therefore, the conversion of CO_2 into value-added chemicals is of paramount importance from a standpoint of green and sustainable chemistry.[4–10] In the last decades, it has been determined that CO_2 is a useful synthon to react with small-membered ring compounds,[11–13] unsaturated compounds,[14] hydrogen,[15] and amines,[16–17] to name a few, for constructing C—C, C—O, C—N, and C—M (metal) bonds on the basis of CO_2 activation. It is worth noting that conversion of CO_2 through C—N bond formation could produce a set of value-added chemicals such as methylamines, formamides, oxazolidinones, quinazolines, carbamates, isocyanates, and polyurethanes, just to name a few.[18] This chapter will cover the recent advances in the synthesis of *N*-formamides and *N*-methylamines using CO_2 as C1 feedstock. Given the three main reducing agents, such as H_2, silanes, and boranes, are used in previously mentioned reactions, the following sections will be divided by the type of reductant used.

5.2 N-METHYL AMINE SYNTHESIS FROM CO_2 AND AMINES

N-Methylated amines are important intermediates in the chemical industry. They are used as platform chemicals for the synthesis of fertilizers, fungicides, synthetic leathers, or polymers, or directly utilized as solvents and formulation agents, to name a few.[19–21] Currently, these compounds are obtained through the *N*-methylation of amines with methanol or formaldehyde,[22–23] methyl iodide, orthoesters, dimethyl carbonate, and formic acid as methyl source.[24–28] Considering the importance of applying CO_2 as the C1 source for the production of carbon-based chemicals, it should be an ideal route to synthesize *N*-methyl or *N,N*-dimethyl amines by the reductive methylation of amines with CO_2.

Catalytic Amination for *N*-Alkyl Amine Synthesis. https://doi.org/10.1016/B978-0-12-812284-6.00005-6

5.2.1 Hydrogen as Reducing Agent

5.2.1.1 Cu Catalyst System

In 1995, Cu/Al_2O_3 as catalyst for the methylation of NH_3 with CO_2 and H_2 was first introduced by Baiker and coworkers.[29] The reaction was performed with a fixed-bed reactor at pressure of 0.6 MPa, in the temperature range of 473–553 K, affording a distribution of mono-, di-, and trimethylamine of 1:0.23:0.07 (Scheme 5.1).[30] It should be noted that monoamine was favored at higher temperature and higher NH_3 concentration. Subsequently, comparative study of synthesis of methylamines from CO_2 and ammonia over M/Al_2O_3 was studied.[31] The highest methylamine production rates were obtained with copper-alumina catalysts. Ni-, Co-, Fe-, and Pt-based catalysts showed little methylamine production, but significant methane formation, whereas only CO, H_2O, and HCN were produced with an Ag catalyst. Cu-Mg-Al mixed oxides were also used as the catalysts for this reaction.[32]

In 2014, Shi and coworkers achieved the *N*-methylation of amines, nitrobenzene, or nitrile derivatives with CO_2 and H_2 in the presence of $CuAlO_x$ which was prepared by a coprecipitation method (Scheme 5.2).[33] It is worth noting that primary amines, secondary amines, nitrobenzene, and aromatic nitrile derivatives can be efficiently and selectively transformed into the corresponding *N*-methyl or *N,N*-dimethyl products with up to 96% yields. Kinetic study suggested that *N*-formamides might be the key intermediates. This work represents one of the most general catalyst systems in the catalytic methylation of amines with CO_2 and H_2.

5.2.1.2 Ru Catalyst System

In 2013, the ruthenium-based catalyst systems for *N*-methylation of the amines with CO_2 and H_2 were individually reported by Leitner[122] and Beller[34] et al., respectively (Scheme 5.3). Compared with methylation of primary and secondary aromatic amines in Leitner's protocol, both aromatic and aliphatic amines could react with CO_2 and H_2 readily to produce *N*-methylamines in Beller's protocol. The proposed mechanism for these processes involves the amidation of the amines by CO_2 hydrogenation, followed by the reduction of the formamide intermediate to the corresponding *N*-methylated product.[35–36]

Following these reports, Leitner and coworkers reported the three-component coupling reaction of primary amines, aldehydes, and CO_2 to produce tertiary *N*-methylamines under similar reaction conditions.[37] Subsequently, this process has been extended to ammonia and ammonium chloride

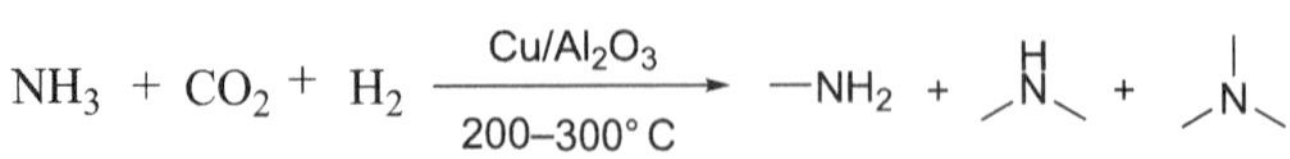

SCHEME 5.1 *N*-Methyl amine synthesis with NH_3, CO_2, and H_2.

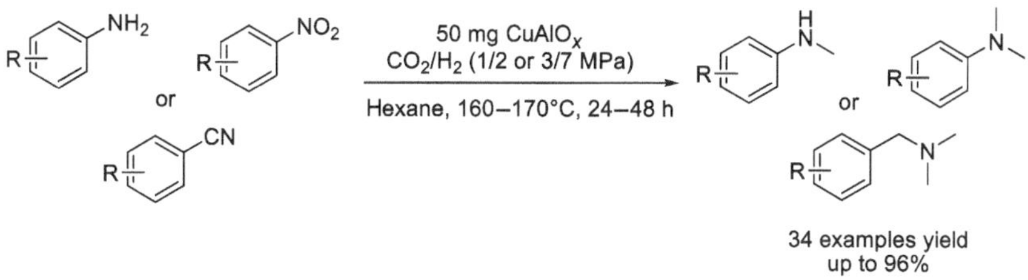

SCHEME 5.2 *N*-Methyl amine synthesis with amine, CO_2, and H_2.

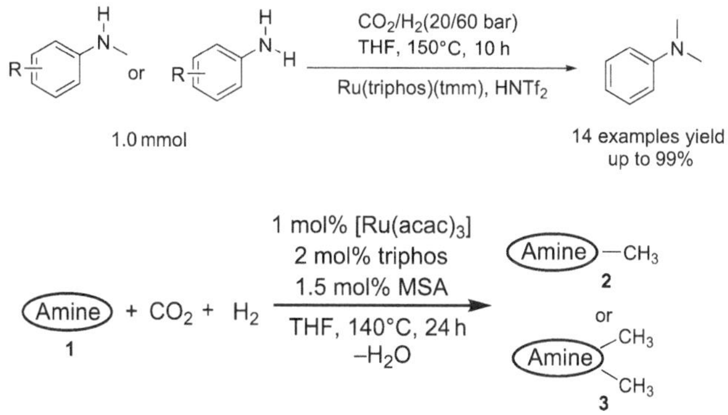

SCHEME 5.3 Ruthenium-catalyzed *N*-methylation of amines with CO_2 and H_2.

for the synthesis of trimethylamine by the use of [Ru(triphos)(tmm)], with and without the addition of $Al(OTf)_3$ as a cocatalyst.[38]

5.2.1.3 *Pd Catalyst System*

In 1997, Baiker described the first example of palladium-catalyzed methylation of the amines, whereby Pd/Al_2O_3 was shown to be the effective catalyst for synthesis of methylamines from CO_2, H_2, and NH_3 in a fixed-bed reactor at 0.6 MPa total pressure in the temperature range of 473–573 K. Nearly two decades later, *N*-methylation of amine and nitro compounds with CO_2/H_2 catalyzed by $Pd/CuZrO_x$ was achieved by Shi and coworkers (Scheme 5.4).[39] It is noteworthy that the primary aromatic amines could be selectively *N*-monomethylated to give *N*-methyl amines.

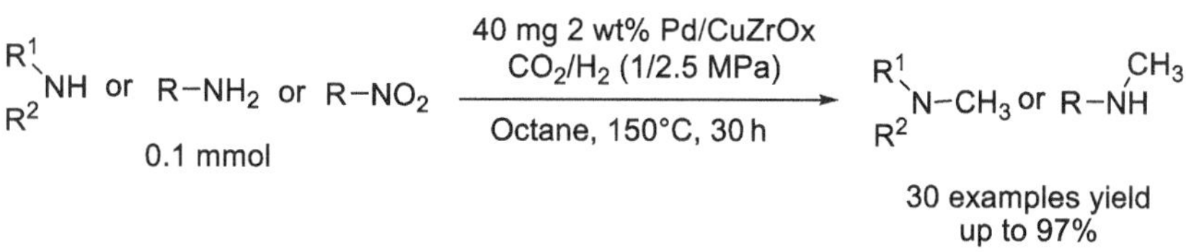

SCHEME 5.4 $Pd/CuZrO_x$-catalyzed methylation of amines using CO_2 and H_2.

5.2.1.4 Pt Catalyst System

In 2014, Shimizu et al. prepared Pt and MoO_x coloaded TiO_2 as effective catalyst for the *N*-methylation of aliphatic and aromatic secondary amines with CO_2/H_2 (5/50 bar).[40] Mechanistic investigations revealed that formic acid and *N*-methylformanilide were key intermediates in the reaction process.

5.2.1.5 Au Catalyst System

In 2015, the methylation of amines with CO_2 and H_2 was reported with very small Au NPs supported on γ-Al_2O_3 as catalyst.[41] The catalyst can be used in the *N*-methylation of both aromatic and aliphatic amines with CO_2/H_2 (2/6 MPa) at 140°C, providing their corresponding tertiary amines in excellent yields.

5.2.2 Hydrosilanes as Reducing Agent

5.2.2.1 Zn Catalyst System

In 2013, Cantat and coworkers[42] describe the *N*-methylation reaction of amines to *N*-methylamines using CO_2 and hydrosilanes catalyzed by a mixture of $ZnCl_2$ (5 mol%) and IPr (5 mol%). The methylation of *N*-methylaniline to *N*,*N*-dimethylaniline was successfully achieved with 95% yield within 20 h at 100°C using 1 bar CO_2 and 2 equiv. phenylsilane.

Two main pathways were proposed for the methylation of N—H bonds using CO_2 and hydrosilanes (paths A and B in Scheme 5.1). Reduction of CO_2 with hydrosilane forms oxysilane or methoxysilane species, which can react with amines to afford formamides and methylamines, respectively. As an alternative, a mechanistic scenario involves formylation of the N—H bond with CO_2 and hydrosilane followed by hydrosilylation of the formamide to the methylamine product (path B) (Scheme 5.5).

Under the similar reaction conditions,[43] the direct [^{11}C]methylation of amines from [^{11}C]CO_2 has been reported for the synthesis of polyethylene

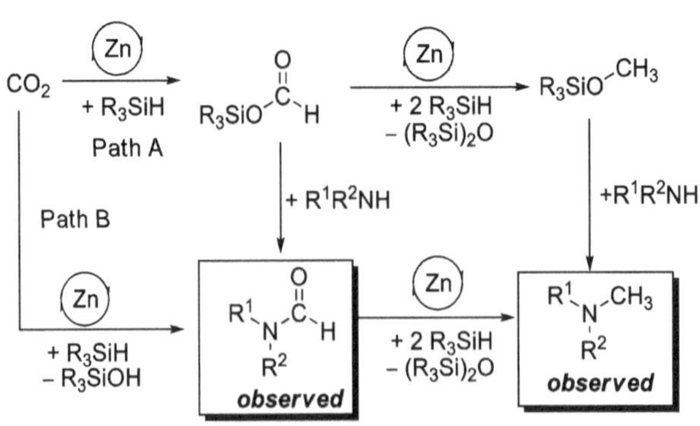

SCHEME 5.5 Proposed pathways for the zinc-catalyzed methylation of amines.

terephthalate radio tracers. In addition, Fluoro-functionalized polymeric NHC-zinc complexes,[44] i.e., [(NHC)H][$ZnXY_2$ (X= Cl, Br)][45] also showed high activity for the reactions of amines with CO_2 and organosilane.

5.2.2.2 Ru Catalyst System

In 2013, Beller and coworkers[46] identified that the mixtures of [$RuCl_2(dmso)_4$] and $nBuPAd_2$ exhibited nice catalytic activity for the methylation of amines with CO_2 (30 bar) and phenyl silane at 100°C. Several kinds of aromatic and aliphatic, secondary, and primary amines were converted into the desired tertiary amines and diamines in 61%–99% yields.

With respect to the mechanism, they proposed the following two reaction pathways for direct methylation with carbon dioxide: (1) formation of formamide occurs through direct reduction of CO_2 to generate A, followed by the reduction to produce the corresponding methylamine product and (2) dehydration of ammonium carbamate under hydrosilylative conditions to generate urea B, followed by reduction to the desired product and one molecule of starting material in the presence of a proton (Scheme 5.6).

5.2.2.3 Cu Catalyst System

In 2015, Cazin et al. developed a protocol for NHC-copper-catalyzed *N*-methylation of amines using CO_2 and silane to provide mixtures of *N*-formylated and *N*-methylated products in reasonable overall yields. On the basis of the DFT calculations, the proposed mechanism suggested that activation of the initial precatalyst to the [Cu(H)(NHC)] species and CO_2 insertion into the Cu—H bond to form the format species I, followed by reaction of the silane to regenerate [Cu(H)(NHC)] and liberate $HC(O)OSiR_3$ (II), which can react with the amine to generate the amide intermediate III, and finally, reduction with hydrosilane to give the mehylation product (Scheme 5.7).[47]

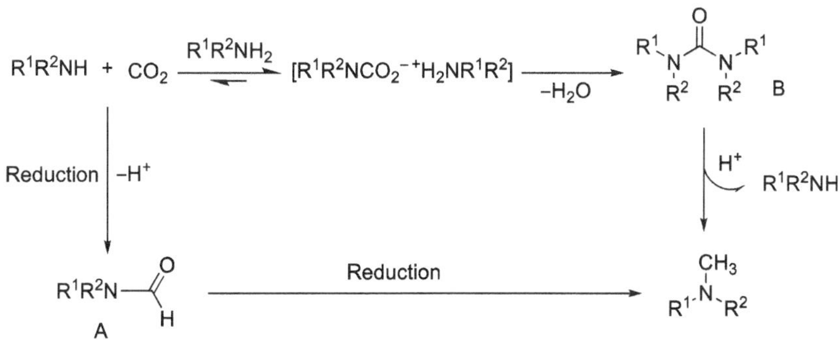

SCHEME 5.6 Proposed pathways for the methylation of amines with CO_2 and silane.

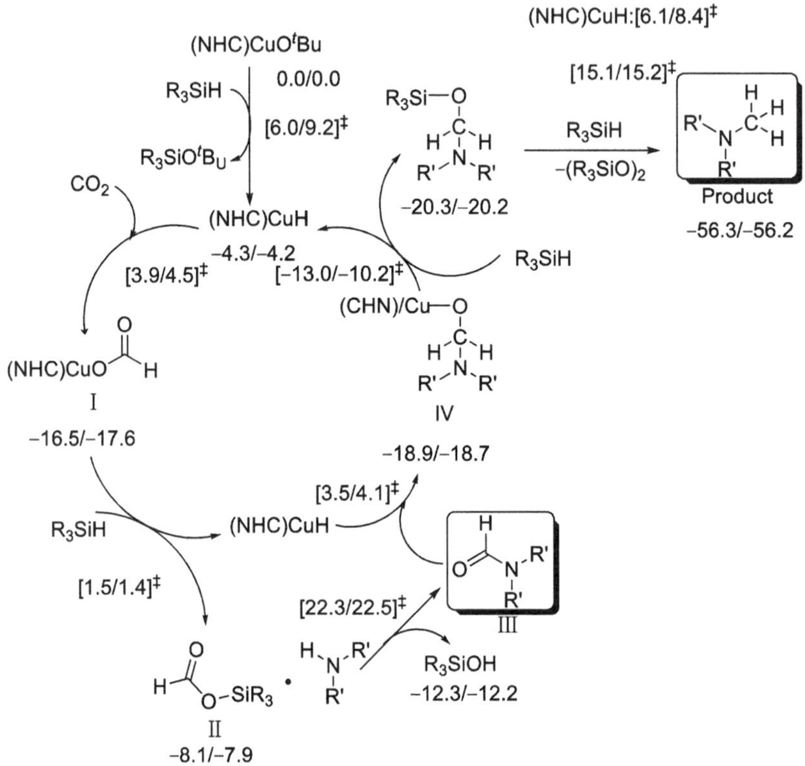

SCHEME 5.7 Pathways for methylation via CO_2 insertion. DFT-calculated ΔG (kcal mol^{-1}) are reported for NHC = IMes and IPr.

5.2.2.4 Ni Catalyst System

In 2015, the Ni(0) complexes catalyzed *N*-methylation of primary and secondary aliphatic amines using CO_2 (1 atm) and $PhSiH_3$ was developed by García and coworkers.[48] With [(dippe)Ni(μ-H)]$_2$ or Ni(COD)$_2$/dcype as catalyst, the reaction gives a mixture of *N*-methylated products and other by-products.

5.2.2.5 Metal-Free Catalyst System

In 2014, Dyson and coworkers employed a metal-free catalyst system for the efficient methylation of various amines with CO_2 (1 atm) and hydrosilane at 50°C.[49] It should be noted that the catalyst system tolerates a broad range of substrates/functional groups such as nitrile and nitro groups, double and triple bonds, providing the corresponding *N*-methylated amines in 63%–91% yields. In addition, $B(C_6F_5)_3$ was also identified as an effective catalyst for *N*-methylation of amines/imines in the presence of $PhSiH_3$ under 0.5 MPa CO_2 pressure, providing the desired products in 26%–93% yields.[50]

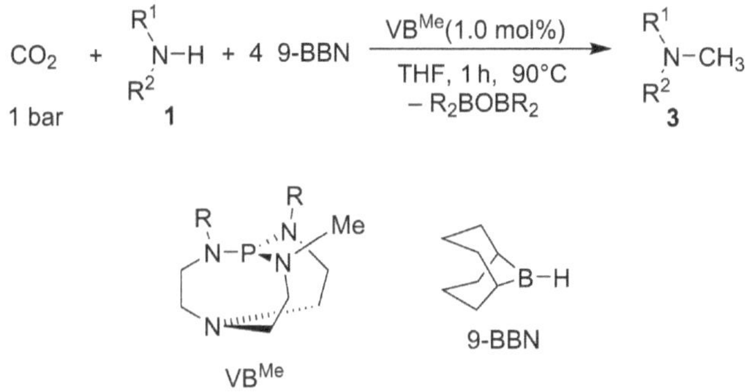

SCHEME 5.8 *N*-Methylation of amines using CO_2 and hydroboranes.

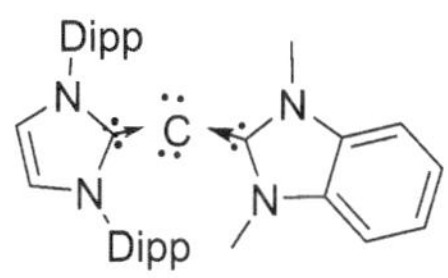

SCHEME 5.9 CDCs 7e.

5.2.3 Boranes as Reducing Agent Under Transition Metal-Free Catalyst System

Cantat et al. describe the first metal-free methylation of amines using CO_2 with hydroboranes as reducing agents.[51] Using proazaphosphatrane as catalyst, a large scope of substrates are transformed into the *N*-methylamines in 33%–99% yields (Scheme 5.8).

The proposed mechanism for this process involved the reduction of CO_2 to a formoxyborane intermediate HCOOBBN, which serves as an electrophile to facilitate the formation of the C—N bond and affords a formamide intermediate. The reduction of the R_2N-COH formamide with 9-BBN affords the methylamine R_2N-CH_3.

Two years later, Ong et al. succeed in constructing CDCs 7e as catalyst for methylation of amines using CO_2 with hydroboranes, providing methylamines in 24%–90% yields (Scheme 5.9).[52]

5.3 CONCLUSIONS AND OUTLOOK

Recent progress in the area of conversion of CO_2 to *N*-methylamines has been reviewed and summarized in this chapter. Molecular H_2 is the desirable reductant because it is the cleanest and atom economical reductive source. However, hydrogenation of CO_2 often requires noble metal catalysts under harsh reaction conditions. In contrast, hydrosilanes and hydroboranes possess a mild reduction potential for CO_2 reduction. With these reductants, the methylation of amines with CO_2 has been achieved under atmospheric pressure of CO_2 at room temperature. Nevertheless, these methodologies are limited by the availability of reductants and the formation of unwanted by-products which required a subsequent workup step for separation. Although significant progress has been made in this field, many challenges still remain to be addressed. Future efforts will focus on developing a catalyst system with high activity that is capable of activating CO_2 under mild conditions. Lastly, detailed mechanistic investigations in the reductive functionalization of CO_2 will continue to play an important role in rational design of more effective catalysts.

REFERENCES

1. Choi, S.; Drese, J. H.; Jones, C. W. *ChemSusChem* **2009**, *2*, 796.
2. Li, L.; Li, Y.; Wen, X.; et al. *Energy Fuels* **2011**, *25*, 3835.
3. Wang, Q.; Luo, J.; Zhong, Z.; Borgna, A. *Energy Environ. Sci.* **2011**, *4*, 42.
4. Arakawa, H.; Aresta, M.; Armor, J. N.; et al. *Chem. Rev.* **2001**, *101*, 953.
5. Aresta, M.; Dibenedetto, A. *Dalton Trans.* **2007**, *28*, 2975.
6. Mikkelsen, M.; Jørgensen, M.; Krebs, F. C. *Energy Environ. Sci.* **2010**, *3*, 43.
7. Riduan, S. N.; Zhang, Y. *Dalton Trans.* **2010**, *39*, 3347.
8. Sakakura, T.; Choi, J.-C.; Yasuda, H. *Chem. Rev.* **2007**, *107*, 2365.
9. Wang, J.-L.; Miao, C.-X.; Dou, X.-Y.; Gao, J.; He, L.-N. *Curr. Org. Chem.* **2011**, *15*, 621.
10. Omae, I. *Catal. Today* **2006**, *115*, 33.
11. North, M.; Pasquale, R.; Young, C. *Green Chem.* **2010**, *12*, 1514.
12. Yang, Z. Z.; He, L. N.; Miao, C. X.; Chanfreau, S. *Adv. Synth. Catal.* **2010**, *352*, 2233.
13. Yang, Z.-Z.; He, L.-N.; Peng, S.-Y.; Liu, A.-H. *Green Chem.* **2010**, *12*, 1850.
14. Gu, Y.; Shi, F.; Deng, Y. *J. Org. Chem.* **2004**, *69*, 391.
15. Zhang, Z.; Xie, Y.; Li, W.; et al. *Angew. Chem. Int. Ed.* **2008**, *47*, 1127.
16. Jiang, T.; Ma, X.; Zhou, Y.; Liang, S.; Zhang, J.; Han, B. *Green Chem.* **2008**, *10*, 465.
17. Shi, F.; Deng, Y.; SiMa, T.; Peng, J.; Gu, Y.; Qiao, B. *Angew. Chem. Int. Ed.* **2003**, *42*, 3257.
18. Yang, Z.-Z.; He, L.-N.; Gao, J.; Liu, A.-H.; Yu, B. *Energy Environ. Sci.* **2012**, *5*, 6602.
19. Jørgensen, K.; Ricci, A. *Modern Animation Methods.* Wiley-VCH: Weinheim, 2000.
20. Rappoport, Z. *The Chemistry of Anilines. 169. John Wiley & Sons: England, 2007.*
21. Salvatore, R. N.; Yoon, C. H.; Jung, K. W. *Tetrahedron* **2001**, *57*, 7785.
22. Bautista, F.; Campelo, J.; Garcia, A.; et al. *J. Catal.* **1997**, *172*, 103.
23. Niphadkar, P.; Joshi, P.; Gurav, H.; Deshpande, S.; Bokade, V. *Catal. Lett.* **2009**, *133*, 175.
24. Clarke, H. T.; Gillespie, H. B.; Weisshaus, S. Z. *J. Am. Chem. Soc.* **1933**, *55*, 4571.
25. Eschweiler, W. *Eur. J. Inorg. Chem.* **1905**, *38*, 880.
26. Farkas, E.; Sunman, C. J. *J. Org. Chem.* **1985**, *50*, 1110.
27. Tundo, P.; Selva, M. *Acc. Chem. Res.* **2002**, *35*, 706.
28. Zhao, Y.; Foo, S. W.; Saito, S. *Angew. Chem. Int. Ed.* **2011**, *50*, 3006.
29. Gredig, S. V.; Koeppel, R. A.; Baiker, A. *J. Chem. Soc. Chem. Commun.* **1995**, *1*, 73.
30. Gredig, S. V.; Koeppel, R. A.; Baiker, A. *Catal. Today* **1996**, *29*, 339.
31. Gredig, S. V.; Koeppel, R.; Baiker, A. *Appl. Catal. A Gen.* **1997**, *162*, 249.
32. Auer, S. M.; Gredig, S. V.; Köppel, R. A.; Baiker, A. *J. Mol. Catal. A Chem.* **1999**, *141*, 193.
33. Cui, X.; Dai, X.; Zhang, Y.; Deng, Y.; Shi, F. *Chem. Sci.* **2014**, *5*, 649.
34. Li, Y.; Sorribes, I.; Yan, T.; Junge, K.; Beller, M. *Angew. Chem. Int. Ed.* **2013**(46)

35. Beydoun, K.; vom Stein, T.; Klankermayer, J.; Leitner, W. *Angew. Chem. Int. Ed.* **2013**, *52*, 9554.
36. Tlili, A.; Frogneux, X.; Blondiaux, E.; Cantat, T. *Angew. Chem. Int. Ed.* **2014**, *53*, 2543.
37. Beydoun, K.; Ghattas, G.; Thenert, K.; Klankermayer, J.; Leitner, W. *Angew. Chem. Int. Ed.* **2014**, *53*.
38. Beydoun, K.; Thenert, K.; Streng, E. S.; Brosinski, S.; Leitner, W.; Klankermayer, J. *ChemCatChem* **2016**, *8*, 135.
39. Cui, X.; Zhang, Y.; Deng, Y.; Shi, F. *Chem. Commun.* **2014**, *50*.
40. Kon, K.; Siddiki, S. M.; Onodera, W.; Shimizu, K. *Chem. A Eur. J.* **2014**, *20*, 6264.
41. Du, X. L.; Tang, G.; Bao, H. L.; et al. *ChemSusChem* **2015**, *8*, 3489.
42. Jacquet, O.; Frogneux, X.; Gomes, C. D. N.; Cantat, T. *Chem. Sci.* **2013**, *4*, 2127.
43. Liger, F.; Eijsbouts, T.; Cadarossanesaib, F.; Tourvieille, C.; Le Bars, D.; Billard, T. *Eur. J. Org. Chem.* **2015**, *2015*, 6434.
44. Yang, Z.-Z.; Yu, B.; Zhang, H.; Zhao, Y.; Ji, G.; Liu, Z. *RSC Adv.* **2015**, *5*, 19613.
45. Santoro, O.; Nahra, F.; Cordes, D. B.; Slawin, A. M.; Nolan, S. P.; Cazin, C. S. *J. Mol. Catal. A Chem.* **2016**, *423*, 85.
46. Li, Y.; Fang, X.; Junge, K.; Beller, M. *Angew. Chem. Int. Ed.* **2013**, *52*, 9568.
47. Santoro, O.; Lazreg, F.; Minenkov, Y.; Cavallo, L.; Cazin, C. S. *Dalton Trans.* **2015**, *44*.
48. González-Sebastián, L.; Flores-Alamo, M.; García, J. J. *Organometallics* **2015**, *34*, 763.
49. Das, S.; Bobbink, F. D.; Laurenczy, G.; Dyson, P. J. *Angew. Chem. Int. Ed.* **2014**, *53*.
50. Yang, Z.; Yu, B.; Zhang, H.; et al. *Green Chem.* **2015**, *17*, 4189.
51. Blondiaux, E.; Pouessel, J.; Cantat, T. *Angew. Chem. Int. Ed.* **2014**, *53*.
52. Chen, W. C.; Shen, J. S.; Jurca, T.; et al. *Angew. Chem. Int. Ed.* **2015**, *54*.

Chapter 6

N-Alkyl Amine Synthesis by Oxidative Amination of Alkane

6.1 INTRODUCTION

N-Alkyl amines are an important class of building blocks which can be used for the synthesis of a variety of biologically active molecules, pharmaceutical, and agrochemical products. Given that inert $C(sp^3)$—H bonds are prevalent in organic compounds, the direct amination of these bonds represents a faster and atom-economical synthetic approach for synthesis of the alkyl amines. However, because of thermodynamic and kinetic stability of the $C(sp^3)$—H bond of the alkanes, catalytic amination of these bonds is a longstanding challenge. Direct $C(sp^3)$—H bond amination reactions are limited by two issues: 1) intrinsic reactivity and 2) the site selectivity. These issues can be overcome by the use of metal catalysts and/or at high temperatures. Over the past years, extensive efforts have been made for the catalytic aminations of $C(sp^3)$—H bonds.[1–19] In this chapter, the catalytic aminations of $C(sp^3)$—H bonds will be summarized. For the scope of this overview, the aminations of $C(sp^3)$—H bonds with azide, CH_3CN, *N*-triflylimino-λ^3-bromane, $TsNBr_2$, among others, will not be considered because it strays from the subject of focus.

In 1955, Saunders described the first $C(sp^3)$—H amination of 1-o-azidophenylpiperidine.[20] The nitrenes were generated by thermal decomposition azido-group in 1-o-azidophenylpiperidine, followed by C—H bond insertion. Since then, a variety of azides, including azidobenzenes,[21–23] azidoformates,[24,25] sulfonyl azides,[23,26,27] and cyanogen azide,[28] have been used for amination of the $C(sp^3)$—H bonds. Given that the high reactivity of free nitrenes is prone to result in numerous side reactions, metal-mediated $C(sp^3)$—H bond aminations with nitrenes have received considerable attention. The first example of metal-mediated C—H amination was reported by Kwart and Kahn in 1967 (Scheme 6.1).[29] Using copper as catalyst, decomposition of benzenesulfonyl azide in cyclohexene solution could afford allylic sulfonamide (3%), enamine (17%), and aziridine (15%). A low yield of 6% was also obtained by Breslow in the amination of cyclohexane with dichloramine-T in the presence of zinc.[30] In 1969, much higher yield of 70% was achieved by Turner in the amination of 1,4-dioxane with chloramine-T in the presence of copper powder at room temperature.[31]

Catalytic Amination for *N*-Alkyl Amine Synthesis. https://doi.org/10.1016/B978-0-12-812284-6.00006-8

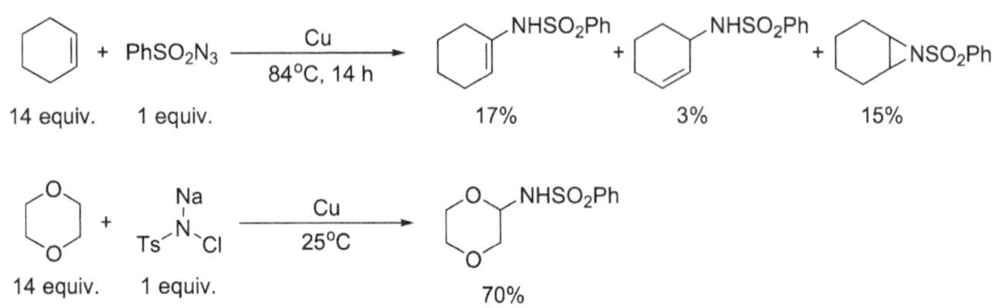

SCHEME 6.1 Early copper-mediated C—H aminations of cyclohexene and 1,4-dioxane.

An important breakthrough for the amination of the C(sp^3)—H bonds was achieved by using iminoiodanes (PhI=NR) function as nitrene precursors. An iron(III)- and manganese(III)-porphyrins catalyst for inter- and intramolecular amination of alkanes with iminoiodanes was first introduced in 1982 by Breslow and colleagues.[32–34] However, these processes are hampered by limitations of iminoiodinane preparation (i.e., scope and stability).

Another advance in this field was made by both Che and Du Bois who developed a protocol for in situ preparation of an iminoiodinane species by simply mixing an amide starting material with PhI(OAc)$_2$.[35,36] This work opened the door for direct using amines as nitrogen source in C(sp^3)—H bonds amination.

6.2 INTERMOLECULAR OXIDATIVE AMINATION OF ALKANES

6.2.1 Cu Catalyst System

The first transition metal-catalyzed C—H amination involving a nitrene was introduced by Kwart and Khan in 1967 (Scheme 6.2).[29] With copper as a catalyst, decomposition of benzenesulfonyl azide for insertion into C—H bond of cyclohexene has been reported to generate a mixture of C—H insertion and aziridination products. This transformation opens the door for the development of C—H amination of the inactivated C(sp^3)—H bonds.

Nearly four decades later, Pérez and colleagues developed copper-homoscorpionate catalyst Tp^{Br3}Cu(NCMe) (Scheme 6.3A) facilitating amination of benzylic C—H bonds and inactive aliphatic alkanes with PhI=NTs in moderate to high yield.[37,38] Following this report, β-diketiminato copper-(II) anilides (Scheme 6.3B),[39,40] and [CuI(TMG$_3$trphen)][PF$_6$] (Scheme 6.3C)[41] were also developed for the amination of unactivated C(sp^3)—H bond with aromatic amines and PhI=NTs, respectively.

In 2007, Zhao and colleagues demonstrated a copper-catalyzed amination of unactivated C(sp^3)—H bonds adjacent to a nitrogen atom with amides (Scheme 6.4).[42] The C—H bond amination reaction proceeds with up to 78% yield utilizing CuBr/^{t}BuOOH catalyst system. Subsequently, this methodology has been extended to copper-catalyzed amination of benzylic and aliphatic alkanes C—H

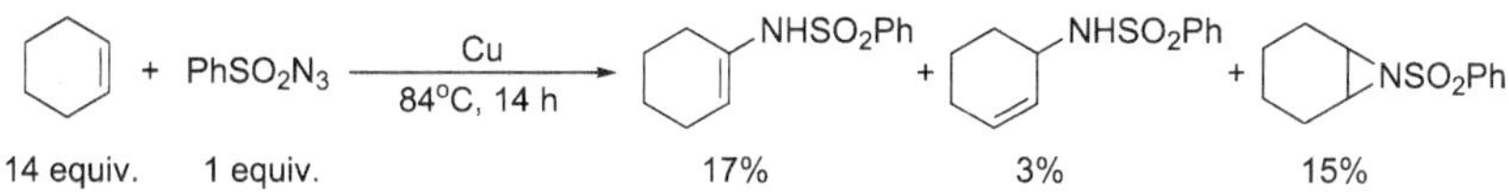

SCHEME 6.2 Copper-catalyzed C—H amination of benzenesulfonyl azide with cyclohexene.

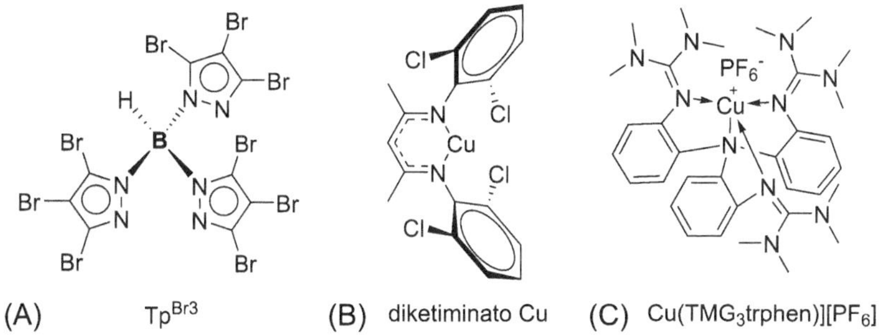

SCHEME 6.3 Copper catalysts of (A) $(Tp^{Br3}Cu(NCMe)$, (B) β-diketiminato copper-(II) anilides, and (C) $[CuI(TMG_3trphen)][PF_6]$.

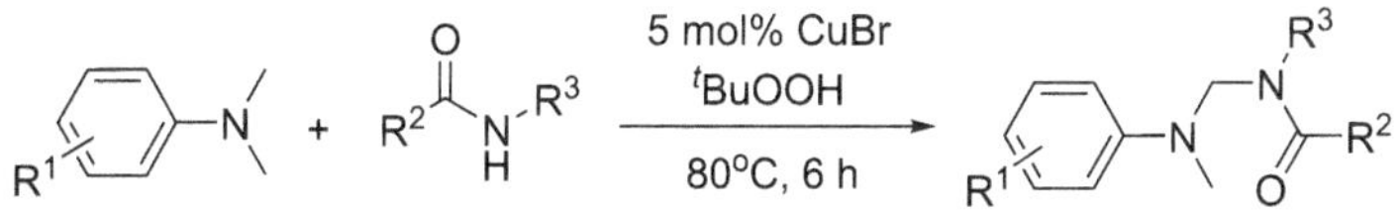

SCHEME 6.4 Copper-catalyzed amination of *N,N*-dimethylaniline derivatives.

bonds with sulfonamides,[43,44] and aromatic and aliphatic amines.[43] In addition, oxidative amination of ethereal with indole/carbazole derivatives could also be realized with this catalyst system.[45]

6.2.2 Fe Catalyst System

In 1982, the first example Fe-catalyzed intermolecular amination of the $C(sp^3)$—H bonds was reported by Breslow and colleagues. Treatment of cyclohexane with tosylimidoiodobenzene (PhI=NTs) in the presence of Fe(III) porphyrins chloride at room temperature affords *N*-cyclohexyltoluene-*p*-sulphonamide in 3.1% yield.[32]

Two decades later, Nicholas and colleagues introduced a photoassisted amination of allylic $C(sp^3)$—H with nitroarenes and carbon monoxide in the presence of $[Cp^*Fe(CO)_2]_2$ to afford *N*-allyl amines in 11%–100% GC yields (Scheme 6.5).[46]

In 2008, Fu and colleagues disclosed a $FeCl_2$/NBS catalyst system for amination of benzylic $C(sp^3)$—H bonds with carboxamides and sulfonamides (Scheme 6.6).[47] The activity order of the benzylic reagents is

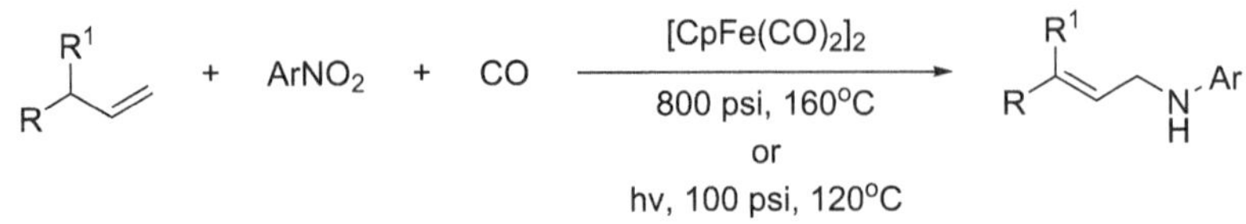

SCHEME 6.5 Fe-catalyzed amination of allylic C—H with nitroarenes and CO.

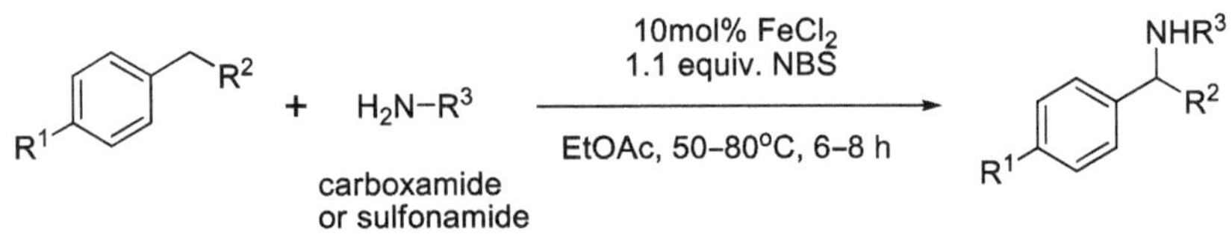

SCHEME 6.6 $FeCl_2$-catalyzed amination of benzylic C—H bonds in the presence of NBS.

diphenylmethane > ethylbenzene > 4-bromoethylbenzene. Subsequent investigation by Chen and colleagues established a protocol for the Fe-catalyzed amination of benzylic $C(sp^3)$—H bonds with imidazoles using a similar $FeCl_2$/DTBP catalyst system.[48] Besides benzylic C—H bonds, $C(sp^3)$—H bond adjacent to a nitrogen atom, such as *N*-substituted pyrrolidin-2-ones and amides, could also be used as reaction partners for $C(sp^3)$—H bond amination with aryl azoles,[49] amides,[50] and arylamines[51] when Fe salt/peroxide is employed as catalyst system.

6.2.3 Mn Catalyst System

In 1982, Breslow and colleagues documented the first example Mn-catalyzed intermolecular amination of the $C(sp^3)$—H bonds. Using Mn(III) porphyrins chloride as catalyst, amination of cyclohexane with tosylimidoiodobenzene (PhI=NTs) at room temperature affords *N*-cyclohexyltoluene-*p*-sulphonamide in 6.8% yield.[32] Subsequent investigation by Mansuy and colleagues found that reaction of alkenes with a high-valent Mn-nitrene favored the allylic amination over the competing aziridination reaction.[52] Mn(TDCPP)(ClO_4) facilitates the allylic amination of cyclohexene with yields up to 70% yield. To demonstrate the synthetic utility of this methodology, the group of Breslow reported manganese porphyrin-catalyzed amination of the aromatic steroid equilenin acetate with PhI=NTs to produce the 11 β-amination product in around 40% yield.[53] Additionally, some striking advances were achieved by Che and coworkers who employed a Mn complex bearing an electron-deficient porphyrin macrocycle (TPFPP) for amination of a set of hydrocarbons, including indane, cyclohexene, or tetrahydrofuran, with good to excellent yields (Scheme 6.7).[35] Noteworthy elements of this process included the use of the alkane as the limiting substrate, the in situ formation of the iminoiodinane, and a high turnover (up to 2600).

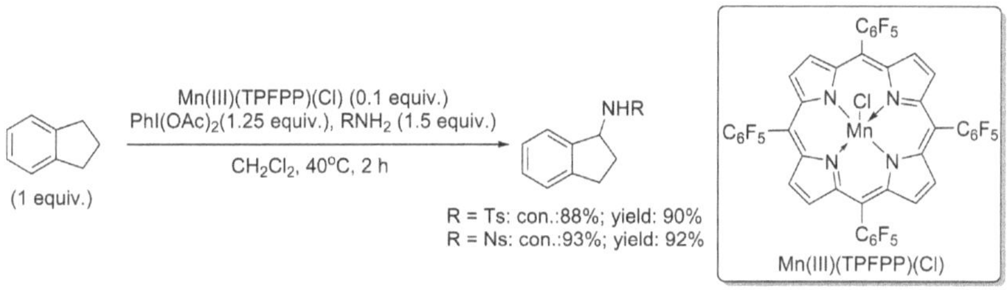

SCHEME 6.7 Amidation of indane catalyzed by manganese(III)-porphyrins.

A manganese-catalyzed asymmetric intermolecular amination of saturated C—H bonds of ethylbenzene and ethylnaphthalenes with PhI=NTs was reported by Che and coworkers in 1999.[54] The amination proceeds up to 85% yield with up to 58% ee utilizing the chiral manganese(iii) porphyrins catalyst system. Subsequently, the same catalyst system was applied for the enantioselective C—H amination of benzylic substrates using $MsNH_2/PhI(OAc)_2$ as an in situ generating system of iminoiodinane.[55] Moderate enantioselectivities of up to 56% ee were observed. A major breakthrough was realized in 2001 by Katsuki who developed a chiral (salen)manganese(III) complex (Scheme 6.8A) that was applied for the enantioselective C—H amination of benzylic and allylic substrates with PhI=NTs.[56] High yields were obtained and in the case of 1,1-dimethylindan, a high enantioselectivity (89% ee) was obtained.

6.2.4 Ru Catalyst System

In 1996, Cenini and colleagues developed a protocol for amination of allylic C—H bond with nitroarene using CO as reductant.[57] Utilizing a $Ru_3(CO)_{12}$/Ph-BIAN catalyst system, allylic amination reaction proceeds with up to 81.9% selectivity when nitroarenes bearing electron-withdrawing substituents.[58]

In 1998, a bis(tosyl)amidoruthenium(III) complex of 1,4,7-trimethyl-1,4,7-triazacyclononane (Me_3tacn) was prepared and applied for amination of saturated C—H bonds.[59] When PhI=NTs was utilized as the nitrogen source, this

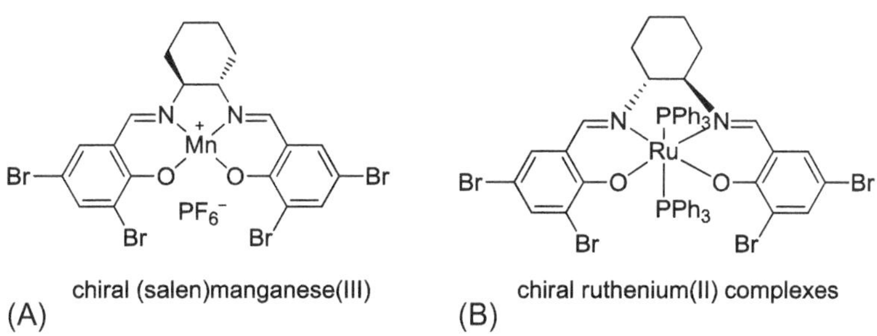

SCHEME 6.8 (A) Chiral (salen)manganese(III) and (B) ruthenium(II) complexes.

reaction could also become catalytic with up to 75 turnovers. Upon further optimization of this reaction, ruthenium cyclic amine or bipyridine complexes were observed to possess significantly improved reactivity, and amination of adamantane, cyclohexene, ethylbenzene, cumene, indane, tetralin, and diphenylmethane afforded *N*-substituted sulfonamides in 80%–93% yields.[60] Using an in situ formation of the iminoiodinane, high yields up to 89% could also be achieved. In addition, [Ru(TPFPP)(CO)] also displays good activity in amination of PhI=NTs with a variety of $C(sp^3)$—H bonds, including benzylic, allylic, and adjacent to ethers, to give the desired products with yields of up to 90%.[35] On the basis of the kinetic and deuterium isotope effect studies, the proposed mechanism for this process involves the rate-limiting formation of a carboradical intermediate.[61]

With respect to the Ru-catalyzed asymmetric amination of $C(sp^3)$—H bond, chiral ruthenium(II) porphyrin complexes are used to catalyze intermolecular C—H amination of benzylic substrates, albeit with up to 43% ees.[54,55] In 2002, chiral ruthenium(II)-salen complexes RuII(salen)$(PPh_3)_2$ (Scheme 6.8B) were identified as an effective catalyst for asymmetric amination of silyl enol ethers, providing the desired products with up to 97% ees.[62]

6.2.5 Rh Catalyst System

In 1997, Müller and colleagues reported the first example of the Rhodium-catalyzed intermolecular amination of $C(sp^3)$—H bonds with NsN=IPh.[63] The $Rh_2(OAc)_4$ displays good activity in decomposition of NsN=IPh to give nitrenes, which can be inserted into a variety of $C(sp^3)$—H bonds, including benzylic, allylic, and adjacent to ethers, to give the desired products with yields of up to 84%. Utilizing the [Rh_2(*R*)-bnp_4] as catalyst, indane can be reacted at room temperature in 71% yield with 31% ee (Scheme 6.9). This work represents the first example of enantioselective C—H amination reaction.

In 2007, Bois and colleagues described the rhodium-catalyzed intermolecular $C(sp^3)$—H amination of benzylic and 3 degrees C—H bonds with $TcesNH_2$[64] (Scheme 6.10). Using the $Rh_2(esp)_2$ as catalyst and $PhI(O_2CtBu)_2$ as the oxidant,

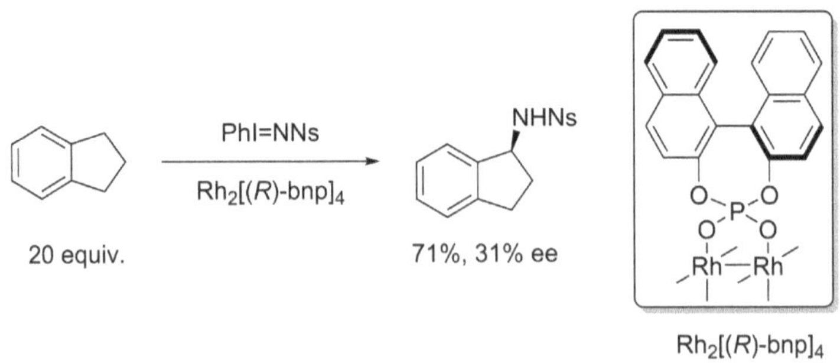

SCHEME 6.9 Rh-catalyzed asymmetric amination of indane with NsN=IPh.

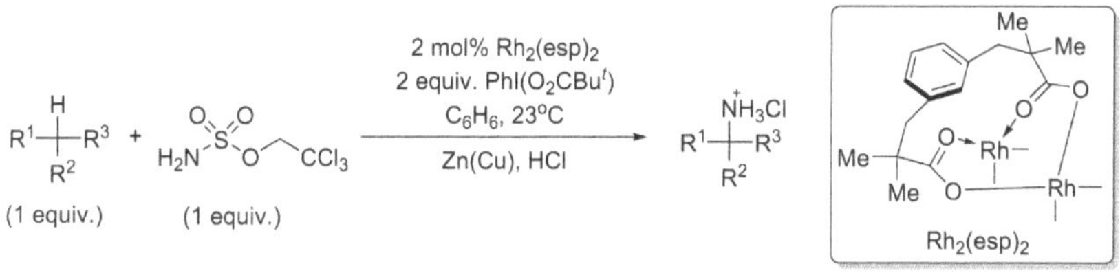

SCHEME 6.10 Rhodium(II)-catalyzed intermolecular amination using $TcesNH_2$.

a range of benzylic amines and tetrasubstituted amines are produced with up to 74% yield. The Tces group is easily removed under mild reductive conditions, which enhances the practicability of this method. The high performance exhibited by $Rh_2(esp)_2$ for this reaction could be rationalized by the kinetic stability of its mixed-valence dimer toward further oxidative damage.[65] Moreover, C—H bonds amination of benzylic methylene groups in the α-position to a stereogenic center can also be realized under the similar catalyst system, providing the desired products with good (dr = 80/20) to excellent (dr > 95/5) diastereoselectivity.[66,67] The DFT calculations suggest a triplet stepwise pathway is favorable for $Rh_2(esp)_2$-catalyzed intermolecular C—H amination.[68] An enantioselective C—H amination of benzylquinolones with $TcesNH_2$ was presented by Bach and colleagues in 2013.[69] Moderate enantioselectivities up to 70% ee were achieved with a chiral C_2-symmetric Rh(II) complex via hydrogen bonds.

In 2006, an important breakthrough was realized by Reddy and Davies who developed chiral $Rh_2(S\text{-TCPTAD})_4$ for the enantioselective C—H amination of benzylic substrates.[70] Much higher enantioselectivities of up to 94% ee may be achieved for amination of the indane using $NsNH_2/PhI(OAc)_2$ as an in situ generating system of iminoiodinane. Unfortunately, these protocols required the use of excess substrate with respect to the nitrogen source. Three years later, the Hashimoto group developed new chiral dirhodiumtetracarboxylates $Rh_2(S\text{-TCPTTL})_4$ with pendent phthalamide arms facilitating enantioselective aminations of allylic C—H bond of silyl enol ethers with enantioselectivities as high as 73%.[71] Highlighting the synthetic utility of this methodology, a new catalytic asymmetric route to (−)-pancracine has been developed.

In 2006, Dauban and colleagues discovered a highly efficient rhodium-catalyzed intermolecular $C(sp^3)$—H amination with a enantiomerically pure sulfonimidamide.[72] Using $[Rh_2(S)\text{-nttl}_4]$ as catalyst, this reaction occurs with excellent diastereoselectivities of up to 99%, particularly at secondary benzylic positions. The high reactivity and selectivity can be rationalized by a pronounced matched effect between the enantiomerically pure sulfonimidamide and the chiral rhodium catalyst. Following this report, they studied the reaction between a sulfur(VI) compound and an iodine(III) oxidant in the presence of $[Rh_2(S)\text{-nta}_4]$, which lead to the formation of a chiral metallanitrene of unprecedented reactivity (Scheme 6.11).[18] The latter allows intermolecular C—H

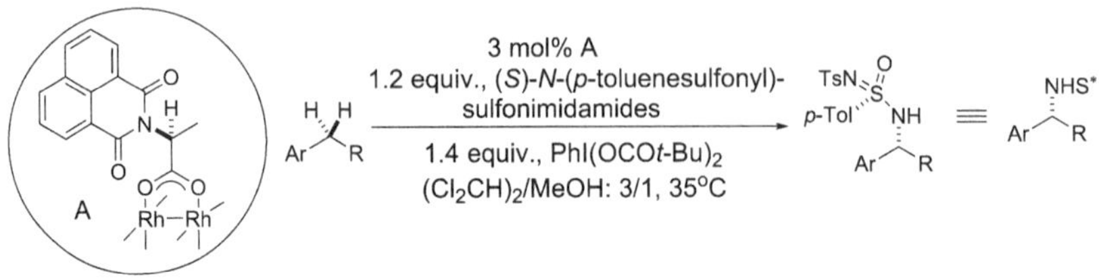

SCHEME 6.11 Catalytic C—H amination of benzylic substrates.

amination to proceed in very high yields up to 92% and excellent diastereoselectivities up to 99%. The scope of this reaction includes benzylic and allylic substrates as well as alkanes. Secondary positions react preferentially, but insertion into activated primary C—H bonds or sterically accessible tertiary sites is also possible. Moreover, using the similar catalyst system, amination of terpenes and enol ethers allylic C—H occurs with excellent yields as well as with high regio- and diastereoselectivity.[73,74] The chemoselectivity can be rationalized by conjugation of allylic C—H bonds with the π-bond. Alkanes are also efficiently functionalized with a net preference for tertiary equatorial C—H bonds. The steric and hyperconjugative effects would explain this.

Noteworthy, Lebel and colleagues developed the *N*-Tosyloxycarbamates as reagents in rhodium-catalyzed $C(sp^3)$—H amination reactions in 2007.[75,76] It is worth noting that metal nitrenes used in C—H insertion reactions were obtained from *N*-tosyloxycarbamates in the presence of an inorganic base and a rhodium(II) dimer complex catalyst (Scheme 6.12). The scope of this reaction includes ethereal, benzylic, tertiary, secondary, and even primary C—H bonds. Moreover, this methodology has been extended to stereoselective $C(sp^3)$—H amination using a chiral *N*-tosyloxycarbamate and *N*-mesyloxycarbamate.[77,78] Utilizing a $Rh_2[(S)\text{-Br-nttl}]_4/K_2CO_3$ catalyst system, chiral carbamate-protected allylic amines, and benzylic and propargylic amines are produced in good yields with up to 99:1 dr.

In addition to showing hypervalent iodine oxidant is used for nitrene-transfer mechanism, Berry and colleagues demonstrate that simple one-electron oxidants such as Ce^{4+} may also be used to achieve catalytic C—H amination via the one-electron mechanistic regime.[79,80]

Besides the nitrene insertion mode, the C—H bond activation pathway could also be used for Rh-catalyzed $C(sp^3)$—H amination. For example, Rh-catalyzed chelation-assisted intermolecular $C(sp^3)$—H amination of 2-(*tert*-butyl)pyridine with sulfonamides at room temperature was achieved by You and colleagues in 2005 (Scheme 6.13).[81] When $PhI(OAc)_2$ was employed, 2-alkylpyridine derivatives reacted with sulfonamides to afford the desired products in good yields. The authors proposed the intermediacy of a five-membered cationic rhodacycle in the catalytic cycle.

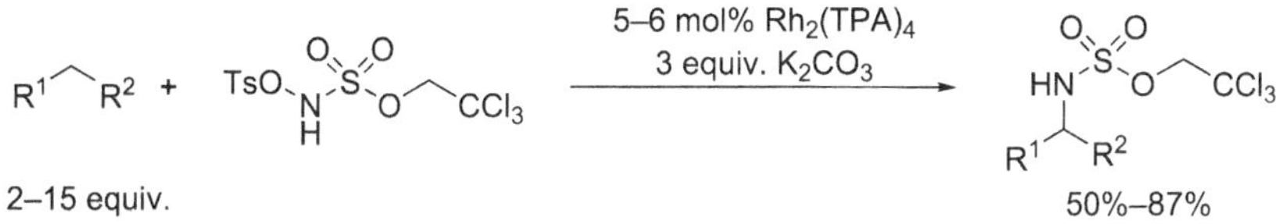

SCHEME 6.12 Rhodium(II)-catalyzed intermolecular amination using *N*-tosyloxycarbamates.

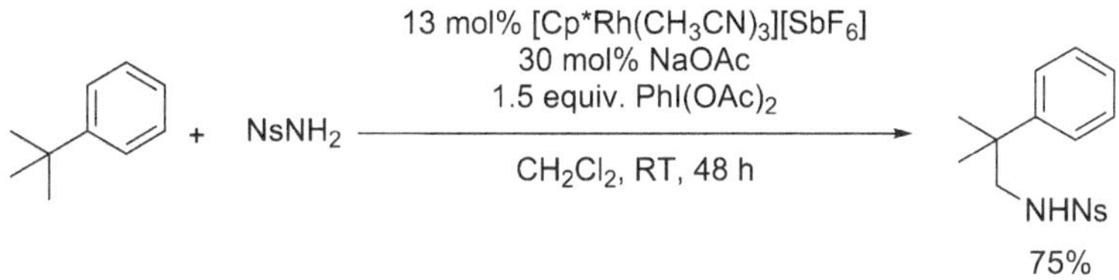

SCHEME 6.13 Rh-catalyzed, chelation-assisted intermolecular $C(sp^3)$—H amination.

6.2.6 Pd Catalyst System

In 2006, Yu and Che reported the first example of Pd-catalyzed intermolecular amination of unactivated $C(sp^3)$—H bonds with the assistance of an oxime group as directing group (Scheme 6.14).[82] Using $Pd(OAc)_2/K_2S_2O_8$ as catalyst system and primary sulfonamides or trifluoroacetamides as the amino source, aliphatic O-methyl oximes could undergo regioselective β-amination of a primary $C(sp^3)$—H bond to give the corresponding monoamidated product in 76%–88% yields. The observed preference for activation of 1 degree C—H bond vs 2 degrees C—H bonds is probably due to steric effect.

In 2008, the White group reported the Pd-catalyzed intermolecular allylic amination of an allylic C—H bond with *N*-(methoxycarbonyl)-*p*-toluenesulfonamide (Scheme 6.15).[83] Utilizing a $(BisSO)PhPd(OAc)_2/Cr(salen)Cl/BQ$ catalyst system, a wide variety of alpha olefins were transformed into the corresponding allylic carbamates with excellent selectivity for the E-linear product. While allyl amine products were obtained in moderate to good yields, linear isomers were highly favored over branched products. Further improvements were made by replacing the Lewis acid additive with a Brønsted base (*N*,*N*-diisopropylethylamine).[84] The role of a base is to increase the effective concentrations of an active nucleophile, which constrains the reaction forward.

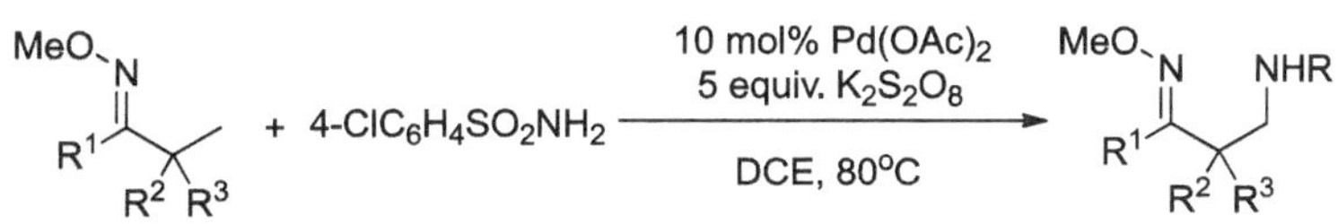

SCHEME 6.14 Pd-catalyzed intermolecular amination of unactivated $C(sp^3)$—H bonds.

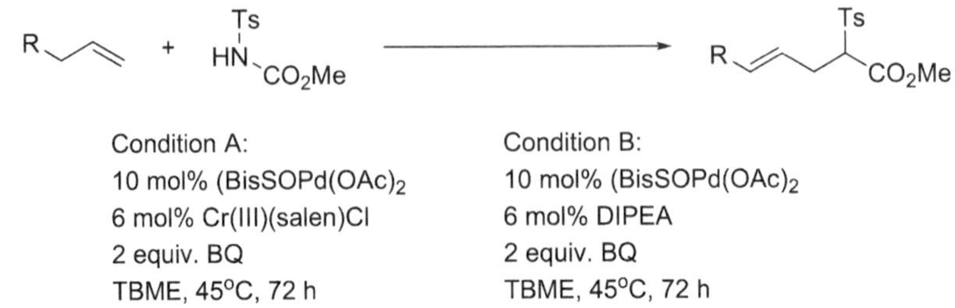

SCHEME 6.15 Pd-catalyzed intermolecular allylic C—H bond amination.

Mechanism studies revealed that proton abstraction by acetate plays a pivotal role as either an internal or external proton abstractor to form the (π-allyl) Pd species and the C—H activation step is the rate-determining step in these reactions.[85] Upon further optimization of reaction conditions, it was observed that an aerobic allylic C—H amination reaction could be operated under BQ/Co(II)/O_2-based redox-rely catalysis.[86] Recently, 1-adamantyl-substituted disulfoxide has identified as an effective ligand for Pd-catalyzed allylic C—H amination with 2,2,2-trichloroethyl tosyl carbamate.[87]

At the same time, Liu and colleagues reported the intermolecular allylic C—H amination under aerobic oxidative conditions in 2008 (Scheme 6.16).[88] Notably, a catalytic amount of maleic anhydride (MA) was used in combination with 6 atm of O_2 to serve as terminal oxidants at ambient temperature. Further improvements were made by utilizing $PhI(OPiv)_2$ in conjunction with 1,4-naphthoquinone.[89] The $Pd(OAc)_2/PhI(OPiv)_2$/1,4-naphthoquinone catalyst system could significantly improve product yield and chemoselectivity and allow use of the olefin as the limiting reagent. Mechanistic studies suggested that the coordination of NQ to an active palladium species is crucial in turnover-limiting allylic C—H bond activation to generate a π-allyl-Pd intermediate.

In addition, Baran and colleagues developed a protocol of *N-tert*-prenylation of indoles based on the allylic C—H amination approach.[90] In contrast to employ electron-deficient amino sources, this reaction offers an opportunity to use indoles directly as the amino coupling partner. Unfortunately, a high loading of the 40 mol% palladium is required in this protocol.

6.2.7 Ag Catalyst System

In 2007, He and colleagues described the first example of silver-catalyzed intermolecular amination of saturated C—H bonds with PhI=NTs.[91] The combination of AgOTf with 4,7-diphenyl-1,10-phenanthroline affords active catalysts for the amination of the activating benzylic C—H bonds at 50°C in 25%–71% yields. However, the cyclic, linear, or branched alkane substrates are amidated in 22%–40% yields.

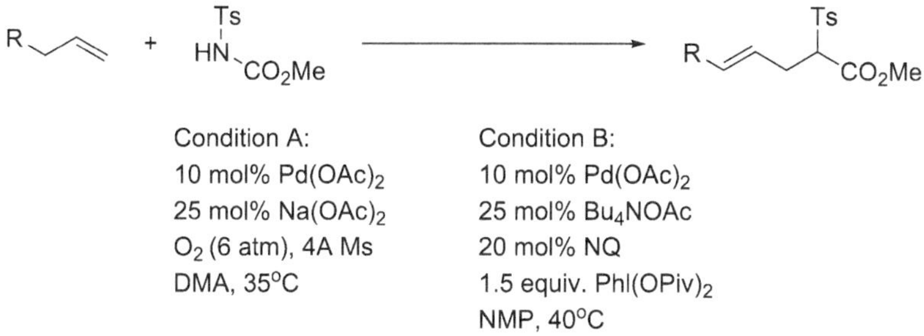

SCHEME 6.16 Pd-catalyzed intermolecular allylic C—H amination.

A year later, the above problem was overcome by using Tp^xAg (Tp^x) hydrotris(pyrazolyl)borate ligand) as catalyst. These substrates can be converted into amino functionalities in 65%–90% yields.[92] The nitrenoid will preferentially insert into tertiary sites over secondary C—H bonds. Subsequently, the same catalyst system has applied for the enantioselective C—H amination of benzylic linear, or branched alkanes substrates using (*S*)-*N*-(*p*-toluenesulfonyl)-*p*-toluenesulfonimidamide and $PhI(OAc)_2$ as an in situ generating system of iminoiodinane.[93] Although much higher yields were obtained in functionalization of various benzylic substrates, the diastereoselectivity remains low in all cases. It is worth noting that the amination of the C—H bonds with yields of up to 70% was obtained when 5 equiv. of alkanes such as pentane and other C6—C8-branched nonactivated hydrocarbons are used.

6.2.8 Co Catalyst System

In 2011, the Chen and Yu group disclosed the first examples of the cobalt-catalyzed benzylic C—H amination reactions.[94] Using $CoBr_2/^tBuOOtBu$ as a catalyst system, various amides including primary or secondary sulfonamides, carboxamides, and carbamates react with benzylic hydrocarbons to afford the C—H amination products in 30%–94% yields.

Subsequent investigation by Ge and colleagues realized cobalt-catalyzed intermolecular dehydrogenative amination of unactivated $C(sp^3)$—H with the assistance of an 8-aminoquinolinyl group as a bidentate directing group

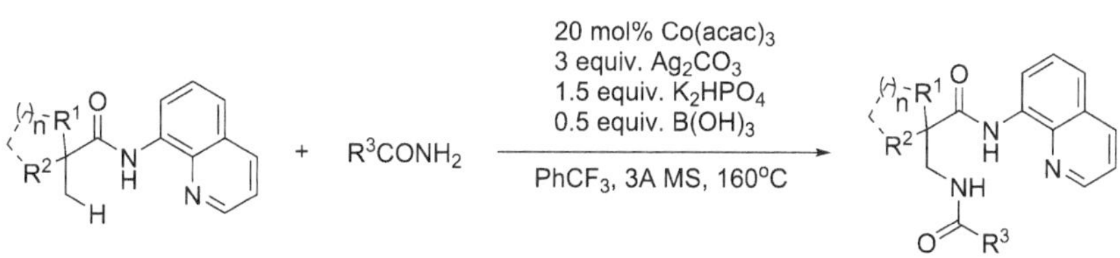

SCHEME 6.17 Cobalt-catalyzed intermolecular dehydrogenative amination of unactivated $C(sp^3)$—H.

(Scheme 6.17).[95] With $Co(acac)_3/Ag_2CO_3$ as the catalyst system, 2,2-disubstituted propanamides reacted with trifluoroacetamide or heptafluorobutanamide to give the corresponding products in 29%–66% yields with a predominant selectivity for C—H bonds of β-methyl groups.

Recently, an important breakthrough for oxidative amination of inert $C(sp^3)$—H bonds was achieved by Shi and colleagues (Scheme 6.18).[96] With the Co-Ni/C as catalyst in the presence of the TBHP and H_2, nitrobenzene and toluene derivatives can be successfully transformed into the corresponding *N*-benzyl aniline derivatives with up to 95% yield. This reaction would be especially attractive given the use of nitroarenes as more straightforward and the step economy source instead of amines. This approach offers a new concept for inert $C(sp^3)$—H bond functionalization.

6.2.9 Re Catalyst System

In 2012, Zhu and colleagues described the amination of the benzylic $C(sp^3)$—H bond with NH_2Ts in the presence of NBS under the catalysis of $ReOI_2(OEt)(PPh_3)_2$ (Scheme 6.19). This reactivity of rhenium catalysts suggests that high-valence rhenium-(V) complexes could catalyze $C(sp^3)$—H bond activation reactions.[97]

6.2.10 Ni Catalyst System

Recently, Cai and colleagues have developed nickel-catalyzed amination of primary benzylic hydrocarbons and sulfonamides (Scheme 6.20).[98] With $Ni(OAc)_2$ as the catalyst and DTBP as the oxidant, this transformation shows excellent tolerance for sulfonamide derivatives and provides the corresponding amination products in good to excellent yields.

6.2.11 Metal-Free System

In 2009, Fan and colleagues described transition-metal-free intermolecular benzylic amination with sulfonamides (Scheme 6.21).[99] With (diacetoxyiodo)benzene (DIB) and iodine as a catalyst system, the reaction produced various alkyl amines in 45%–98% yields. Moreover, this catalyst system could also facilitate the amination of purine nucleoside derivatives with $C(sp^3)$—H bonds adjacent to ethers.[100] In addition, amination of $C(sp^3)$—H bonds adjacent to ethers with a

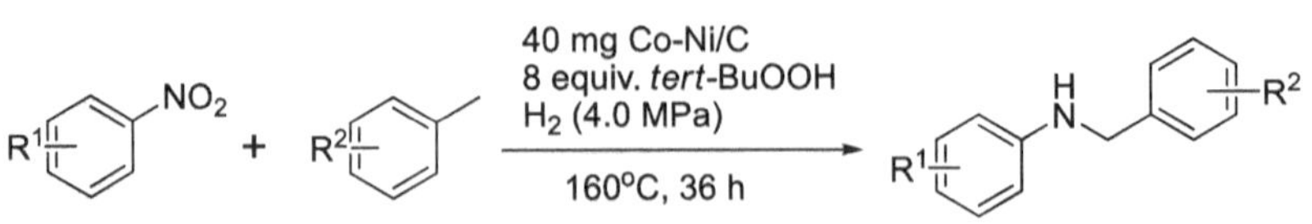

SCHEME 6.18 Co-Ni/C-catalyzed oxidative amination of toluene derivatives with nitrobenzene derivatives.

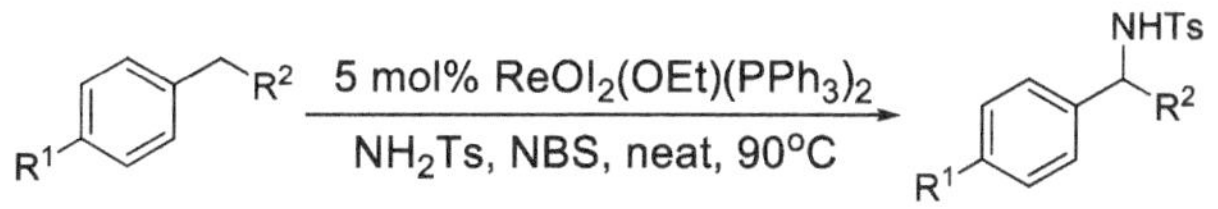

SCHEME 6.19 $ReOI_2(OEt)(PPh_3)_2$-catalyzed amination of benzylic $C(sp^3)$—H bonds.

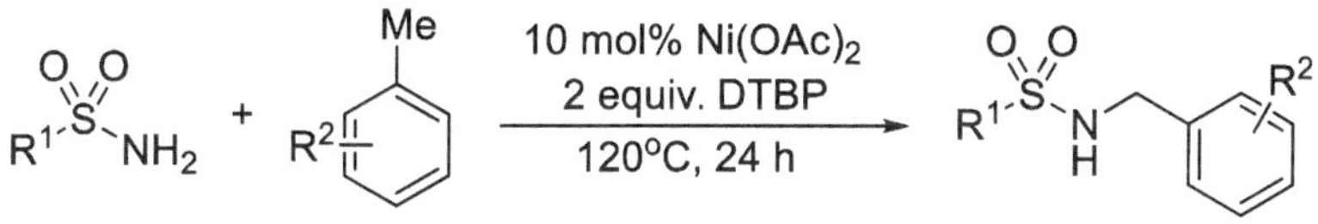

SCHEME 6.20 Nickel-catalyzed amination of primary benzylic C—H bonds with sulfonamides.

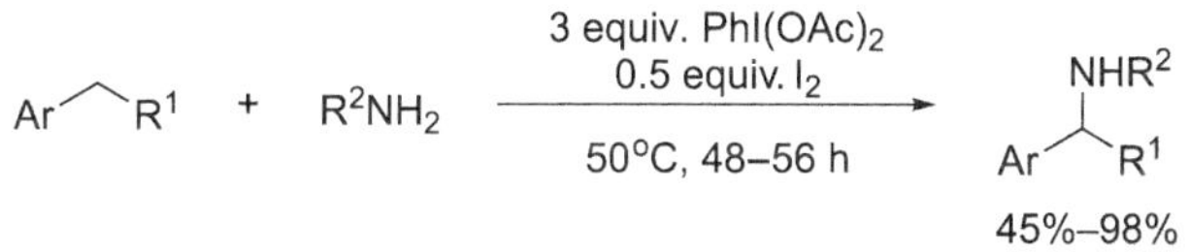

SCHEME 6.21 Metal-free amination of $C(sp^3)$—H bonds of ethers with purine nucleoside.

wide range of sulfonamides, amides, imide, and secondary amines could be realized by utilizing a NaH/Ph_2IPF_6 catalyst system.[101]

In 2011, Chang and colleagues demonstrated control of chemoselective amination between aryl $C(sp^2)$—H and benzylic $C(sp^3)$—H bonds with PIDA as oxidant.[102] When phthalimide was used as nitrogen source, aryl $C(sp^2)$—H bond amination took place, whereas a change in the nitrogen source to dibenzenesulfonimide could afford benzylic $C(sp^3)$—H bond amination.

In the next year, Loh and colleagues developed procedure for α-amination of aldehydes (Scheme 6.22).[103] Hypoiodite formed in situ from substoichiometric molecular iodine and stoichiometric amounts of sodium percarbonate has been shown to catalyze the $C(sp^3)$—H bonds amination for the synthesis of α-amino acetals. Following this report, the α-C—H amination of ketones with amine substrates undergo cross-coupling to generate synthetically useful α-amino ketones has also been developed using similar catalyst system.[104]

On the basis of several control experiments, plausible mechanistic details were proposed (Scheme 6.23). Initially, oxidation of iodine or iodide with hydrogen peroxide could form the active cationic iodine species hypoiodite

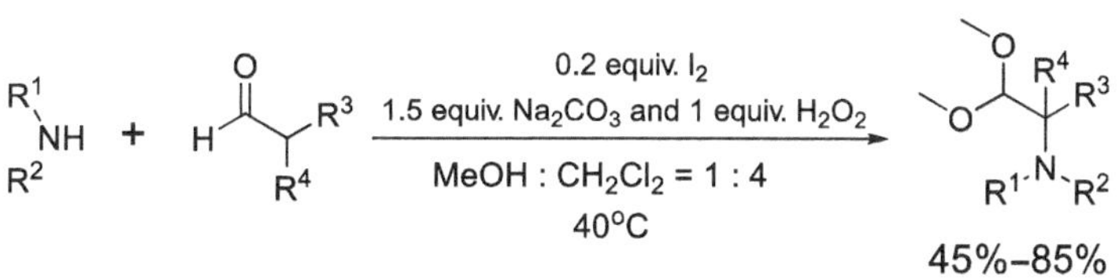

SCHEME 6.22 Iodinine-catalyzed α-amination of aldehydes.

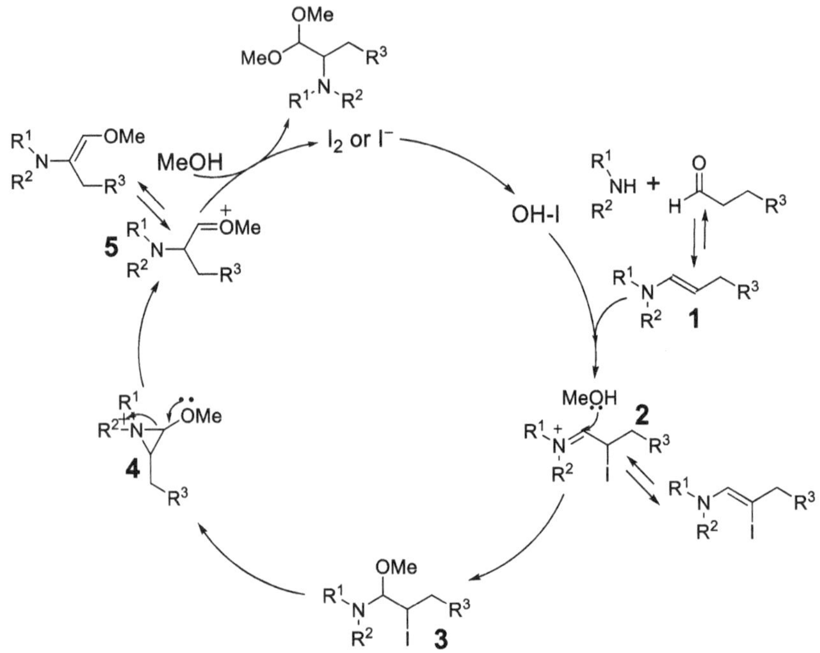

SCHEME 6.23 Plausible mechanism for iodine-catalyzed α-amination of aldehydes.

acid, which reacts with enamine **1** to give iminium ion **2.** Then, the intermediate **2** could be attacked by methanol to provide iodo-substituted intermediate **3**, which could undergo intramolecular cyclization to afford aziridinium ion **4**. Subsequent attack of one more methanol molecule on the ring-opened intermediate **5** could afford the desired product.

Moreover, DDQ could also be used as an oxidant for amination of benzylic and allylic $C(sp^3)$—H bond with sulfonamides in 60%–87% yields.[105] In addition, other catalyst systems, such as TBAI/TBHP[106] and Ph_2SO/Tf_2O,[107] have also been developed for allylic $C(sp^3)$—H bond amination.

6.3 INTRAMOLECULAR OXIDATIVE AMINATION OF ALKANES

6.3.1 Cu Catalyst System

In 2011, Nicholas and colleagues reported Cu-catalyzed intramolecular amination of $C(sp^3)$—H bonds (Scheme 6.24).[108] Using CuI-diimine complexes as catalyst and PhI=O as oxidant, the cyclization of carbamates and sulfamates to the corresponding five- and six-membered *N*,*O*-heterocycles resulted in good yields. Furthermore, this methodology was attempted for the asymmetric $C(sp^3)$—H amination by using various chiral diimine ligands, but the products were isolated with only moderate enantioselectivity.

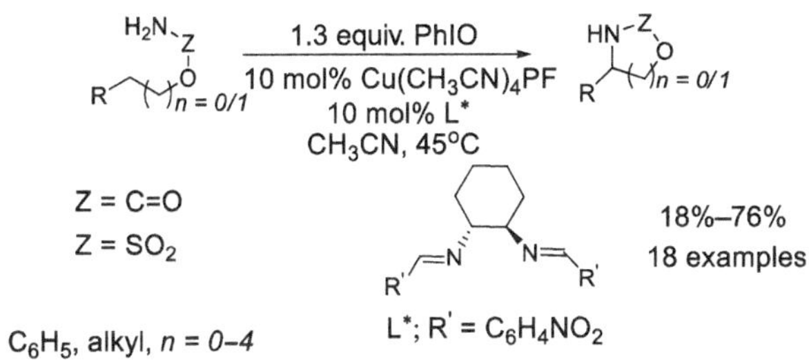

SCHEME 6.24 $C(sp^3)$—H amination of carbamates.

Subsequently, Kanai and Kuninobu disclosed a Cu-catalyzed intramolecular $C(sp^3)$—H and $C(sp^2)$—H amination using 8-aminoquinoline as bidentate directing group.[109] The reaction proceeded at a terminal methyl group, as well as at the internal benzylic position of an alkyl chain (Scheme 6.25).[110] Almost at the same time, this reaction was realized by the Ge group with a CuCl/duroquinone/$PhCO_2Na$ catalyst system under an air atmosphere.[111]

Recently, Shi and coworkers successfully realized an intramolecular amination of unactivated $C(sp^3)$—H bonds to construct indoline derivatives catalyzed by a $Cu(OAc)_2/Ag_2CO_3$ system. The amination proceeded well at primary C—H bonds. Preliminary mechanistic studies suggest this process involves a radical pathway (Scheme 6.26).[112] Moreover, a $Cu(OAc)_2/MnO_2$ catalyst system was used to deliver isoindilinones from benzylic C—H of 2-methylbenzamides via intramolecular amination.[113]

In addition, Fu et al. reported a one-pot approach to synthesize tetrahydroisoquino[2,1-a]-quinazolinone derivatives by a sequential Cu-catalyzed *N*-arylation, intramolecular amination of unactivated $C(sp^3)$—H bonds adjacent to a nitrogen atom (Scheme 6.27).[114]

6.3.2 Fe Catalyst System

In 1983, Gellman et al. reported an intramolecular directed metal-nitrene insertion reaction with Fe^{III}(TPP)Cl catalyst (Scheme 6.28).[115]

In 2012, an important breakthrough was achieved by White and coworkers who reported a highly selective and general C—H amination by using [Fe^{III}Pc] as catalyst and $PhI(OPiv)_2$ as an oxidant. The [Fe^{III}Pc] catalyst displays a strong preference for allylic C—H amination over aziridination and all other C—H bond types (Scheme 6.29).[116]

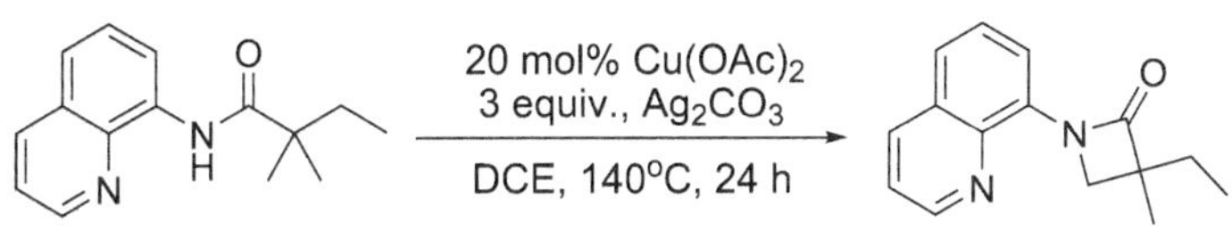

SCHEME 6.25 Copper-catalyzed intramolecular $C(sp^3)$—H amination.

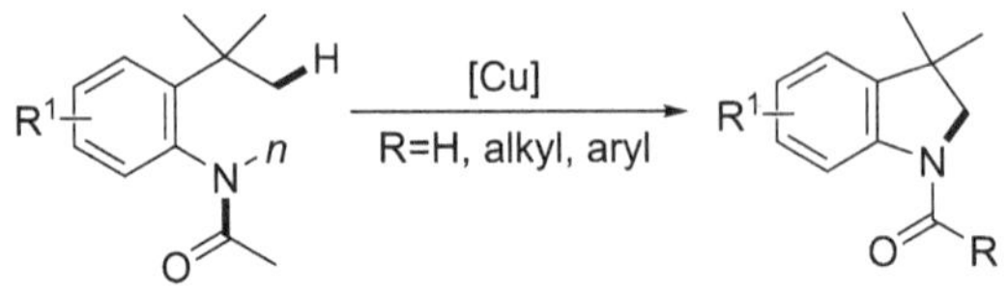

SCHEME 6.26 Copper-catalyzed C—H amination.

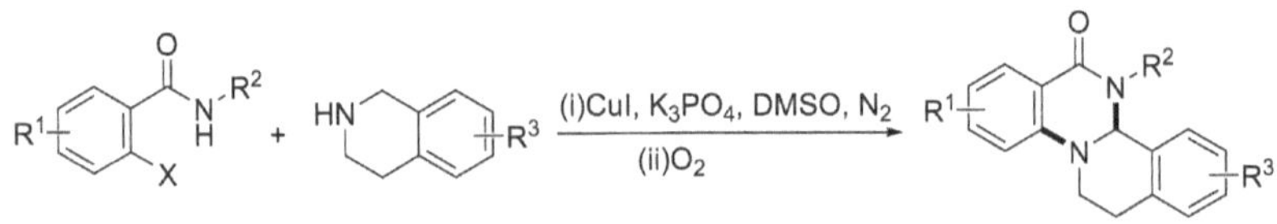

SCHEME 6.27 Copper-catalyzed one-pot synthesis of tetrahydroisoquinolino[2,1-a]-quinazolinone derivatives.

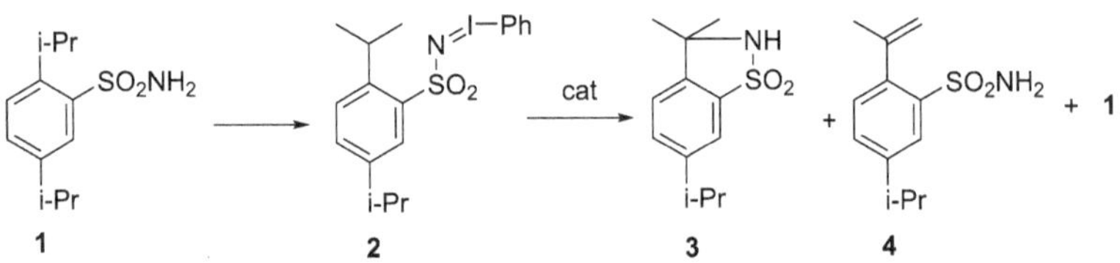

SCHEME 6.28 Intramolecular nitrogen functionalization.

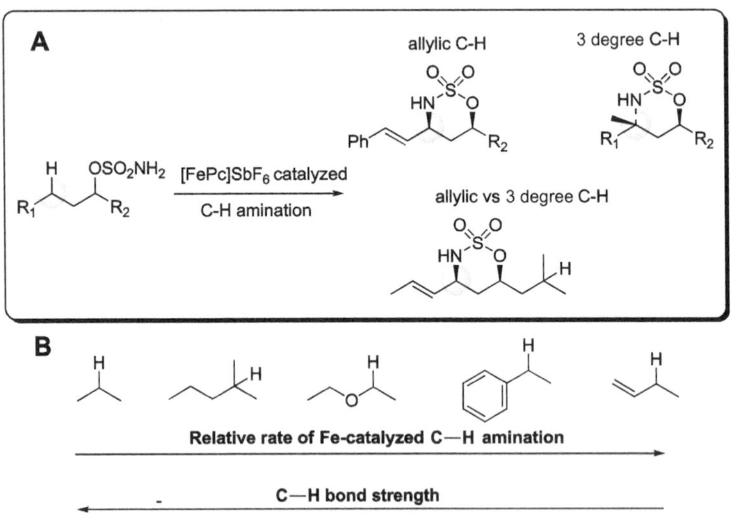

SCHEME 6.29 Reactivity trends for Fe-catalyzed intramolecular C—H amination.

Nonheme iron complex $[Fe(qpy)(MeCN)_2](ClO_4)_2$ catalyzed the intra- and intermolecular amination of various sulfamate esters $C(sp^3)$—H bonds were reported by Che[117] and colleagues in 2013 (Scheme 6.30). Note that the substrate scope spans cycloalkanes, benzylic/allylic hydrocarbons, and natural products α-pinene and β-pinene. Experimental studies (including ESI-MS analysis) and DFT calculations suggested that 7-coordinate iron-imide/nitrene intermediate $[Fe(qpy)(NR)(X)]^{n+}$ (CX, X=NR, solvent, or anion) is a key intermediate in this reaction.

Recently, Wong reported an intramolecular α-amination of 2-aminoacetamides for synthesis of imidazolidin-4-ones with evolved variants of cytochrome P450 BM3 from Bacillus megaterium.[118]

6.3.3 Mn Catalyst System

In 1983, Gellman et al. reported an intramolecular directed metal-nitrene insertion reaction with Mn^{III}(TPP)Cl catalyst,[115] while it showed inferior activity compared to Fe^{III}(TPP)Cl and led to a low total yield of the three analyzed products.

More than two decades later, Che and coworkers reported the chiral manganese(III) Schiff-base complex-catalyzed enantioselective intramolecular amination of sulfamate esters (Scheme 6.31).[119]

6.3.4 Ru Catalyst System

In 2002, Che et al. for the first time achieved the ruthenium porphyrins-catalyzed intramolecular amination reactions of saturated C—H bonds via a "$PhI(OAc)_2$+ sulfamate esters" procedure.[120] In the presence of the electron-deficient ruthenium porphyrin [Ru(tpfpp)(CO)] (H_2tpfpp=*meso-tetra* (pentafluorophenyl)porphyrin), various cyclic sulfamidates were synthesized in 52%–88% yields with high regioselectivity. Of note, the optically active

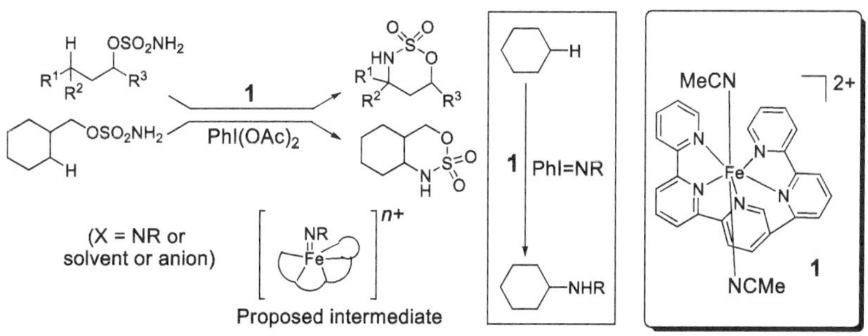

SCHEME 6.30 Intramolecular amination of $C(sp^3)$—H Bond.

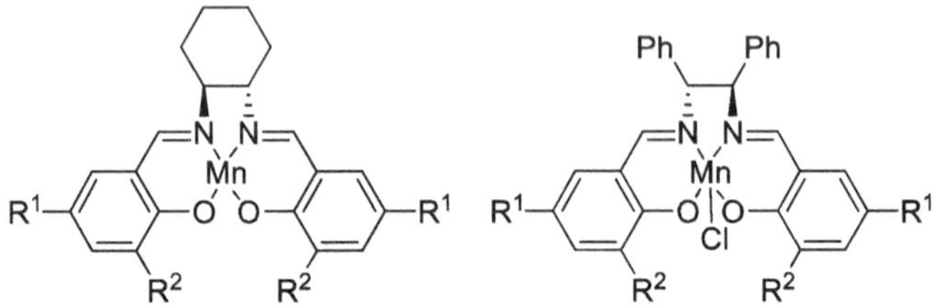

SCHEME 6.31 Enantioselective intramolecular amination catalyzed by manganese(III) Schiff-bases.

cyclic sulfamidates were also obtained in up to 87% ee in the presence of the chiral ruthenium porphyrin [Ru(por*)(CO)] (H_2por* = 5,10,15,20-tetrakis {(1*S*,4*R*,5*R*,8*S*)-1,2,3,4,5,6,7,8-octahydro-1,4:5,8-dimethanoanthracene-9-yl}-porphyrin) when prochiral sulfamate esters were used as starting materials. However, 20%–77% yields and the use of equivalent amount of additive (Al_2O_3) were not satisfactory.

A similar protocol was shown by Blakey and colleagues in 2008.[121] Far higher enantioselectivities of up to 90% ee and yields of up to 84% were obtained in the asymmetric amination of benzylic C—H bonds (Scheme 6.32) via a cationic ruthenium(II)-pybox catalyst. It is worth noting that two examples of amination of allylic C—H bonds were also contained and up to 92% ee were afforded. However, 2.3 equiv. of MgO is still required for this catalyst system. Later, Blakey et al. revealed that the ruthenium(II) 2,6-bis (imino)pyridyl complexes were also active for the intramolecular C—H amination of sulfamate esters under a similar condition.[122]

Efficient diruthenium complexes-catalyzed allylic C—H amination with bis-(homoallylic) sulfamate esters was developed by Musaev and Bois et al. in 2011.[9] A range of products containing oxathiazinane heterocycle were afforded with good to excellent selectivities in the presence of [Ru_2(hp)$_4$Cl] (2.5 mol%) and Å molecular sieves. A mechanism involving hydrogen-atom abstraction/radical recombination and the intermediacy of a discrete, albeit short-lived, diradical species is proposed for [Ru_2(hp)$_4$Cl]-promoted C—H amination on the basis of the computational and experimental results.

6.3.5 Rh Catalyst System

In 2001, the first example of Rh-catalyzed intramolecular amination of carbamates bonds was reported by Du Bois and colleagues.[123] With $Rh_2(OAc)_4$, PhI $(OAc)_2$, and MgO as catalyst system, the carbamates containing benzylic and tertiary C—H centers were cyclized to oxazolidinone products in 74%–84% yields. In the case of less reactive substrates such as *n*-butylcarbamate, however,

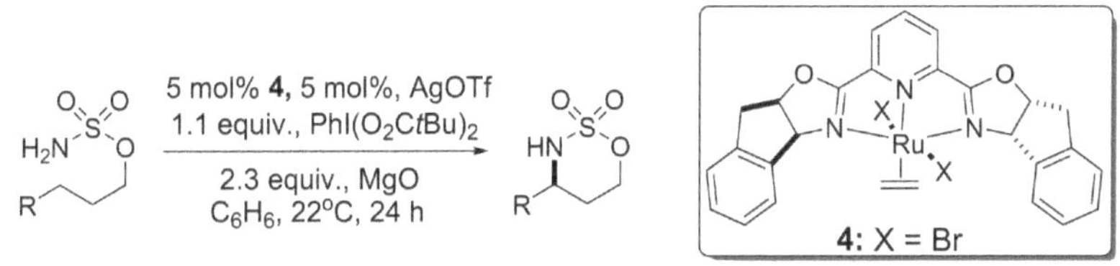

SCHEME 6.32 Asymmetric amination of benzylic and allylic C—H bonds.

the more effective triphenylacetate (tpa) complex [$Rh_2(tpa)_4$] (5 mol%) is required. It should be noted that C—N bond formation in all the examples reported is stereospecific and the cyclic carbamates are produced with *syn*-diastereoselectivity, with retention of stereochemistry at the insertion site (Scheme 6.33). Subsequently, this methodology has been extended to the amination of sulfonamides.[123] A similar phenomenon was observed that [$Rh_2(OAc)_4$] (2 mol%) afforded good to excellent yields of six-membered ring insertion products only with the substrates containing benzylic, allylic, and tertiary C—H centers and the exceptional case required the use of $Rh_2(oct)_4$. Furthermore, the synthesis of *N,O*-acetal oxathiazinane heterocycles could also be achieved with the above catalyst system. Of note, the generated *N,O*-acetal oxathiazinane heterocycles could be used as iminium ion equivalents and further react with the nucleophiles such as allylsilanes, silyl enol ethers, and ketene acetals,[124] as well as alkynylzinc reagents,[125] to afford the desired oxathiazinane products. Lastly, it is found that by changing the Rh-complex to $Rh_2(oct)_4$, the diastereoselective C—H amination of chiral sulfamate esters could be also efficiently promoted, and high yields and diastereoselectivity were obtained.[126]

Highlighting the synthetic utility of this methodology, the Rh-catalyzed intramolecular saturated C—H amination reaction has applied for the preparation of various valuable chemicals such as the bromopyrrole alkaloids, Manzacidin

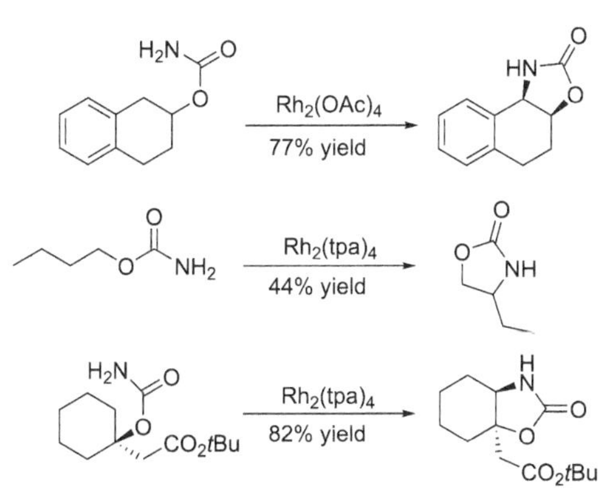

Reaction comditions: ctalyst (5 mol %), $PhI(OAc)_2$ (1.4 equiv.), MgO (2.3 equiv.), CH_2Cl_2, 40°C

SCHEME 6.33 The Rh-catalyzed oxidative cyclization of carbamates.

A and C,[127] carbamate-protected glycols,[128] benzene-fused cyclic sulfamates,[129] seven-membered ring sulfamates,[130] an immunomodulator (+)-conagenin,[131] aconitine,[132] (−)-cytoxazone and (+)-*epi*-cytoxazone,[133] five-membered sulfamidates,[134] (−)-kaitocephalin,[135] substituted pyrrolidines and piperidines,[136] functional epiandrosterone,[137] 1,2-disubstituted diamondoids,[138] myriocin,[139,140] and iminosugars.[141,142]

In addition to the widely used [$Rh_2(OAc)_4$], some other Rh-complexes such as calix[4]arene-derived tetracarboxylate dirhodium(II) inclusion complexes,[143] $R_2(esp)_2$ (esp = α,α,α′,α′-tetramethyl-1,3-benzenedipropanoate),[144–149] $R_2(tpa)_4$[150] and $Rh_2(oct)_4$[151] have also been explored to catalyze the intramolecular saturated C-H amination with the aid of MgO and $PhI(OAc)_2$ derivatives.

In 2008, Du Bois et al. showed the intramolecular enantioselective C—H amination with a chiral rhodium carboxamidate catalyst.[152] Under the optimized conditions (1.2 equiv. PhI=O, 3 Å MS and CH_2Cl_2), the $Rh_2(Snap)_4$ (2 mol%) excellent exhibited enantioselectivity in cyclization reactions of sulfamate esters and >80% enantiomeric excesses were generally obtained. Additionally, a strong preference for allylic C—H bond insertion was also observed.

In 2012, Berry et al. developed a new mixed-valent $Rh_2^{II,III}$ dimer, [$Rh_2(espn)_2Cl$] ($espn^{2-}$ = α,α,α′,α′-tetramethyl-1,3-benzenedipropanoate), which is able to efficiently catalyze the intramolecular C—H amination of sulfamates to afford the cyclic sulfamates with TONs > 1400 and 72% yield.[153] It is worth noting that catalyst loading is as low as 0.05 mol% and the use of equivalent amount of MgO could also be avoided.

In the intramolecular saturated C—H amination, the iodine(III) oxidants have been widely used. In fact, it can be replaced with other terminal oxidants. For example, in 2009, Du Bois et al. revealed that the halogenating agents such as NaOCl as an oxidant are also effective for converting sulfamate substrates to the corresponding [1,2,3]-oxathiazinane-2,2-dioxide heterocycles when combined with $Rh_2(oct)_4$ (3 mol%) under biphasic alkaline conditions (CH_2Cl_2, aq Na_2HPO_4).[154] A mechanism likely involving the intermediacy of an *N*-centered radical is proposed, which is distinct from that induced by hypervalent iodine reagents.

Different from the classical combination of Rh-complexes, $PhI(OAc)_2$ and MgO, Lebel and colleagues revealed a new combination of $Rh_2(tpa)_4$ (6 mol%) and K_2CO_3 (3 equiv.), which is able to efficiently promote the intramolecular C—H amination reaction of selected *N*-tosyloxycarbamates to afford the corresponding oxazolidinones in moderate to good yields.[155] Of note, the stereospecific C—N bond formation is also tolerated when a chiral enantioenriched *N*-tosyloxycarbamate was used as the starting material and the corresponding oxazolidinone product was provided with complete retention of configuration (Scheme 6.34).[76] Further reaction optimization revealed that when with *N*-mesyloxycarbamates as a precursor to synthesize oxazolidinones, the reaction was allowed to perform under greener and milder conditions (3 mol%

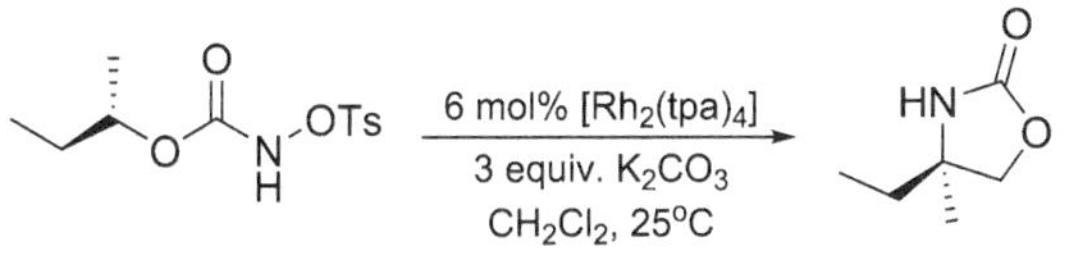

SCHEME 6.34 Rhodium-catalyzed intramolecular C—H amination.

$Rh_2(tpa)_4$, 1.5 equiv. Aq. sat. K_2CO_3, EtOAc, 22°C).[156] It is found that conformationally restricted cyclic secondary *N*-mesyloxycarbamates furnish *cis*-oxazolidinones in excellent selectivity, but *trans*-oxazolidinones were prepared from acyclic secondary *N*-mesyloxycarbamates using the combination of $Rh_2(oct)_4$ (1 mol%) and KOAc (3 equiv.).

In addition to the continuing attempts to develop different catalyst systems, the exploration about the reaction mechanism of $Rh_2^{II,II}$- and $Rh_2^{II,III}$-catalyzed intramolecular amination of carbamates and sultamates has been done. For example, in 2007, Che et al. performed a DFT study on the mechanism of $Rh_2^{II,II}$-catalyzed intramolecular C—H aminations of carbamates. Two kinds of reaction mechanisms, concerted singlet insertion and stepwise triplet recombination, were proposed.[157] A similar result was suggested by Du Bois in the intramolecular C—H aminations of sulfamates catalyzed by $Rh_2(O_2CR)_4$.[158] Further, the mechanisms of the dirhodium (Rh_2L_4, L = formate, *N*-methylformamide, *S*-nap)-catalyzed intramolecular C—H aminations of 3-phenylpropylsulfamate ester have also been investigated in detail by Zhao et al. with BPW91 density functional theory computations.[159] It was revealed that in the $Rh_2(formate)_4$-promoted reaction, a concerted mechanism was observed but $Rh_2(N\text{-methylformamide})_4$ and $Rh_2(S\text{-nap})_4$ catalyzed the reaction via a stepwise process. Zhao et al. also explored the reaction mechanisms on the intramolecular allylic C—H amination of 4-pentenylsulfamate, promoted by selected dirhodium paddlewheel complexes with a DFT approach.[160] A predominant singlet concerted, highly asynchronous pathway and an alternate triplet stepwise pathway are obtained when catalyzed by $Rh_2(OAc)_4$. The process is similar in the case of $Rh_2(CHCOCF_3)_4$. But a mixed singlet-triplet pathway is preferred in the C—H insertion catalyzed by $Rh_2(NCH_3CHO)_4$.

In the dirhodium-catalyzed C—H amination, the existence of Rh_2-nitrene intermediates is widely accepted. Recently, a detailed investigation of the ground electronic state (GES), redox potentials, and C—H amination of $[Rh_2^{II,III}(O_2CCH_3)_4(L)_n]^+$ (L = none, Cl^-, and H_2O), $[Rh_2(esp)_2]^+$, and $Rh_2(espn)_2Cl$ (esp = α,α,α′,α′-tetramethyl-1,3-benzenedipropanoate and espn = α,α,α′,α′-tetramethyl-1,3-benzenedipropanamidate) has been reported by Berry and Musaev.[161] DFT calculations suggest a doublet GES for $[Rh_2(esp)_2[NSO_3(CH_2)_3Ph](L)]^+$ and a quartet GES for $[Rh_2(espn)_2[NSO_3(CH_2)_3Ph](L)]^+$. CASSCF calculations describe the GES of $[Rh_2(esp)_2[NSO_3(CH_2)_3Ph]$ as Rh_2(II,II) with a coordinated nitrene radical cation. Conversely, the GES of $[Rh_2(espn)_2[NSO_3(CH_2)_3Ph]$ is Rh_2(II,III) with a coordinated triplet nitrene. It is found that quartet transition states (^{4}TSs) react via a stepwise radical mechanism, whereas doublet transition states (^{2}TSs) react via a concerted mechanism that is lower in energy.

6.3.6 Pd Catalyst System

In 2007, White and colleagues reported palladium-catalyzed intramolecular diastereoselective allylic C—H amination to furnish cyclic carbamates using their Pd^{II}-sulfoxide catalyst system (Scheme 6.35).[162] The proposed mechanism for allylic C—H amination involves a Pd^{II}/bis(sulfoxide)-promoted allylic C—H cleavage to form a π-allylpalladium intermediate, followed by acetate-mediated deprotonation, and finally resulting functionalization. The DFT study showed that either the C—H activation or the Pd(0) oxidation may be the step determining the energy span of the reaction.[163] Moreover, this methodology has been extended to preparation of *syn*-1,3-amino alcohol motif,[164,165] α- and β-homophenylalanine precursors[166] using the similar catalyst system.

In 2009, Liu and colleagues developed palladium-catalyzed intramolecular aerobic oxidative allylic amination (Scheme 6.36).[167] It is worth noting that the regioselectivity could be switched from five- to seven-membered by adding Brønsted base. This methodology provides an efficient route for the synthesis of seven-membered cyclic amides.

In addition to allylic C—H bond, other inactivated C(sp^3)—H bonds could also undergo intramolecular amination reaction by using a directing group. In 2008, Yu and colleagues developed a Pd-catalyzed C(sp^3)—H amination procedure for the preparation of β-, γ-, and δ-lactams with a $Pd(OAc)_2/CuCl_2/AgOAc$ catalyst system (Scheme 6.37).[168] Subsequent independent investigations by Chen[169,170] and Daugulis[171] using picolinamide as directing group established Pd-catalyzed intramolecular C(sp^3)—H amination of the γ and δ positions of picolinamide protected amine substrates for the synthesis of aziridines, pyrrolidines, and indolines with $Pd(OAc)_2/PhI(OAc)_2$.[169] In addition, 8-aminoquinoline could also be used as a directing group for Pd-catalyzed unactivated β-C(sp^3)—H bond amination of carboxamides to synthesize β-lactams with high regioselectivity. C_6F_5I finely tuned the electronic and steric property

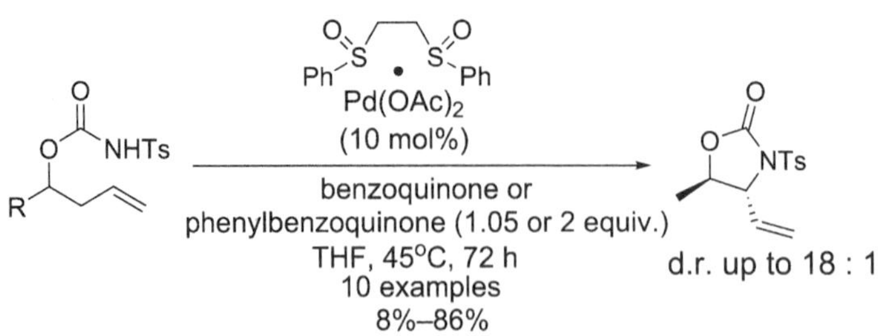

SCHEME 6.35 White's diastereoselective allylic C—H amination.

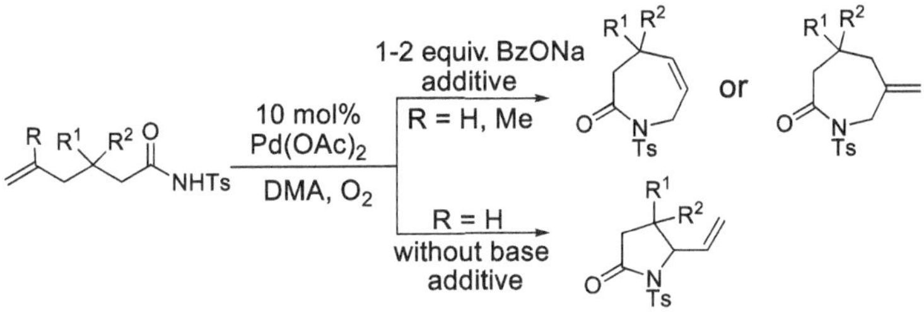

SCHEME 6.36 Palladium-catalyzed intramolecular aerobic oxidative allylic amination.

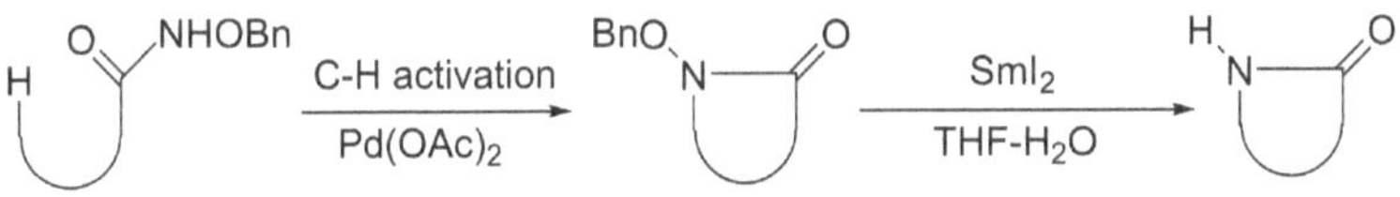

SCHEME 6.37 Pd-catalyzed C—H Activation reaction.

of reaction intermediates to favor the C—N bond forming process, which is deduced from the experiment results and related references.[51,107,111,172,173]

A major breakthrough was realized in 2014 by Gaunt who used an unprotected aliphatic secondary amine to direct a palladium-catalyzed C—H amination of an adjacent methyl group. This reaction proceeds through an unusual four-membered ring cyclopalladation pathway. The development of this method leads to the synthesis of aziridines and β-lactams (Scheme 6.38).[174] Recently, a Pd-catalyzed enantioselective C—H amination to aziridines in high enantiomeric was successfully developed by using a chiral BINOL phosphoric acid ligands.[175]

6.3.7 Ag Catalyst System

In 2004, He and colleagues reported an efficient amination reaction of saturated C—H bonds catalyzed by a unique disilver(I) complex (Scheme 6.39A). With $AgNO_3$/$t$$Bu_3$tpy as a catalyst system, carbamates and sulfamates underwent amination to afford corresponding five-membered ring and six-membered ring insertion products in good to excellent yields (Scheme 6.39B). The mechanism studies suggested that a silver-nitrene intermediate was probably involved in the reaction.[176] Subsequent investigation by Shi and colleagues extended the

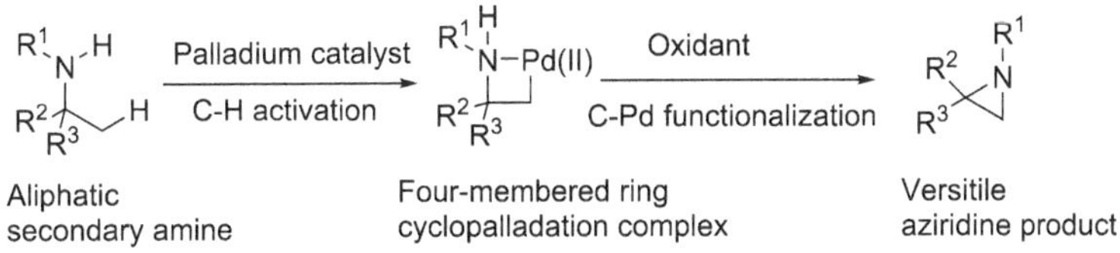

SCHEME 6.38 Palladium-catalyzed directed C—H activation of aliphatic secondary amines.

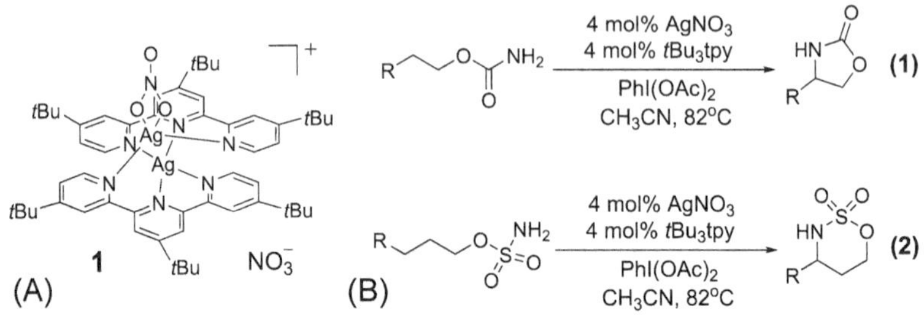

SCHEME 6.39 (A) A unique disilver(I) catalyst $[Ag_2(^tBu_3tpy)_2(NO_3)](NO_3)$ (1). (B) Oxidative cyclization of carbamates [Eq. (1)] and sulfamates [Eq. (2)].

Ag-catalyzed site-selective amination reaction to linear amines using a similar $Ag(OAc)/dtbpy/PhI(OAc)_2$ catalyst system. This method prefers primary $C(sp^3)$—H bonds and exhibits good chemo- and regioselectivity compared to Hofmann-Löffler-Freytag reaction and nitrene insertion.[177]

This catalyst system was further optimized through investigation of the effects of ligand on site-selectivity. Notably, Ag catalysts supported by ^{t}Bubipy appear to prefer amination of the most electron-rich C—H bond, while silver supported by a tpa ligand is more sensitive to the steric environment around the C—H bond (Scheme 6.40).[178,179]

6.3.8 Co Catalyst System

In 2015, the cobalt-catalyzed amination of unactivated $C(sp^3)$—H bonds using 8-aminoquinoline as bidentate directing group was established by Ge and colleagues (Scheme 6.41).[95] With $Co(OAc)_2$ as catalyst and Ag_2CO_3 as oxidant, this protocol enables an efficient access to β-lactam, γ-lactam, or β-amino amide derivatives. Deuterium-labeling experiments indicate that the $C(sp^3)$—H bond cleavage of amide is the rate-limiting step in the catalytic process.[95]

6.3.9 Ni Catalyst System

In 2014, Ge and colleagues developed Ni-catalyzed intramolecular cyclization reaction of 2,2-disubstituted propionamides (Scheme 6.42).[180] The regioselectivity will react preferentially for β-methyl groups over the γ-methyl or β-methylene C—H bonds. Furthermore, $C(sp^3)$—H bond functionalization of β-methyl groups was favored over the aromatic $C(sp^2)$—H bonds.

6.3.10 Bimetal Catalyst System

In 2014, Chuang developed the first Co/Pd-cocatalyzed intramolecular C—H amination and aziridination reactions with $CoPd(OAc)_4$ as catalyst and $PhI(OAc)_2$ as oxidant. A preference for tertiary and benzylic C—H bonds over

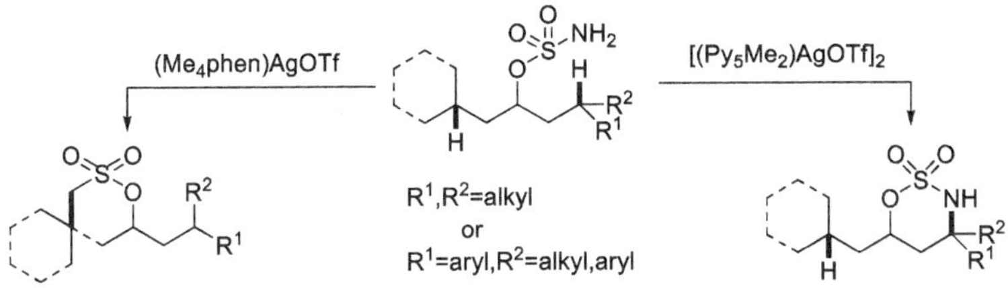

SCHEME 6.40 Tunable regioselectivity between similar 3 degrees alkyl C(sp^3)—H bonds.

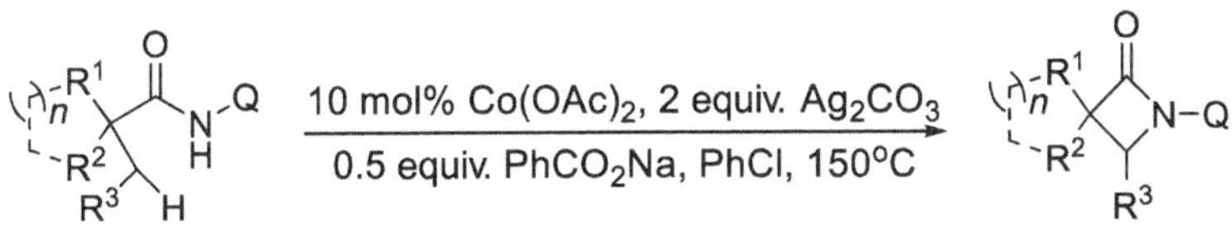

SCHEME 6.41 Scope of α,α-disubstituted propanamides.

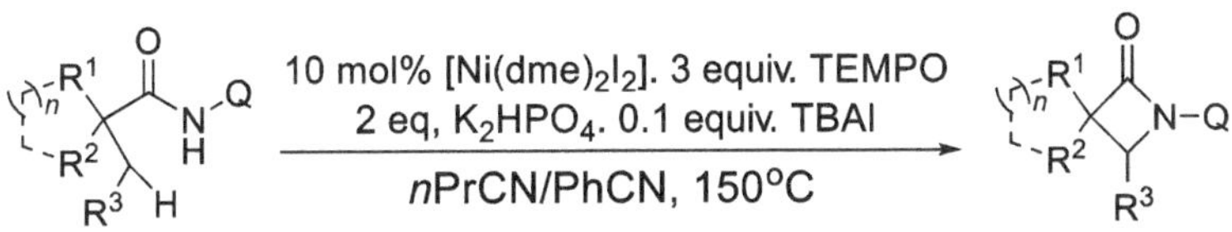

SCHEME 6.42 C—H amination on primary and secondary C(sp^3)—H carbon atoms.

secondary C—H bonds in the formation of six-membered cyclic oxathiazinane products was observed (Scheme 6.43).[181]

Zhao investigated the bimetallic paddlewheel complexes-catalyzed intramolecular benzylic nitrene insertion by using DFT calculations. It is found that the imido N species were found to be predominant in the Ru-Ru and Pd-Mn cases whereas a mixed singlet-triplet nitrene and imido species were observed in the Rh-Rh and Pd-Co cases. In addition, the triplet nitrene N species are

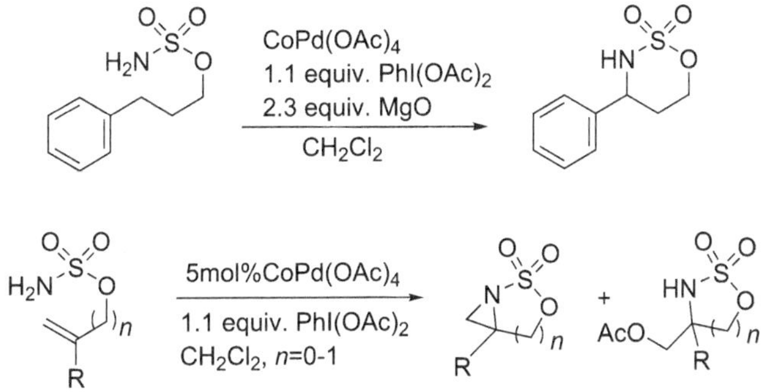

SCHEME 6.43 $CoPd(OAc)_4$-catalyzed intramolecular amination and aziridination.

preferred in the Pd-Ni and Pd-Zn systems. The calculations suggest the Pd-Mn, Pd-Co, and Pd-Ni paddlewheel complexes could be used as economical alternatives to expensive dirhodium/diruthenium complexes in C—H amination catalysis.[172]

6.3.11 Metal-Free Catalyst System

In 2013, Zhang developed an efficient synthesis of tetramic acid derivatives via intramolecular C(sp^3)—H aminations of 1-acetyl *N*-aryl cyclopropane/cyclopentanecarboxamides in the presence of $PhI(OPiv)_2$ and CH_3CH_2COOH (Scheme 6.44).[182]

In 2015, Shi succeeded in establishing intramolecular C(sp^3)—H amination under mild reaction conditions using iodoarene as catalyst and *m*-CPBA as oxidant (Scheme 6.45).[183] The C—H functionalization involving iodine(III) reagents-generated in situ occurs readily at sterically hindered tertiary C—H bonds. Importantly, this leads to the experimentally observed amination at a chiral center without loss of stereochemical information.

Subsequently, Muñiz and colleagues developed an NIS promoted Hofmann-Löffler reaction under visible light (Scheme 6.46A).[184] The employed NIS is more economic than related reagent combinations that include the requirement for hypervalent iodine reagents. In investigating further the metal-free sp^3 C—H amination reaction, they discovered that I_2/NBS catalyst system was capable of the regioselective formation of the piperidines. The reaction overrides the common preference for pyrrolidine formation within the Hofmann-Löffler manifold and significantly enlarges position-selective C—H amination reactions (Scheme 6.46B).[185]

6.4 CONCLUSIONS AND OUTLOOK

The direct catalytic amination of inert C(sp^3)—H bonds has emerged as a powerful and atom efficient transformation for the synthesis of alkyl amines. In principle, mainly two strategies have been developed for the amination of C(sp^3)—H bonds. One attractive method is metal nitrenoids insertion into the C—H bond for the amination of C(sp^3)—H bonds. Another effective method involves transition-metal-catalyzed direct C—H bond activation-based amination, in which aliphatic C—H first cleavaged to produce C—M bond followed by formation of C—N bond. The secondary or even tertiary C—H bond

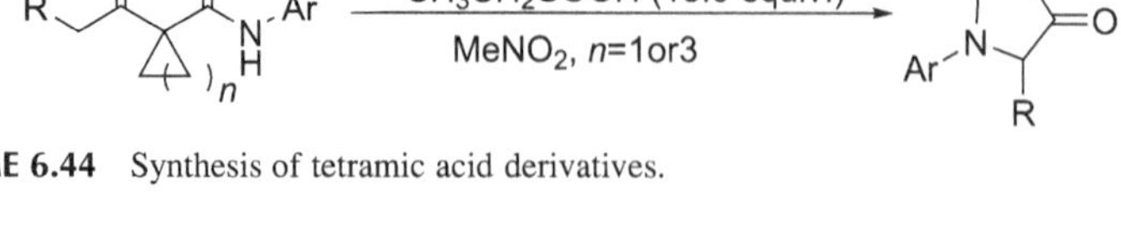

SCHEME 6.44 Synthesis of tetramic acid derivatives.

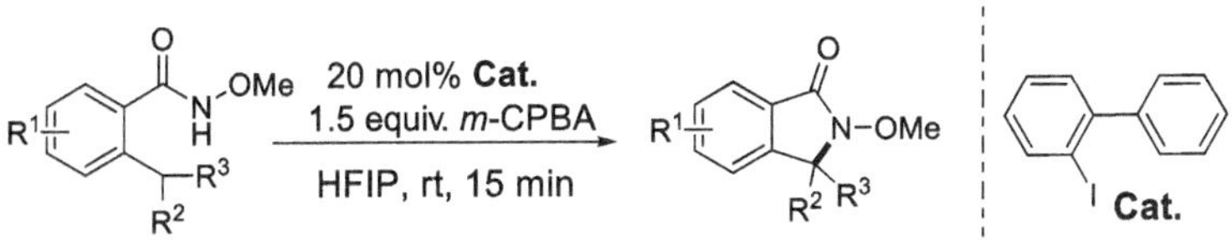

SCHEME 6.45 Metal-free stereospecific sp^3 C—H amination.

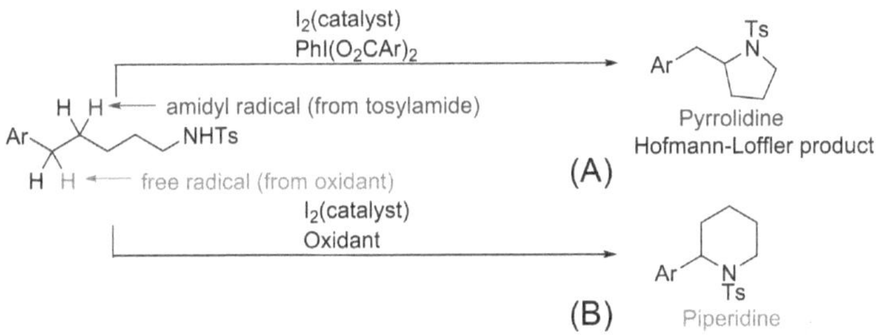

SCHEME 6.46 The position-selective intramolecular C—H amination of aliphatic groups.

sites are generally favored in C—H insertions resulting from compromise between electronic and steric factors, while primary positions are preferentially functionalized in the C—H activation-based amination owing to steric effects.

Although significant progress has been made in direct amination of $C(sp^3)$—H bonds, many challenges still remain to be addressed. First of all, the majority of C—H bond activation-based catalyst systems are confined to use the directing group and no general catalyst system has been reported for the selective amination of simple alkanes. Moreover, design of ligands for the asymmetric $C(sp^3)$—H bond activation-based amination and C—H bond insertion amination is still a challenge. The development of new methods for the efficient amination of $C(sp^3)$—H bond utilizing O_2 or air as the greener oxidant is still in high demand. Finally, further mechanistic studies are required to design more efficient protocol for $C(sp^3)$—H bonds catalytic amination reaction.

REFERENCES

1. Zhao, J.; Zhang, Q. *Acta Chim. Sin.* **2015**, *73*, 1235.
2. Shi, Z.; Mu, Y.; Zhu, C. *Synlett* **2015**, *27*, 486.
3. Fu, H.; Yu, J. *Rep. Org. Chem.* **2015**, *5*, 1.
4. Chen, G.; Nack, W. *Synlett* **2015**, *26*, 2505.
5. Zatolochnaya, O. V.; Gevorgyan, V. *Nat. Chem.* **2014**, *6*, 661.
6. Uchida, T.; Katsuki, T. *Chem. Rec.* **2014**, *14*, 117.
7. Louillat, M. L.; Patureau, F. W. *Chem. Soc. Rev.* **2014**, *43*, 901.
8. Zhang, L.; Deng, L. *Chin. Sci. Bull.* **2012**, *57*, 2352.
9. Roizen, J. L.; Harvey, M. E.; Du Bois, J. *Acc. Chem. Res.* **2012**, *45*, 911.
10. Ramirez, T. A.; Zhao, B.; Shi, Y. *Chem. Soc. Rev.* **2012**, *41*, 931.

11. Gephart, R. T.; Warren, T. H. *Organometallics* **2012**, *31*, 7728.
12. Collet, F.; Lescot, C.; Dauban, P. *Chem. Soc. Rev.* **2011**, *40*, 1926.
13. Bois, J. D. *Org. Process Res. Dev.* **2011**, *15*, 758.
14. Marder, T. B.; Dyer, P. W.; Fairlamb, I. J.; Gibson, S.; Scott, P. *Dalton Trans.* **2010**, *39*, 10335.
15. Mancheno, D. E.; Thornton, A. R.; Stoll, A. H.; Kong, A.; Blakey, S. B. *Org. Lett.* **2010**, *12*, 4110.
16. Thansandote, P.; Lautens, M. *Chem. Eur. J.* **2009**, *15*, 5874.
17. Collet, F.; Dodd, R. H.; Dauban, P. *Chem. Commun.* **2009**, 5061.
18. Liang, C.; Collet, F.; Robert-Peillard, F.; Muller, P.; Dodd, R. H.; Dauban, P. *J. Am. Chem. Soc.* **2008**, *130*, 343.
19. Davies, H. M.; Long, M. S. *Angew. Chem. Int. Ed.* **2005**, *44*, 3518.
20. Saunders, K. H. *J. Chem. Soc.* **1955**, 3275.
21. Smolinsky, G. *J. Am. Chem. Soc.* **1960**, *82*, 4717.
22. Smolinsky, G. *J. Org. Chem.* **1961**, *26*, 4108.
23. Smolinsky, G.; Feuer, B. I. *J. Am. Chem. Soc.* **1964**, *86*, 3085.
24. Lwowsk, W.; Mattingly, T. W. *Tetrahedron Lett.* **1962**, *3*, 277.
25. Lwowski, W.; Mattingly, T. W. *J. Am. Chem. Soc.* **1965**, *87*, 1947.
26. Sloan, M. F.; Renfrow, W. B.; Breslow, D. S. *Tetrahedron Lett.* **1964**, *5*, 2905.
27. Breslow, D. S.; Sloan, M. F.; Newburg, N. R.; Renfrow, W. B. *J. Am. Chem. Soc.* **1969**, *91*, 2273.
28. Anastassiou, A. G.; Simmons, H. E. *J. Am. Chem. Soc.* **1967**, *89*, 3177.
29. Kwart, H.; Khan, A. A. *J. Am. Chem. Soc.* **1967**, *89*, 1951.
30. Breslow, D.; Sloan, M. *Tetrahedron Lett.* **1968**, *9*, 5349.
31. Carr, D.; Seden, T.; Turner, R. *Tetrahedron Lett.* **1969**, *10*, 477.
32. Breslow, R.; Gellman, S. H. *J. Chem. Soc. Chem. Commun.* **1982**, 1400.
33. Breslow, R.; Gellman, S. H. *J. Am. Chem. Soc.* **1983**, *105*, 6728.
34. Mahy, J.-P.; Bedi, G.; Battioni, P.; Mansuy, D. *N. J. Chem.* **1989**, *13*, 651.
35. Yu, X. Q.; Huang, J. S.; Zhou, X. G.; Che, C. M. *Org. Lett.* **2000**, *2*, 2233.
36. Espino, C. G.; Du Bois, J. *Angew. Chem. Int. Ed.* **2001**, *40*, 598.
37. Diaz-Requejo, M. M.; Belderrain, T. R.; Nicasio, M. C.; Trofimenko, S.; Perez, P. J. *J. Am. Chem. Soc.* **2003**, *125*, 12078.
38. Fructos, M. R.; Trofimenko, S.; Diaz-Requejo, M. M.; Perez, P. J. *J. Am. Chem. Soc.* **2006**, *128*, 11784.
39. Jang, E. S.; McMullin, C. L.; Kass, M.; Meyer, K.; Cundari, T. R.; Warren, T. H. *J. Am. Chem. Soc.* **2014**, *136*, 10930.
40. Gephart, R. T., 3rd; Huang, D. L.; Aguila, M. J.; Schmidt, G.; Shahu, A.; Warren, T. H. *Angew. Chem. Int. Ed.* **2012**, *51*, 6488.
41. Bagchi, V.; Paraskevopoulou, P.; Das, P.; Chi, L.; Wang, Q.; Choudhury, A.; Mathieson, J. S.; Cronin, L.; Pardue, D. B.; Cundari, T. R.; Mitrikas, G.; Sanakis, Y.; Stavropoulos, P. *J. Am. Chem. Soc.* **2014**, *136*, 11362.
42. Zhang, Y.; Fu, H.; Jiang, Y.; Zhao, Y. *Org. Lett.* **2007**, *9*, 3813.
43. Zeng, H. T.; Huang, J. M. *Org. Lett.* **2015**, *17*, 4276.
44. Powell, D. A.; Fan, H. *J. Org. Chem.* **2010**, *75*, 2726.
45. Yang, Q.; Choy, P. Y.; Fu, W. C.; Fan, B.; Kwong, F. Y. *J. Org. Chem.* **2015**, *80*, 11193.
46. Srivastava, R. S.; Kolel-Veetil, M.; Nicholas, K. M. *Tetrahedron Lett.* **2002**, *43*, 931.
47. Wang, Z.; Zhang, Y.; Fu, H.; Jiang, Y.; Zhao, Y. *Org. Lett.* **2008**, *10*, 1863.
48. Xia, Q.; Chen, W.; Qiu, H. *J. Org. Chem.* **2011**, *76*, 7577.
49. Xia, Q.; Chen, W. *J. Org. Chem.* **2012**, *77*, 9366.

50. Mao, X.; Wu, Y.; Jiang, X.; Liu, X.; Cheng, Y.; Zhu, C. *RSC Adv.* **2012**, *2*, 6733.
51. Sun, M.; Zhang, T.; Bao, W. *Tetrahedron Lett.* **2014**, *55*, 893.
52. Mahy, J. P.; Bedi, G.; Battioni, P.; Mansuy, D. *Tetrahedron Lett.* **1988**, *29*, 1927.
53. Yang, J.; Weinberg, R.; Breslow, R. *Chem. Commun.* **2000**, 531.
54. Zhou, X. G.; Yu, X. Q.; Huang, J. S.; Che, C.-M. *Chem. Commun.* **1999**, 2377.
55. Liang, J. L.; Huang, J. S.; Yu, X. Q.; Zhu, N. Y.; Che, C. M. *Chem. Eur. J.* **2002**, *8*, 1563.
56. Kohmura, Y.; Katsuki, T. *Tetrahedron Lett.* **2001**, *42*, 3339.
57. Cenini, S.; Ragaini, F.; Tollari, S.; Paone, D. *J. Am. Chem. Soc.* **1996**, *118*, 11964.
58. Ragaini, F.; Cenini, S.; Tollari, S.; Tummolillo, G.; Beltrami, R. *Organometallics* **1999**, *18*, 928.
59. Au, S.-M.; Zhang, S.-B.; Fung, W.-H.; Yu, W.-Y.; Che, C.-M.; Cheung, K.-K. *Chem. Commun.* **1998**, 2677.
60. Au, S. M.; Huang, J. S.; Che, C. M.; Yu, W. Y. *J. Org. Chem.* **2000**, *65*, 7858.
61. Au, S. M.; Huang, J. S.; Yu, W. Y.; Fung, W. H.; Che, C. M. *J. Am. Chem. Soc.* **1999**, *121*, 9120.
62. Liang, J.-L.; Yu, X.-Q.; Che, C.-M. *Chem. Commun.* **2002**, 124.
63. Nägeli, I.; Baud, C.; Bernardinelli, G.; Jacquier, Y.; Moraon, M.; Müllet, P. *Helv. Chim. Acta* **1997**, *80*, 1087.
64. Fiori, K. W.; Du Bois, J. *J. Am. Chem. Soc.* **2007**, *129*, 562.
65. Zalatan, D. N.; Du Bois, J. *J. Am. Chem. Soc.* **2009**, *131*, 7558.
66. Norder, A.; Herrmann, P.; Herdtweck, E.; Bach, T. *Org. Lett.* **2010**, *12*, 3690.
67. Norder, A.; Warren, S. A.; Herdtweck, E.; Huber, S. M.; Bach, T. *J. Am. Chem. Soc.* **2012**, *134*, 13524.
68. Wang, J.; Zhao, C.; Weng, Y.; Xu, H. *Catal. Sci. Technol.* **2016**, *6*, 5292.
69. Höke, T.; Herdtweck, E.; Bach, T. *Chem. Commun.* **2013**, *49*, 8009.
70. Reddy, R. P.; Davies, H. M. *Org. Lett.* **2006**, *8*, 5013.
71. Anada, M.; Tanaka, M.; Shimada, N.; Nambu, H.; Yamawaki, M.; Hashimoto, S. *Tetrahedron* **2009**, *65*, 3069.
72. Liang, C.; Robert-Peillard, F.; Fruit, C.; Muller, P.; Dodd, R. H.; Dauban, P. *Angew. Chem. Int. Ed.* **2006**, *45*, 4641.
73. Lescot, C.; Darses, B.; Collet, F.; Retailleau, P.; Dauban, P. *J. Org. Chem.* **2012**, *77*, 7232.
74. Dauban, P.; Darses, B.; Jarvis, A.; Mafroud, A.-K.; Estenne-Bouhtou, G.; Dargazanli, G. *Synthesis* **2013**, *45*, 2079.
75. Lebel, H.; Huard, K. *Org. Lett.* **2007**, *9*, 639.
76. Huard, K.; Lebel, H. *Chem. Eur. J.* **2008**, *14*, 6222.
77. Lebel, H.; Spitz, C.; Leogane, O.; Trudel, C.; Parmentier, M. *Org. Lett.* **2011**, *13*, 5460.
78. Lebel, H.; Trudel, C.; Spitz, C. *Chem. Commun.* **2012**, *48*, 7799.
79. Kornecki, K. P.; Berry, J. F. *Chem. Eur. J.* **2011**, *17*, 5827.
80. Perry, R. H.; Cahill, T. J., 3rd; Roizen, J. L.; Du Bois, J.; Zare, R. N. *Proc. Natl. Acad. Sci. U. S. A.* **2012**, *109*, 18295.
81. Huang, X.; Wang, Y.; Lan, J.; You, J. *Angew. Chem. Int. Ed.* **2015**, *54*, 9404.
82. Thu, H. Y.; Yu, W. Y.; Che, C. M. *J. Am. Chem. Soc.* **2006**, *128*, 9048.
83. Reed, S. A.; White, M. C. *J. Am. Chem. Soc.* **2008**, *130*, 3316.
84. Reed, S. A.; Mazzotti, A. R.; White, M. C. *J. Am. Chem. Soc.* **2009**, *131*, 11701.
85. Engelin, C.; Jensen, T.; Rodriguez-Rodriguez, S.; Fristrup, P. *ACS Catal.* **2013**, *3*, 294.
86. Pattillo, C. C.; Strambeanu, I. I.; Calleja, P.; Vermeulen, N. A.; Mizuno, T.; White, M. C. *J. Am. Chem. Soc.* **2016**, *138*, 1265.
87. Li, Y. G.; Li, L.; Yang, M. Y.; He, G.; Kantchev, E. A. B. *J. Org. Chem.* **2017**, *82*, 4907.

88. Liu, G.; Yin, G.; Wu, L. *Angew. Chem. Int. Ed.* **2008**, *120*, 4811.
89. Yin, G. Y.; Wu, Y. C.; Liu, G. S. *J. Am. Chem. Soc.* **2010**, *132*, 11978.
90. Luzung, M. R.; Lewis, C. A.; Baran, P. S. *Angew. Chem. Int. Ed.* **2009**, *48*, 7025.
91. Li, Z.; Capretto, D. A.; Rahaman, R.; He, C. *Angew. Chem. Int. Ed.* **2007**, *119*, 5276.
92. Gomez-Emeterio, B. P.; Urbano, J.; Diaz-Requejo, M. M.; Perez, P. J. *Organometallics* **2008**, *27*, 4126.
93. Beltrán, Á.; Lescot, C.; Mar Díaz-Requejo, M.; Pérez, P. J.; Dauban, P. *Tetrahedron* **2013**, *69*, 4488.
94. Ye, Y.-H.; Zhang, J.; Wang, G.; Chen, S.-Y.; Yu, X.-Q. *Tetrahedron* **2011**, *67*, 4649.
95. Wu, X.; Yang, K.; Zhao, Y.; Sun, H.; Li, G.; Ge, H. *Nat. Commun.* **2015**, *6*, 6462.
96. Pang, S.; Shi, F. *Tetrahedron Lett.* **2016**, *57*, 5872.
97. Peng, H.; Lin, A.; Zhang, Y.; Jiang, H.; Zhou, J.; Cheng, Y.; Zhu, C.; Hu, H. *ACS Catal.* **2012**, *2*, 163.
98. Li, Z. L.; Jin, L. K.; Cai, C. *Org. Biomol. Chem.* **2017**, *15*, 1317.
99. Fan, R.; Li, W.; Pu, D.; Zhang, L. *Org. Lett.* **2009**, *11*, 1425.
100. Guo, H.-M.; Xia, C.; Niu, H.-Y.; Zhang, X.-T.; Kong, S.-N.; Wang, D.-C.; Qu, G.-R. *Adv. Synth. Catal* **2011**, *353*, 53.
101. Buslov, I.; Hu, X. *Adv. Synth. Catal.* **2014**, *356*, 3325.
102. Kim, H. J.; Kim, J.; Cho, S. H.; Chang, S. *J. Am. Chem. Soc.* **2011**, *133*, 16382.
103. Tian, J. S.; Ng, K. W.; Wong, J. R.; Loh, T. P. *Angew. Chem. Int. Ed.* **2012**, *51*, 9105.
104. Jiang, Q.; Xu, B.; Zhao, A.; Jia, J.; Liu, T.; Guo, C. *J. Org. Chem.* **2014**, *79*, 8750.
105. Ramesh, D.; Ramulu, U.; Mukkanti, K.; Venkateswarlu, Y. *Tetrahedron Lett.* **2012**, *53*, 2904.
106. Sun, J.; Wang, Y.; Pan, Y. *J. Org. Chem.* **2015**, *80*, 8945.
107. Zhang, Z.; Du, H.; Xu, J.; Li, P. *Chem. Commun.* **2016**, *52*, 11547.
108. Barman, D. N.; Nicholas, K. M. *Eur. J. Org. Chem.* **2011**, *2011*, 908.
109. Wang, Z.; Ni, J.; Kuninobu, Y.; Kanai, M. *Angew. Chem. Int. Ed.* **2014**, *53*, 3496.
110. Wang, H.; Li, Y.; Wang, Z.; Lou, J.; Xiao, Y.; Qiu, G.; Hu, X.; Altenbach, H.-J.; Liu, P. *RSC Adv* **2014**, *4*, 25287.
111. Sun, W. W.; Cao, P.; Mei, R. Q.; Li, Y.; Ma, Y. L.; Wu, B. *Org. Lett.* **2014**, *16*, 480.
112. Pan, F.; Wu, B.; Shi, Z. J. *Chem. Eur. J.* **2016**, *22*, 6487.
113. Yamamoto, C.; Takamatsu, K.; Hirano, K.; Miura, M. *J. Org. Chem.* **2016**, *81*, 7675.
114. Tian, H.; Qiao, H.; Zhu, C.; Fu, H. *RSC Adv.* **2014**, *4*, 2694.
115. Breslow, B.; Gellman, S. H. *J. Am. Chem. Soc.* **1983**, *105*, 6728.
116. Paradine, S. M.; White, M. C. *J. Am. Chem. Soc.* **2012**, *134*, 2036.
117. Liu, Y.; Guan, X.; Wong, E. L.-M.; Liu, P.; Huang, J.-S.; Che, C.-M. *J. Am. Chem. Soc.* **2013**, *135*, 7194.
118. Ren, X.; O'Hanlon, J. A.; Morris, M.; Robertson, J.; Wong, L. L. *ACS Catal.* **2016**, *6*, 6833.
119. Zhang, J.; Hong Chan, P. W.; Che, C.-M. *Tetrahedron Lett.* **2005**, *46*, 5403.
120. Liang, J. L.; Yuan, S. X.; Huang, J. S.; Yu, W. Y.; Che, C. M. *Angew. Chem. Int. Ed.* **2002**, *41*, 3465.
121. Milczek, E.; Boudet, N.; Blakey, S. *Angew. Chem. Int. Ed.* **2008**, *47*, 6825.
122. Blakey, B. S.; Bon, L. J. *Heterocycles* **2012**, *84*, 1313.
123. Espino, C. G.; Wehn, P. M.; Chow, J.; Du Bois, J. *J. Am. Chem. Soc.* **2001**, *123*, 6935.
124. Williams Fiori, K.; Fleming, J. J.; Du Bois, J. *Angew. Chem. Int. Ed.* **2004**, *43*, 4349.
125. Fleming, J. J.; Fiori, K. W.; Du Bois, J. *J. Am. Chem. Soc.* **2003**, *125*, 2028.
126. Wehn, P. M.; Lee, J.; Du Bois, J. *Org. Lett.* **2003**, *5*, 4823.
127. Wehn, P. M.; Du Bois, J. *J. Am. Chem. Soc.* **2002**, *124*, 12950.
128. Parker, K. A.; Chang, W. *Org. Lett.* **2005**, *7*, 1785.

129. Wehn, P. M.; Du Bois, J. *Org. Lett.* **2005**, *7*, 4685.
130. Toumieux, S.; Compain, P.; Martin, O. R.; Selkti, M. *Org. Lett.* **2006**, *8*, 4493.
131. Yakura, T.; Yoshimoto, Y.; Ishida, C.; Mabuchi, S. *Synlett* **2006**, *2006*, 930.
132. Conrad, R. M.; Du Bois, J. *Org. Lett.* **2007**, *9*, 5465.
133. Narina, S. V.; Kumar, T. S.; George, S.; Sudalai, A. *Tetrahedron Lett.* **2007**, *48*, 65.
134. Wyszynski, F. J.; Thompson, A. L.; Davis, B. G. *Org. Biomol. Chem.* **2010**, *8*, 4246.
135. Takahashi, K.; Yamaguchi, D.; Ishihara, J.; Hatakeyama, S. *Org. Lett.* **2012**, *14*, 1644.
136. Hatakeyama, S.; Takahashi, K.; Yamaguchi, D.; Ishihara, J. *Synlett* **2013**, *25*, 133.
137. Yamashita, S.; Himuro, M.; Hayashi, Y.; Hirama, M. *Tetrahedron Lett.* **2013**, *54*, 1307.
138. Hrdina, R.; Metz, F. M.; Larrosa, M.; Berndt, J.-P.; Zhygadlo, Y. Y.; Becker, S.; Becker, J. *Eur. J. Org. Chem* **2015**, *2015*, 6231.
139. Nambu, H.; Noda, N.; Niu, W.; Fujiwara, T.; Yakura, T. *Asian J. Org. Chem.* **2015**, *4*, 1246.
140. Noda, N.; Nambu, H.; Ubukata, K.; Fujiwara, T.; Tsuge, K.; Yakura, T. *Tetrahedron* **2017**, *73*, 868.
141. Nocquet, P. A.; Hensienne, R.; Wencel-Delord, J.; Wimmer, E.; Hazelard, D.; Compain, P. *Org. Biomol. Chem.* **2015**, *13*, 9176.
142. Nocquet, P. A.; Hensienne, R.; Wencel-Delord, J.; Laigre, E.; Sidelarbi, K.; Becq, F.; Norez, C.; Hazelard, D.; Compain, P. *Org. Biomol. Chem* **2016**, *14*, 2780.
143. Brodsky, B. H.; Du Bois, J. *Chem. Commun.* **2006**, 4715.
144. Kim, M.; Mulcahy, J. V.; Espino, C. G.; Du Bois, J. *Org. Lett.* **2006**, *8*, 1073.
145. Yakura, T.; Yoshimoto, Y.; Ishida, C. *Chem. Pharm. Bull.* **2007**, *55*, 1385.
146. Tanino, T.; Ichikawa, S.; Matsuda, A. *Org. Lett.* **2011**, *13*, 4028.
147. Olson, D. E.; Roberts, D. A.; Du Bois, J. *Org. Lett.* **2012**, *14*, 6174.
148. Kono, M.; Harada, S.; Nemoto, T. *Chem. Eur. J.* **2017**, *23*, 7428.
149. St Denis, J. D.; Lee, C. F.; Yudin, A. K. *Org. Lett.* **2015**, *17*, 5764.
150. Yakura, T.; Sato, S.; Yoshimoto, Y. *Chem. Pharm. Bull.* **2007**, *55*, 1284.
151. Olson, D. E.; Du Bois, J. *J. Am. Chem. Soc.* **2008**, *130*, 11248.
152. Zalatan, D. N.; Du Bois, J. *J. Am. Chem. Soc.* **2008**, *130*, 9220.
153. Kornecki, K. P.; Berry, J. F. *Chem. Commun.* **2012**, *48*, 12097.
154. Zalatan, D. N.; Du Bois, J. *Synlett* **2009**, *2009*, 143.
155. Lebel, H.; Huard, K.; Lectard, S. *J. Am. Chem. Soc.* **2005**, *127*, 14198.
156. Lebel, H.; Mamani Laparra, L.; Khalifa, M.; Trudel, C.; Audubert, C.; Szponarski, M.; Leduc, C. D.; Azek, E.; Ernzerhof, M. *Org. Biomol. Chem* **2017**, *15*, 4144.
157. Lin, X.; Zhao, C.; Che, C. M.; Ke, Z.; Phillips, D. L. *Chem. Asian J.* **2007**, *2*, 1101.
158. Fiori, K. W.; Espino, C. G.; Brodsky, B. H.; Du Bois, J. *Tetrahedron* **2009**, *65*, 3042.
159. Zhang, X.; Ke, Z.; DeYonke, N. J.; Xu, H.; Li, Z.-F.; Xu, X.; Zhang, X.; Su, C.-Y.; Phillips, D. L.; Zhao, C. *J. Org. Chem.* *78*, 12460.
160. Zhang, X.; Xu, H.; Zhao, C. *J. Org. Chem.* **2014**, *79*, 9799.
161. Varela-Alvarez, A.; Yang, T.; Jennings, H.; Kornecki, K. P.; Macmillan, S. N.; Lancaster, K. M.; Mack, J. B. C.; Bois, J. D.; Berry, J. F.; Musaev, D. G. *J. Am. Chem. Soc* **2016**, *138*, 2327.
162. Fraunhoffer, K. J.; White, M. C. *J. Am. Chem. Soc.* **2007**, *129*, 7274.
163. Duarte, F. J. S.; Poli, G.; Calhorda, M. J. *ACS Catal.* **2016**, *6*, 1772.
164. Rice, G. T.; White, M. C. *J. Am. Chem. Soc.* **2009**, *131*, 11707.
165. Qi, X. B.; Rice, G. T.; Lall, M. S.; Plummer, M. S.; White, M. C. *Tetrahedron* **2010**, *66*, 4816.
166. Jiang, C.; Covell, D. J.; Stepan, A. F.; Plummer, M. S.; White, M. C. *Org. Lett.* **2012**, *14*, 1386.
167. Wu, L.; Qiu, S.; Liu, G. *Org. Lett.* **2009**, *11*, 2707.
168. Wasa, M.; Yu, J.-Q. *J. Am. Chem. Soc.* **2008**, *130*, 14058.

169. He, G.; Zhao, Y.; Zhang, S.; Lu, C.; Chen, G. *J. Am. Chem. Soc.* **2012**, *134*, 3.
170. He, G.; Zhang, S. Y.; Nack, W. A.; Li, Q.; Chen, G. *Angew. Chem. Int. Ed.* **2013**, *52*, 11124.
171. Nadres, E. T.; Daugulis, O. *J. Am. Chem. Soc.* **2012**, *134*, 7.
172. Zhang, S. J.; Sun, W. W.; Cao, P.; Dong, X. P.; Liu, J. K.; Wu, B. *J. Org. Chem.* **2016**, *81*, 956.
173. Zhang, X.; Xu, H.; Liu, X.; Phillips, D. L.; Zhao, C. *Chem. Eur. J.* **2016**, *22*, 7288.
174. MeNally, M.; Haffemayer, B.; Collins, B. L.; Gaunt, M. J. *Nature* **2014**, *510*, 129.
175. Smalley, A. P.; Cuthbertson, J. D.; Gaunt, M. J. *J. Am. Chem. Soc.* **2017**, *139*, 1412.
176. Cui, Y.; He, C. *Angew. Chem. Int. Ed.* **2004**, *116*, 4306.
177. Zhu, X.; Chiba, S. *Org. Biomol. Chem.* **2014**, *12*, 4567.
178. Alderson, J. M.; Phelps, A. M.; Scamp, R. J.; Dolan, N. S.; Schomaker, J. M. *J. Am. Chem. Soc.* **2014**, *136*, 16720.
179. Corbin, J. R.; Schomaker, J. M. *Chem. Commun.* **2017**, *53*, 4346.
180. Wu, X.; Zhao, Y.; Ge, H. *Chem. Eur. J.* **2014**, *20*, 9530.
181. Huang, G. H.; Li, J. M.; Huang, J. J.; Lin, J. D.; Chuang, G. J. *Chem. Eur. J.* **2014**, *20*, 5240.
182. Mao, L.; Li, Y.; Xiong, T.; Sun, K.; Zhang, Q. *J. Org. Chem.* **2013**, *78*, 733.
183. Zhu, C.; Liang, Y.; Hong, X.; Sun, H.; Sun, W.-Y.; Houk, K. N.; Shi, Z. *J. Am. Chem. Soc* **2015**, *137*, 7564.
184. O'Broin, C. Q.; Fernandez, P.; Martinez, C.; Muniz, K. *Org. Lett.* **2016**, *18*, 436.
185. Zhang, H.; Muñiz, K. *ACS Catal.* **2017**, *7*, 4122.

Chapter 7

N-Alkyl Amine Synthesis With Nitroarene/Benzonitrile and Alcohol

7.1 INTRODUCTION

It is well known that primary amines can be efficiently produced by the hydrogenation of nitro- and nitrile-containing molecules.[1–3] Thus, it would be an ideal reaction if substituted amines could be synthesized in one step from nitro or nitrile compounds and alcohols. By doing this, a multistep reaction would be realized in one pot. Generally, the nitro and nitrile groups can be reduced by alcohol, H_2, and glycerol, followed by the *N*-alkylation reaction with alcohol to synthesize *N*-alkyl amine.

7.2 SYNTHESIS OF N-ALKYL AMINES FROM NITROARENES AND ALCOHOLS

7.2.1 Alcohol as Reducing Agent

7.2.1.1 Ru Catalyst System

In 2010, Deng and Li et al. reported the $Ru(CO)(H)_2(PPh_3)_3$/NHC-catalyzed reductive *N*-alkylation of nitroarene with an excessive amount of primary alcohols at 150°C under an argon atmosphere that afforded the corresponding secondary amines.[4] Such a process combined two catalytic cycles, i.e., the reduction of nitroarene to amine, and the *N*-alkylation of the resultant amine with the alcohol through a borrowing hydrogen pathway. The alcohol oxidation, nitro reduction, and imine reduction were realized in a cascade (Scheme 7.1). It should be noted that an excessive amount of alcohol (up to 7.5 equiv.) was required to get decent yields as both the reactant and the solvent. In this reaction, tertiary amines are the typical products and no secondary amine can be generated. Of note, secondary amines can be synthesized efficiently with the addition of suitable base as cocatalyst. For example, secondary amines can be synthesized with good yields with Ru $(acac)_3$/dppe/$KHCO_3$ as catalyst.[5]

Almost at the same time, Shi and colleagues reported the reaction of nitroarenes with primary alcohols catalyzed by [Ru(*p*-cymene)Cl_2]/dppb/K_2CO_3 at

Catalytic Amination for N-Alkyl Amine Synthesis. https://doi.org/10.1016/B978-0-12-812284-6.00007-X

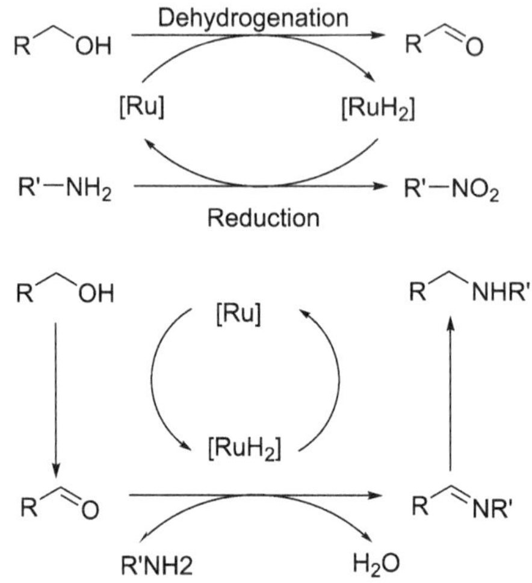

SCHEME 7.1 Mechanism of the reductive *N*-alkylation of nitroarene with alcohol.

130°C under Argon atmosphere to obtain secondary *N*-alkyl amines with 73%–95% isolated yields.[6] Noteworthy, methanol, ethanol, and other aliphatic alcohols can also behave as the reducing agent and alkylation agent, producing yields of the desired products at 70%–93%.

Liu et al. also reported the synthesis of secondary amines from various nitroarenes and primary alcohols catalyzed by a phosphine-amine ruthenium(II) complex.[7] Interestingly, the yields of amine compounds increased as the reaction proceeded under ambient pressure of molecular hydrogen (Table 7.1). By carrying out the reaction under atmospheric D_2, less than 5% of product with D-labeled at the benzylic position was obtained, indicating that the reduction of imines into amines was accomplished primarily by the in situ generated metal-hydrides. Presumably, the hydrogen atmosphere slows down the reductive elimination of hydrides from the metal center, which then assists the reduction of imine.

TABLE 7.1 Synthesis of Secondary Amine From Nitroarene and Alcohol

Entry	Nitro-Substrate	Alcohol	H_2	2degree-Amine	Imine
1	*p*-$BrC_6H_4NO_2$	$C_6H_5CH_2OH$	–	86%	14%
2	*p*-$BrC_6H_4NO_2$	$C_6H_5CH_2OH$	1 atm	100%	0
3	*p*-$ClC_6H_4NO_2$	$C_6H_5CH_2OH$	–	60%	6
4	*p*-$ClC_6H_4NO_2$	$C_6H_5CH_2OH$	1 atm	100%	0

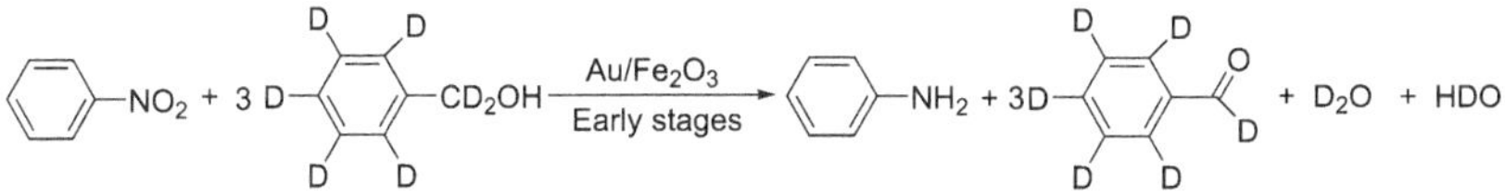

SCHEME 7.2 Isotopic studies in the alkylation of aniline with benzyl d_7-alcohol catalyzed by Au/ Fe_2O_3.

It is worth mentioning that the secondary amine derivatives can be transferred into tertiary amines by using the same catalyst under 1 atm H_2 at 150°C.

7.2.1.2 Au Catalyst System

In 2011 it was reported that heterogeneous Au/Fe_2O_3 can be an active catalyst for the reductive *N*-alkylation of nitroarenes. Of note, the synthesis of secondary and tertiary amines can be controlled by tuning the reaction conditions.[8] By monitoring the reaction between nitrobenzene and d_7-benzyl alcohol at the initial stage through GC-MS, <5% of deuterium incorporation was found in the —NH_2 group of aniline (the intermediate product of the process), and only d_6-benzaldehyde could be detected, which suggests that the protons in the NH_2 group of aniline derives from the proton of the hydroxyl group of benzyl alcohol, while the benzylic hydrides are transferred into water (Scheme 7.2).

Almost at the same time, it was found that dialkyl amines can be synthesized with nitroarenes and alcohols catalyzed by Au/TiO_2-VS without external hydrogen source.[9] This methodology was also proposed to proceed through transfer hydrogenation form the alcohol (forming the aldehyde), formation of an intermediate imine by coupling of aldehyde and the newly formed amine, and finally, reduction of the imine by Au-H species formed during the oxidation of the alcohol. If the reaction temperature was increased from 120°C to 150°C, further alkylation of the dialkyl amines may occur, thus leading to the formation of trialkyl amines.

7.2.1.3 Ni Catalyst System

With Raney Ni as catalyst,[10,11] nitroarenes could be efficiently reduced and alkylated by methanol/ethanol to the corresponding secondary/tertiary amines with up to 98% yields under N_2 atmosphere at 413/443 K. The generation of *N*-alkyl amines from nitroarenes and alcohols was carried out by multistep reaction pathways in one reactor over a multifunctional catalyst. The essential features of the one-pot synthesis of *N*-alkyl amines involve first formation of aniline from the in-situ and/or transfer hydrogenation of nitrobenzene. These steps followed by stepwise *N*-alkylation of aniline to produce *N*-alkyl amine/ *N,N*-dialkyl amine. Methanol/ethanol in these reaction systems could be versatile, since it acted simultaneously as a hydrogen source, alkylating reagent and solvent.

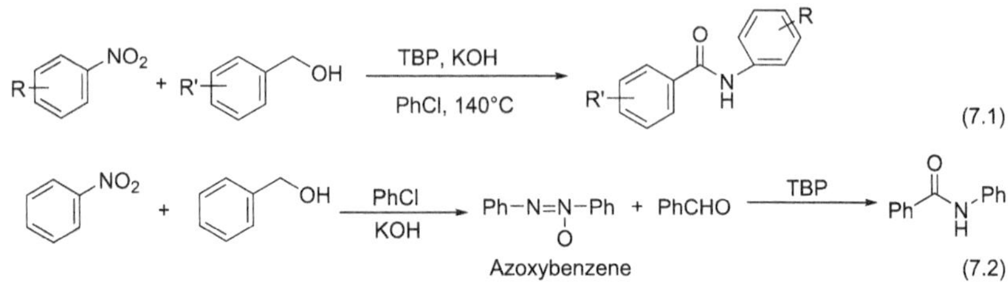

SCHEME 7.3 The synthesis of amide from nitroarenes and aromatic alcohols (7.1) and control experiment (7.2).

7.2.1.4 Peroxide/TBP Catalyst System

Deng et al. developed a direct amination of alcohols with nitroarenes to form aromatic amides in moderate to good yields mediated by peroxide and TBP in the absence of transition-metal catalyst (Scheme 7.3, Eq. 7.1).[12] Although the mechanism for the amidation of alcohols with nitroarenes is not completely clear, the control experiments (Eq. 7.2) revealed that hydrogen generated from alcohol oxidation acted as the nitro reducing reagent to form azoxybenzene and then reacted with aldehyde generated in situ to obtain amides. In this way, the alcohol oxidation, nitro reduction, and amide formation were realized in a cascade.

7.2.1.5 Photocatalytic System

Although heterogeneous catalysts are easily recycled, the previously mentioned systems need to be performed at relatively high reaction temperatures (>363 K). Therefore, improved methods capable of producing secondary amines at room temperature are desired. Au-TiO_2 and Pd/TiO_2 catalysts were used to the photocatalytic *N*-alkylation of nitroarenes in 2013 and 2015, respectively. Both of the reactions have similar reaction mechanism, and the proposed reaction mechanism is illustrated in Scheme 7.4. The reaction is initiated by photoexcitation of TiO_2, producing the electron (e^-) and positive hole (h^+) pairs. The h^+ oxidizes

$$TiO_2 + h\nu \longrightarrow e^- + h^+$$

$$R^1\text{-}CH_2OH + 2h^+ \longrightarrow R^1\text{-}CHO + 2H^+$$

$$H^+ + e^- + \text{Pd/Au} \longrightarrow \text{H-Pd or H-Au}$$

$$R^2\text{-}NO_2 + 6\text{H-Pd/Au} \longrightarrow R^2\text{-}NH_2 + 2H_2O + 6\text{Pd}$$

$$R^1\text{-}CHO + R^2\text{-}NH_2 \rightleftharpoons R^1\text{-}{=}N\text{-}R^2 + H_2O$$

$$R^1CH{=}NR^2 + 2\text{H-Pd} \longrightarrow R^1\text{-}CH_2\text{-}NH\text{-}R^2 + 2\text{Pd}$$

SCHEME 7.4 Proposed pathway for secondary amine formation from alcohols and nitroarenes on the photoactivated Pd/TiO_2 catalyst.

alcohol and produces aldehyde and H^+. Photoexcited electrons reduce H^+ on the Pd particles and produce hydrides (H-Pd or H-Au species). These H-Pd or H-Au species may reduce nitroarene and produce aniline. Catalytic condensation of the formed aniline and aldehyde by the Lewis acid sites on the surface of TiO_2 may produce imine. The imine is then hydrogenated by the H-Pd or H-Au species and is converted to produce secondary amine. Therefore, these photocatalytic and catalytic sequences may facilitate one-pot synthesis of secondary amine from alcohols and nitroarenes at room temperature.[13,14]

7.2.2 Glycerol as Reducing Agent

Renewable biomass has gained extensive attention because the derivation of fuels and chemicals from petroleum faces tremendous challenges for environmental and economic sustainability. Hydrogen generation from glycerol was reported recently. That means glycerol would be ideal if it could be used directly as a reducing agent in the synthesis of fine chemicals without thermal cracking. In the *N*-alkyl amine synthesis using nitroarenes and alcohols as starting materials, except for its role as an alkylation reagent, alcohol was employed as the hydrogen source to reduce nitrobenzene. Thus, a large excess of alcohol was needed, in which the ratio of alcohol to nitrobenzene is in the range of 6:1 to 10:1. It would be an improvement if the coupling reaction could be realized with an equivalent amount of nitrobenzene and alcohol.

A highly efficient Au/Ag-Mo nanorods catalyst was reported by Shi et al. for the one-pot synthesis of amine using equal molar ratio of nitrobenzene and alcohol as starting materials, and bio-based glycerol as the hydrogen source (Scheme 7.5).[15] Good to excellent yields of *N*-alkyl amines and imines were achieved under mild conditions with 1:1 molar ratio of alcohols and nitroarenes. To investigate the reaction mechanism, the reaction mixture was characterized by LC-MS, which suggested the oxidation of glycerol to generate glyceraldhyde, and benzyl alcohol to benzaldehyde. Thus, benzylalcohol and glycerol were all responsible for the reduction of *p*-methyl nitrobenzene at the initial stage. The further step is the formation of *N*-benzylidene-4-methylaniline and its further reduction to *N*-benzyl-*p*-methyl aniline by glycerol. Interestingly, the glycerol aldehyde generated in situ did not react with aniline, possibly because of its specific structure.

$RuCl_3/PPh_3/K_2CO_3$ could also be an effective catalyst system in the one-pot synthesis of secondary amines from nitroarenes with an equivalent amount of alcohols using glycerol as the reducing agent at 130°C.[16] And tertiary amines can be synthesized by simply increasing the reaction temperature to 150°C, and good to excellent results are achieved. An improved $RuCl_3$-catalyst system for the construction of aryl amines from nitroarenes and alcohols was reported in 2016.[17]

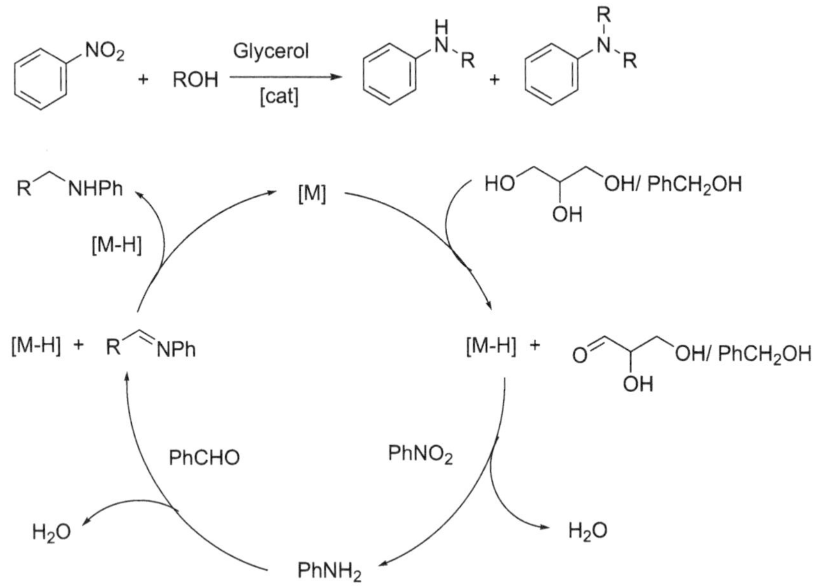

SCHEME 7.5 The plausible mechanism of the synthesis of *N*-alkyl aniline using glycerol as the reducing agent.

7.2.3 Hydrogen as Reducing Agent

Molecular hydrogen is the cheapest reducing agent in industry; therefore its use for the reductive *N*-alkylation of nitroarenes would be beneficial. Shimizu and colleagues provide the first example of the heterogeneously catalyzed one-step synthesis of *N*-substituted anilines from nitroaromatics and benzyl alcohols under 0.2 MPa H_2 atmosphere using Ag/Al_2O_3 as catalyst in the presence of KF (Scheme 7.6).[18] Systematic studies on the effects of the metals, Ag particle size, and supporting oxides reveal that the use of metals with weak metal-hydrogen bond energy, smaller particle size, and the support having both acidic and basic sites were important features of the catalyst.

In the presence of a Ru/C-$NaNO_2$ catalyst, alicyclic amines and aromatic amines were produced in high yield.[19] The hydrogenation reaction undergoes nitro group reduction, ring hydrogenation, and continues to *N*-alkylation

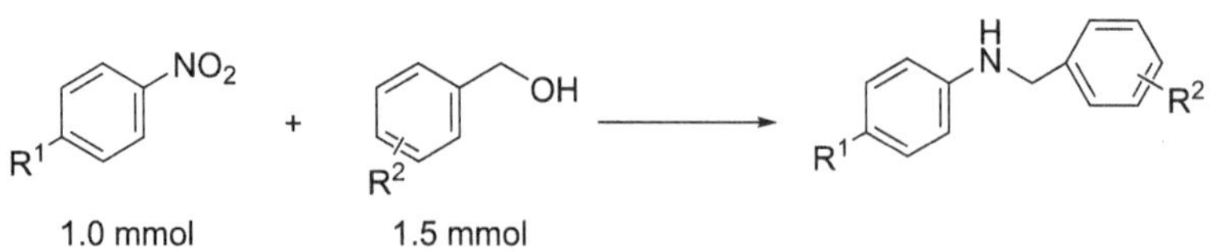

SCHEME 7.6 Reductive *N*-alkylation of nitroarenes and aromatic alcohols with H_2 as reducing agent.

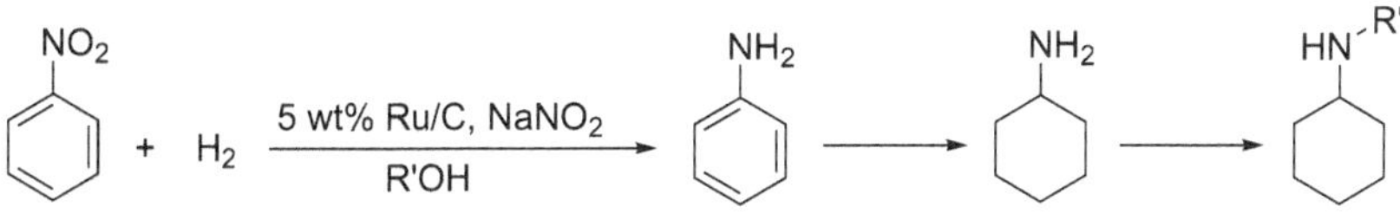

SCHEME 7.7 Reaction pathways for hydrogenation of nitrobenzene.

depending on the reaction conditions used. More severe reaction conditions resulted in reactions that proceeded further. Thus, aniline, cyclohexylamine, and *N*-alkyl cyclohexlamine could be produced almost quantitatively by simply altering the reaction conditions. Among the reaction parameters, temperature showed highest dependency on the product composition. The nitrobenzene conversion and aniline selectivity were 79.6% and 100%, respectively, at 80°C while 100% of nitrobenzene was converted to cyclohexylamine at 90°C. As temperature increased above 90°C, the cyclohexylamine decreased gradually and *N*-alkyl-cyclohexylamine started to generate. Eventually the cyclohexylamine was completely converted to *N*-alkyl-cyclohexylamine at 170°C (Scheme 7.7).

$Cu_{30}Cr_5/\gamma\text{-}Al_2O_3$ was used in a fixed-bed tubular reactor for the reductive *N*-alkylation of nitrobenzene with benzyl alcohol.[20] The XRD, H_2-TPR, XPS characterization results indicate that addition of chromium to $Cu_{30}Cr_5/\gamma\text{-}Al_2O_3$ made the Cu^0 particles smaller and better dispersed, and Cu^0 species were assumed to be the active sites of the catalysts. The catalyst system is highly stable and the yield of *N,N*-benzyl aniline still remained above 85% while the catalyst was run 150h under optimum conditions.

7.3 SYNTHESIS OF *N*-ALKYL AMINES FROM BENZONITRILES AND ALCOHOLS

7.3.1 Alcohol as Reducing Agent

Shi and colleagues reported using benzonitriles as the substrates to produce the same secondary amine products with $RuCl_3/PPh_3/K_2CO_3$ as catalyst at 140°C[6] (Scheme 7.8). In this reaction, secondary amines were normally synthesized with up to 95% yield. However, *N,N,N*-diethyl-benzyl amine was produced with 90% yield if ethanol was applied.

Recently, Beller et al. investigated the domino ruthenium-catalyzed transfer hydrogenation of nitriles with subsequent *N*-monoalkylation by alcohols.[21] A variety of aromatic, heteroaromatic, aliphatic, and disubstituted nitriles were converted with isopropanol into the secondary amines with excellent yields up

$$R^1CN + R^2OH \xrightarrow[140°C,\ Ar]{RuCl_3/PPh_3/K_2CO_3} R^1CH_2NHR^2$$

SCHEME 7.8 Reductive *N*-alkylation of nitriles with alcohols.

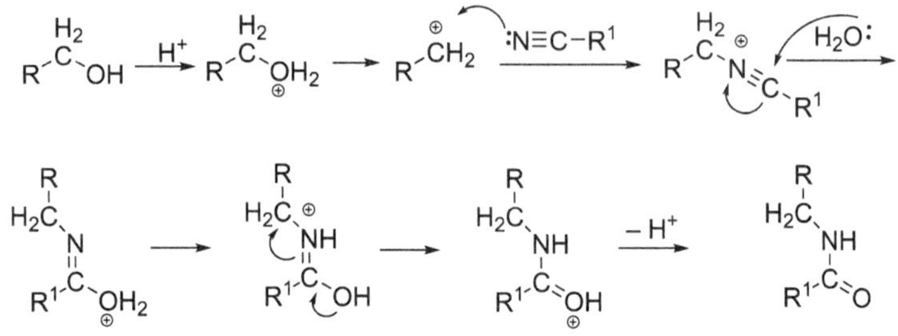

SCHEME 7.9 Amide synthesis via Ritter reaction.

to 99%. The effects of different alcohols such as ethanol, 2-butanol, and 2-pentanol were studied, and the corresponding secondary amines in low to excellent yields were obtained. The proposed mechanism for the *N*-monoalkylation reaction proceeds with hydrogenation of nitrile by using $RuCl_3/PPh_3$ with isopropanol and sodium hydroxide at the base to give primary amine. While isopropanol is dehydrogenated to acetone, the acetone further reacts with the primary amine produced in situ to give the desired imine by releasing water as a side product. Finally, imine is subsequently hydrogenated to furnish the desired secondary amine.

7.3.2 Ritter Reaction

The Ritter reaction (reported by Ritter et al. in 1948) is a classical route to synthesize amide from nitriles and alcohols (Scheme 7.9).[22,23] In this reaction, a carbocation generated in situ from an alcohol, and an alcohol derivative or an olefin is trapped by a nitrile to produce a nitrilium species, which generate amide after hydrolysis. Typically, this reaction was catalyzed by acid catalysts. At that point various catalysts like Brønsted acids (solid acids),[24–30] Lewis acids,[31–36] acid functionalized ionic liquids,[37] organic acids,[38–41] zeolites,[42] supported acid catalysts,[43–48] metal complexes,[49–51] montmorillonite,[52,53] anhydrides,[54] and I_2[55] were used in Ritter reactions involving various alcohols and nitriles. Because recent progress in this area has been adequately summarized, it will not be reviewed in this chapter.

7.4 CONCLUSIONS AND OUTLOOK

Diverse hydrogen sources as well as the synthesis of *N*-alkyl amines from nitroarenes or benzonitriles and alcohols have been presented and summarized in this chapter. Among all the reducing agents, H_2 is most ideal because it is environmental friendly, plentiful, and easily industrialized. However, the present systems face disadvantages such as strict operation conditions, low selectivity, and low catalyst efficiency. It is also of concern that hydrogenative hydroxyl group cleavage severely inhibits the development of this reaction.

Therefore, new work should focus on the development of new catalysts with high activity in nitroarene/nitrile reduction and the following *N*-alkylation reaction, but reduced activity for the hydroxyl group cleavage.

REFERENCES

1. He, L.; Wang, L. C.; Sun, H.; et al. *Angew. Chem. Int. Ed.* **2009**, *48*, 9538.
2. Reguillo, R.; Grellier, M.; Vautravers, N.; et al. *J. Am. Chem. Soc.* **2010**, *132*, 7854.
3. Samuelsen, G. S.; Garik, V. L.; Smith, G. B. L. *J. Am. Chem. Soc.* **1950**, *72*, 3872.
4. Feng, C.; Liu, Y.; Peng, S.; Shuai, Q.; Deng, G.; Li, C.-J. *Org. Lett.* **2010**, *12*, 4888.
5. Liu, Y.; Chen, W.; Feng, C.; Deng, G. *Chem. Asian J.* **2011**, *6*, 1142.
6. Cui, X.; Zhang, Y.; Shi, F.; Deng, Y. *Chem. Eur. J.* **2011**, *17*, 2587.
7. Lee, C.-C.; Liu, S.-T. *Chem. Commun.* **2011**, *47*, 6981.
8. Peng, Q.; Zhang, Y.; Shi, F.; Deng, Y. *Chem. Commun.* **2011**, *47*, 6476.
9. Tang, C. H.; He, L.; Liu, Y. M.; Cao, Y.; He, H. Y.; Fan, K. N. *Chem. Eur. J.* **2011**, *17*, 7172.
10. Li, X.; Zhang, J.; Xiang, Y.; et al. *Sci. Chin. Ser. B Chem.* **2008**, *51*, 248.
11. Xu, L.; Li, X.; Zhu, Y.; Xiang, Y. *N. J. Chem.* **2009**, *33*, 2051.
12. Xiao, F.; Liu, Y.; Tang, C.; Deng, G.-J. *Org. Lett.* **2012**, *14*, 984.
13. Stíbal, D.; Sá, J.; Bokhoven, J. A. V. *Catal. Sci. Technol.* **2013**, *3*, 94.
14. Selvam, K.; Sakamoto, H.; Shiraishi, Y.; Hirai, T. *N. J. Chem.* **2015**, *39*, 2467.
15. Cui, X.; Zhang, C.; Shi, F.; Deng, Y. *Chem. Commun.* **2012**, *48*, 9391.
16. Cui, X.; Deng, Y.; Shi, F. *ACS Catal.* **2013**, *3*, 808.
17. Tan, D.-W.; Li, H.-X.; Young, D. J.; Lang, J.-P. *Tetrahedron* **2016**, *72*, 4169.
18. Shimizu, K.; Shimura, K.; Nishimura, M.; Satsuma, A. *ChemCatChem* **2011**, *3*, 1755.
19. Oh, S. G.; Mishra, V.; Cho, J. K.; et al. *Catal. Commun.* **2014**, *43*, 79.
20. Song, G.; Tian, J.; Chen, L.; Li, Y. *Res. Chem. Intermediat.* **2015**, *41*, 5399.
21. Werkmeister, S.; Bornschein, C.; Junge, K.; Beller, M. *Eur. J. Org. Chem.* **2013**, *2013*, 3671.
22. Ritter, J. J.; Kalish, J. *J. Am. Chem. Soc.* **1948**, *70*, 4048.
23. Ritter, J. J.; Minieri, P. P. *J. Am. Chem. Soc.* **1948**, *70*, 4045.
24. Firouzabadi, H.; Iranpoor, N.; Khoshnood, A. *Catal. Commun.* **2008**, *9*, 529.
25. Okuhara*, T.; Chen, X.; Matsuda, H. *Appl. Catal. A Gen.* **2000**, *200*, 109.
26. Khaksar, S.; Fattahi, E.; Fattahi, E. *Tetrahedron Lett.* **2011**, *52*, 5943.
27. Olah, G. A.; Yamato, T.; Iyer, P. S.; Trivedi, N. J.; Singh, B. P.; Prakash, G. K. S. *Mater. Chem. Phys.* **1987**, *17*, 21.
28. Jiang, D.; He, T.; Ma, L.; Wang, Z. *RSC Adv.* **2014**, *4*, 64936.
29. Kukushkin, V. Y.; Pombeiro, A. J. L. *Inorg. Chim. Acta* **2005**, *358*, 1.
30. Qin, Y.; Qingfu, W.; Zhengkun, Y. *Chem. Soc. Rev.* **2015**, *44*, 2305.
31. Callens, E.; Burton, A. J.; Barrett, A. G. M. *Tetrahedron Lett.* **2006**, *47*, 8699.
32. Firouzabadi, H.; Sardarian, A. R.; Badparva, H. *Synth. Commun.* **1994**, *24*, 601.
33. Guérinot, A.; Reymond, S.; Cossy, J. *Eur. J. Org. Chem.* **2012**, *2012*, 19.
34. Mukhopadhyay, M.; Iqbal, J. *J. Org. Chem.* **1997**, *62*, 1843.
35. Ramon, R. S.; Marion, N.; Nolan, S. P. *Chem. Eur. J.* **2009**, *15*, 8695.
36. Mukhopadhyay, M.; Reddy, M. M.; Maikap, G. C.; Iqbal, J. *J. Org. Chem.* **1995**, *60*, 2670.
37. Feng, J.; Yingjie, L.; Haifeng, D.; Zonghan, L.; Jungang, C.; Dapeng, L. *Chem. Res. Chin. Univ.* **2010**, *26*, 384.
38. Barbero, M.; Bazzi, S.; Cadamuro, S.; Dughera, S. *Eur. J. Org. Chem.* **2009**, *2009*, 430.

39. Jiang, S.; Wang, Z.; Jiang, Z.; Li, J.; Zhou, S.; Pu, L. *Lett. Org. Chem.* **2012**, *9*, 24.
40. Sanz, R.; Martínez, A.; Guilarte, V.; Álvarez-Gutiérrez, J. M.; Rodríguez, F. *Eur. J. Org. Chem.* **2007**, *2007*, 4642.
41. Shi, M.; Tian, G.-Q. *Tetrahedron Lett.* **2006**, *47*, 8059.
42. Chen, X.; Okuhara, T. *J. Catal.* **2002**, *207*, 194.
43. Gawande, M. B.; Rathi, A. K.; Nogueira, I. D.; Varma, R. S.; Branco, P. S. *Green Chem.* **2013**, *15*, 1895.
44. Lakouraj, M. M.; Mokhtary, M. *Monatsh. Chem.* **2008**, *140*, 53.
45. Nasr-Esfahani, M.; Montazerozohori, M.; Karami, Z. *Org. Prep. Proced. Int.* **2016**, *48*, 321.
46. Salehi, P.; Motlagh, A. R. *Synth. Commun.* **2000**, *30*, 671.
47. Tamaddon, F.; Khoobi, M.; Keshavarz, E. *Tetrahedron Lett.* **2007**, *48*, 3643.
48. Tamaddon, F.; Tavakoli, F. *J. Mol. Catal. A Chem.* **2011**, *337*, 52.
49. Goto, A.; Endo, K.; Saito, S. *Angew. Chem. Int. Ed.* **2008**, *47*, 3607.
50. Ibrahim, N.; Hashmi, A. S. K.; Rominger, F. *Adv. Synth. Catal.* **2011**, *353*, 461.
51. Top, S.; Jaouen, G. *J. Org. Chem.* **1981**, *46*, 78.
52. Kumar, H. M. S.; Reddy, B. V. S.; Anjaneyulu, S.; Reddy, E. J.; Yadav, J. S. *N. J. Chem.* **1999**, *23*, 955.
53. Lakouraj, M. M.; Movassagh, B.; Fasihi, J. *Synth. Commun.* **2000**, *30*, 821.
54. Martíneza, A. G.; Alvareza, R. M.; Vilara, E. T.; Frailea, A. G.; Hanackb, M.; Subramanianb, L. R. *Tetrahedron Lett.* **1989**, *30*, 581.
55. Theerthagiri, P.; Lalitha, A.; Arunachalam, P. N. *Tetrahedron Lett.* **2010**, *51*, 2813.

Index

Note: Page numbers followed by *s* indicate schemes.

B

C

D

E

G

H

I

Kolibri 360 GmbH
Ferdinand-Jühlke-Straße 7
99095 Erfurt, Deutschland
produktsicherheit@kolibri360.de